The Minerals, Metals & Materials Series

The Minerals, Metals & Materials Series publications connect the global minerals, metals, and materials communities. They provide an opportunity to learn about the latest developments in the field and engage researchers, professionals, and students in discussions leading to further discovery. The series covers a full range of topics from metals to photonics and from material properties and structures to potential applications.

More information about this series at http://www.springer.com/series/15240

Muammer Kaya

Electronic Waste and Printed Circuit Board Recycling Technologies

TMS

🐴 Springer

Muammer Kaya
Mining Engineering Department
Eskisehir Osmangazi University
Eskisehir, Turkey

ISSN 2367-1181 ISSN 2367-1696 (electronic)
The Minerals, Metals & Materials Series
ISBN 978-3-030-26592-2 ISBN 978-3-030-26593-9 (eBook)
https://doi.org/10.1007/978-3-030-26593-9

© The Minerals, Metals & Materials Society 2019
This work is subject to copyright. All rights are reserved by the Publisher, whether the whole or part of the material is concerned, specifically the rights of translation, reprinting, reuse of illustrations, recitation, broadcasting, reproduction on microfilms or in any other physical way, and transmission or information storage and retrieval, electronic adaptation, computer software, or by similar or dissimilar methodology now known or hereafter developed.
The use of general descriptive names, registered names, trademarks, service marks, etc. in this publication does not imply, even in the absence of a specific statement, that such names are exempt from the relevant protective laws and regulations and therefore free for general use.
The publisher, the authors, and the editors are safe to assume that the advice and information in this book are believed to be true and accurate at the date of publication. Neither the publisher nor the authors or the editors give a warranty, express or implied, with respect to the material contained herein or for any errors or omissions that may have been made. The publisher remains neutral with regard to jurisdictional claims in published maps and institutional affiliations.

This Springer imprint is published by the registered company Springer Nature Switzerland AG
The registered company address is: Gewerbestrasse 11, 6330 Cham, Switzerland

Memoriam

In the memory of decedent my mother, Asiye Kaya; my father, Hayrullah Kaya; and my Ph.D. supervisor, Prof. Dr. Andre Laplante (Mining and Metallurgical Engineering Department, McGill University, Canada).

Prof. Dr. Andre Laplante (1953–2006)

After his engineering studies in Montreal and in Toronto, he entered the Department of Mining, Metals, and Materials Engineering at McGill University in 1980 and was named Associate Professor in 1986. He was a specialist in gold recovery. Recognized worldwide for his research, he was named Honorary Visiting Researcher at AJ Parker Research Center for Hydrometallurgy at Murdoch University in Western Australia. Avid sports lover, he also took part in the cross-country ski team of McGill University. All his colleagues and students will remember his wonderful teaching abilities, his exemplary honesty, and his enormous joie de vivre.

Preface

Globally, e-waste is the most traded hazardous waste on the planet.

This book provides an up-to-date review of recycling technology of electronic waste (e-waste) and waste printed circuit boards (WPCBs), which are rich in base and precious metals, and the essential component of end-of-life (EoL) electrical and electronic equipment (EEE). From the economic and environmental perspectives, the efficient recycling of WPCBs is of great importance. For the extraction of metals from WPCBs, a large amount of research study has been done to establish an environmentally friendly and economic way to recover metals from WPCBs based on physico-mechanical, pyrometallurgical, and hydrometallurgical processes.

Nowadays, population increases, rapid economic growths, technological innovations, living standard improvements, shortened life spans of EEE, and consumer attitude changes have resulted in significant increases in the amount of waste electrical and electronic equipment (WEEE) that needs to be safely managed, which contains a considerable amount of metal resource with higher metal contents than found in primary ores. The handling of e-waste, including combustion in incinerators, disposing in landfill, or exporting overseas, is no longer permitted due to environmental pollution and global legislations. The WPCB is not only a dangerous toxic waste but at the same time a rich secondary resource for polymetallic and nonmetallic materials. Developing new, clean, and economical technologies for recycling of WPCBs is a significant challenge. Processing of the waste should both optimize recovery of metals and nonmetals and minimize final waste volume and processing emissions.

This book deals with the outline of e-waste problem; its diverse categories, composition, and management; and various recycling processes especially the green eco-friendly ones with unique attention toward extraction of both valuable metals and cheap nonmetals. Here, various WPCB treatment methods and WPCB recycling techniques, which can be classified into conventional and advanced recycling technologies, are comprehensively revised and analyzed based on their advantages and disadvantages. At present, there are a lot of techniques of recycling

WPCBs in the world. Researchers have proposed many methods, including physico-mechanical separation, pyrometallurgical method, hydrometallurgical processing, biotechnology, microwave treatment, supercritical fluid technology, etc., to recycle WPCBs from different point of view. However, physico-mechanical processing + hydrometallurgical processing is still the most competitive technology for WPCB recycling, and it will be the most commonly used technology to recycle WPCBs globally. This book also indicates the future research direction of WPCB recycling should focus on a combination of several techniques or in series recycling to maximize the benefits of process.

Keywords: Electronic waste (e-waste); WEEE; Printed circuit boards; Recycling techniques; Pyrometallurgy; Hydrometallurgy; Mechanical separation; Precious metals extractions; Base metal extraction

Eskisehir, Turkey   Muammer Kaya

Introduction

Twentieth-Century Waste Management Policy
  "How do we get rid of our waste efficiently with minimum damage to public health and the environment?"
  Twenty-First-Century Resource Management: Circular Economy Policy
  "How do we handle our discarded resources in ways which do not deprive future generations of some, if not all, of their value?"

The Green and Circular Economy (GCE) Initiative, set up by the United Nations Environment Programme (UNEP), spearheads the transformations needed to create a viable future for generations to come. Its aim is to shape an economy "that results in improved human weel-being and social equity, while significantly reducing environmental risks and ecological scarcities. In its simplest expression, a green economy can be thought of as one which is low carbon, resource efficient and socially inclusive". The circular economy (CE) concept was developed as a strategy for reducing the demand of its economy upon natural resources, as well as for mitigating environmental damage. It refers to reducing, reusing, and recycling (3Rs) activities conducted in the process of production, circulation, and consumption. The CE approach to efficient resource use integrates cleaner production and industrial ecology in broader systems. The production and use of metals play a crucial role in creating that future economy. According to the Green Economy Initiative Report, the main sources of future economic development and growth will be renewable energy technologies, resource- and energy-efficient buildings and equipment, low-carbon public transport systems, infrastructure for fuel-efficient and clean-energy vehicles, and waste management and recycling facilities. All of these rely heavily on metals. Metals within products are selected to fulfill specific functions. The increasing complexity of this functional demand has led to the use of an increasing number of elements. Metal/element use intensity in products increased significantly in the twentieth century. Metals are used as chemical compounds, as components in alloys, or in some cases as pure metal. Global metal needs will increase three to nine times than all the metal used in the world today. The growth of demand for some metals is much faster than the others. There is an accelerated demand for and potential scarcity of some elements and resources. As a global demand for many metals continues to rise, more low-quality ores are mined, leading to an overall decrease in ore grades. Recycling is in fact one of the most immediate, tangible, and low-cost investments available for decoupling economic growth from environmental degradation and escalating resource use [1]. This is one of the main reasons for writing this book.

The development of the electronic industry is associated with an increased demand on metals, especially more precious metals (PMs) and more scarce/rare metals, and leads to increased growth of WEEE. The demand of EEE has increased dramatically with the rapid advancement in information and communication technology (ICT). Drastic innovations on the electrical and electronic technologies further shortened the life and thus enhanced the generation of waste from WEEE. E-waste consists of WEEEs or goods which are not fit for their originally intended use. Half of the e-waste is coming from electrical appliances and the rest from electronic goods [2]. There are four main sources of e-waste: small/large home appliances, hospital medical equipment, office machines (ICT and telecom equipment), and industrial equipment/machines. Consumer and lighting equipment, electrical and electronic tools, entertainment devices, toys and sports equipment, and monitoring and controlling equipment are also important source of e-waste. In the last three decades, ICT restructures societies and economies. EEEs can become e-waste due to rapid advancement in technology; development in society; change in style, fashion, and status; greater demands on EEE; nearing the end of their useful life span, and not taking precaution while handling them. The replacement of EEE becomes more frequent, which results in large quantities of e-waste that need to be disposed [3]. E-waste covers a wide range of compositions depending on its resources, processing, and management procedures. According to the Association of Plastics Manufacturers in Europe (APME), the material composition of EEE is 38% ferrous metal, 28% nonferrous metal, 19% plastics, 4% glass, 1% wood, and 10% others [4]. E-waste processing and handling cause immediate impacts on the environment and human health.

EEE contains a range of components made of a wide variation of metals, plastics, and other substances. Modern electronics can contain up to 60 different elements; many are valuable (precious metals (Au, Ag, Pd, Pt, Ru, Rh, and Ir), special metals (Co, In and Sb) (SMs), and base metals (Cu and Sn) (BMs)), some are hazardous (halogens, Hg and Cd), and some are both (Pb) [5]. Mobile phones contain more than 40 elements [1]. The most complex mix of substances is usually present in PCBs. Metal fraction (MF) constitutes the most valuable and easiest-to-recycle materials from WEEE. There are ample capacities and markets available for recycled metals. Current recycling processes are capable of recovering <95% of the in-feed metals. Additionally, the presence of PMs makes e-waste recycling attractive economically. A major challenge for metal recovery is the heterogeneity and complexity of WPCBs, which includes metal diversity and their liberation. Generally, WPCB recycling process involves sorting, disassembly, upgrading, and refining. Disassembly is used for separation of target electronic components (ECs), particularly metals, from their organic substance. Upgrading includes physico-mechanical, pyrometallurgical, and hydrometallurgical treatments. Refining is a final step to get high-purity metals. Despite all legislative efforts to establish recycle and environment-friendly waste management efforts, the majority of resources today are lost. Several causes can be identified as follows: first, insufficient collection efforts; second, partly inappropriate recycling technologies; and third, and above all, large and often illegal exports of waste into regions with no or inappropriate recycling infrastructures in place. Large emissions of hazardous substances are associated with this.

Either WPCB can be directly treated, which includes landfill and incineration, or the recycling of WPCB by various kinds of technologies is also an alternative. Compared to direct treatments, recycling is more favorable due to both environmental and economic consideration given the enrichment of materials in WPCB. It is regarded as a secondary resource since the concentration of PMs and organic resins/polymers in WPCB is normally ten times higher than rich nonrenewable (sometimes even hundred times higher). Therefore, recycling is not simply a reduction of waste but a reuse of resources with better economic feasibility and less environmental impact [6].

The three e-waste recycling techniques that are most commonly used are chemical, metallurgical, and physical processes. Chemical recycling methods include pyrolysis, gasification, combustion, supercritical fluid depolymerization, and hydrogenolytic degradation. However, the acids and gases produced during hydrometallurgical and pyrometallurgical techniques make them unsuitable for eco-friendly recycling (green technology). The physical and mechanical recycling techniques, which are easier to perform and are more friendly to the environment, do not use any chemicals and eliminate also the problems of secondary pollution. Among these techniques, the mechanical processing of WPCBs is a recycling technique that may involve comminution, size separation, magnetic separation, air classification, density separation, and electrostatic separation to concentrate metals, being an acceptable economically and environmental option. These techniques focus on separating the metallic fractions (MF) from the nonmetallic fractions (NMF) based on the differences in the intrinsic properties of metallic and nonmetallic components such as density, size, electrical conductivity, response to magnetic field, surface properties, etc.

In this book, various WPCB treatment methods and WPCB recycling techniques, which can be classified into conventional recycling technologies (i.e., direct treatment (landfill and incineration) and physico-mechanical separation) and advanced recycling technologies (i.e., pyrometallurgy, hydrometallurgy, biometallurgy, electrometallurgy, vacuum metallurgy, supercritical fluids, full recovery of nonmetallic fraction (NMF), etc.), are comprehensively reviewed and analyzed based on their advantages and disadvantages from industrial application point of view. Unfortunately, despite the fact that many efforts to develop recycling technologies have been endeavored, these technologies are still rather exclusive and inadequate because of the intricacy of the e-waste system. Hence, the demerits of each process are debated and discussed from the viewpoint of technical advancement and environmental protection. Also, industrial application possibilities (i.e., commercial extent) and the recycling technique evaluation criteria are discussed including economic, environmental, and marketability point of views.

Recycling has long been known to be an environmentally friendly strategy and an appropriate way to manage WEEE streams. The proper technologies to recycle metals and nonmetals from e-waste without negative effects to the environment and human health are urgent and essential. The diverse types of secondary resources recycled from WEEE can potentially replace an equivalent quantity of materials that would otherwise need to be produced from primary resources. Moreover, with the development of eco-design, extended/individual producer responsibility (EPR) system, and sustainable supply chain management, resources and components contained in WEEE are likely to be used in the electronics industry, forming a closed-loop system of resource supply. Therefore, it is possible to achieve self-sufficiency of resources in the electronics industry using secondary resources recycled from WEEE. E-waste recycling can effectively provide a solution to the new norm of resource constraint faced in the world [7].

Recycling is primarily an economic industrial activity with a strong environmental and social implication. Metal recycling is increasingly promoted as an effective way to address resource scarcity and mitigate environmental impacts associated with metal production and use; but, there is little systemic information available regarding recycling performance and still less on the true recycling rates that are possible and how to do better considering the system in its totality. The development of the recycling technologies for WEEE has entered a new stage. The WEEE disposing technologies have evolved from simple disassembly, classification, and sorting to high value-added utilization technologies. Among all these WEEE, WPCBs are considered as the most valuable components due to PM contained. Previous studies found that the presence of PMs are richer in WPCBs than in typical metal mines, which are driven recycling PMs from WPCBs to a profitable business without proper pollution controls in developing countries. However, recovering PMs from WPCBs is a challenge because WPCBs are both valuable and harmful simultaneously, which are caused by their complex materials makeup. Hence, the proper technologies to recycle metals from WPCBs without negative effects to the environment and human health are urgent and essential.

In this book, the revised and discussed technologies are current, state-of-the-art, and best available technologies (BAT) that are currently used in the e-waste/WPCB recycling industry. Then, the integrated technological flowsheets, including MFs and NMFs enrichment and PM recovery which are WPCB techniques, are put forward. Recycling transforms WEEE into a resource allowing the recovery and reuse of the metallic and nonmetallic components and mitigating the environmental impacts. This book reveals opportunities for research and policy to reduce the risks from accumulating e-waste and ineffective recycling.

In Chap. 1, the definitions, compositions, and classifications of e-wastes are described. In Chap. 2, PCB structures, characteristics, sources, and assembly structure, methods of fastening ECs and joining ECs in PCBs, and solders, soldering methods, and desoldering are covered. Chapter 3 covers WPCB recycling chain and treatment options, dismantling, size reduction, and separation and purification processes. In Chap. 4, dismantling, desoldering, sensing, and active disassembly systems are extensively covered. Chapter 5 contains traditional and advanced WPCB recycling technologies. The separation of metals from e-waste using physico-mechanical processing and hydrometallurgical and pyrometallurgical routes is critically analyzed. Pyrometallurgical routes are comparatively economical and eco-efficient if the hazardous emissions are controlled. In industrial e-waste recycling, currently, pyrometallurgical routes are used initially for the segregation and upgrading of PMs (Au and Ag) into BMs (Cu, Pb, and Ni) followed by hydrometallurgical and electrometallurgical processing for the recovery of pure BMs and PMs. But, hydrometallurgy is a promising treatment due to its low capital cost, high selectivity, and lower environmental impact. Chapter 6 involves WPCB size reduction and classification equipment. In Chap. 7, physical properties of WEEE are given. WPCB sorting and separation technologies and equipment are covered. In Chap. 8, current industrial-scale e-waste/WPCB recycling lines are included. Chapter 9 involves recycling of NMF technologies from WPCBs. In Chap. 10, state-of-the-art hydrometallurgical recovery of metals is covered. This book emphasizes the recycling of WPCBs by physico-mechanical and hydrometallurgical treatments. Chapter 11 comprises hydrometallurgical recovery of critical REEs and special metals from WEEE. In Chap. 12, the perspectives of WPCB recycling are extensively covered. Finally, conclusions are given in tabulated form in Chap. 13.

To develop a clean and environment-friendly green technology for resource recovery from WPCBs as soon as possible is not only to raise the utilization rate of natural resources but also to reduce pollution, to protect environment, and to achieve energy saving and emission reduction. Therefore, this work is of great significance for national advantages, and at the same time, it can bring economic benefits to the enterprise.

References

1. http://wedocs.unep.org/bitstream/handle/20.500.11822/8423/-Metal%20Recycling%20Opportunities %2c%20 Limits%2c%20Infrastructure-2013Metal_recycling.pdf?sequence=3&isAllowed=y
2. Kaya M (2016) Recovery of metals and nonmetals from electronic waste by physical and chemical recycling processes. Waste Manag 57:64–90. https://doi.org/10.1016/j.wasman.2016.08.004
3. Zhou Y, Qiu K (2010) A new technology for recycling materials from waste printed circuit boards. J Hazard Mater 175(1–3):823–828. https://doi.org/10.1016/j.jhazmat.2009.10.083
4. Association of Plastics Manufacturers in Europe (APME) (2004) Plastics—a material of choice for the electrical and electronic industry-plastics consumption and recovery in Western Europe 1995. APME: Brussels. p 1
5. Mesker CEM, Hagelüken C, Van Damme G (2009) TMS 2009 annual meeting & exhibition, San Francisco, California, USA, EPD Congress 2009 proceedings

Ed. by. SH Howard, P Anyalebechi, L Zhang, pp 1131–1136, ISBN No: 978-0-87339-732-2
6. Ning C, Lin CSK, Hui DCW (2017) Waste printed circuit board (PCB) recycling techniques. Top Curr Chem (Z) 375:43. https://doi.org/10.1007/s41061-017-0118-7
7. Gu Y, Wu Y, Xu Ming, Mu X, Zuo T (2016) Waste electrical and electronic equipment (WEEE) recycling for a sustainable resource supply in the electronics industry in China. J Cleaner Prod 127:331–338. https://doi.org/10.1016/j.jclepro.2016.04.041

Contents

1 E-Waste and E-Waste Recycling 1
    1.1 Definition and Classification of E-Waste 1
    1.2 Fundamentals of E-Waste and E-Waste Recycling 2
        1.2.1 Significance and Characteristics of WEEE and E-Waste Recycling (Amounts and Compositions) 3
    1.3 History and Cornerstones of E-Waste Problem and E-Waste Recycling 10
    1.4 Representative Sample Preparation and Analysis 16
    1.5 Life Cycle of Electronics and E-Waste 19
    1.6 Life Cycle Assessment (LCA) and Life Cycle Management (LCM) 23
        1.6.1 LCA Principles 23
        1.6.2 The Several Steps of LCA 24
        1.6.3 LCA Tool 25
    1.7 Objectives of WPCB/E-Waste Recycling Opportunities 26
    1.8 General Driving Force for E-Waste Processing 26
    1.9 E-Waste Concerns and Challenges 29
    References 30

2 Printed Circuit Boards (PCBs) 33
    2.1 Printed Circuit Boards (PCBs) 33
    2.2 Structure and Contents of PCBs 34
    2.3 Sources and Value of WPCBs 37
    2.4 Characterization of WPCBs 41
    2.5 Impact on Metal Resources 43
    2.6 Methods of Fastening Electronic Components on WPCBs 47

	2.7	Methods of Joining Components in PCBs: Soldering	48
		2.7.1 Solder	48
		2.7.2 Pb-Free Solder	48
		2.7.3 Solder Paste	49
		2.7.4 Solder Flux	49
		2.7.5 Solder Alloys	50
	2.8	Soldering Methods of Electronic Products	51
		2.8.1 Dip Soldering	51
		2.8.2 Wave Soldering	52
		2.8.3 Reflow Soldering	52
	2.9	Solder Mask	53
	2.10	Characterization of Wastes from PCB Manufacturing	53
	References	55
3	**WPCB Recycling Chain and Treatment Options**	59
	3.1	Recycling Chain	59
	3.2	Dismantling/Disassembly	62
		3.2.1 Selective Disassembly	62
		3.2.2 Simultaneous Disassembly	62
	3.3	Size Reduction	63
	3.4	Separation/Upgrading/Extraction	65
	3.5	Purification and Refining Endprocesses	66
	References	68
4	**Dismantling and Desoldering**	69
	4.1	Dismantling/Disassembly Process/System	69
	4.2	Desoldering for Disassembly	72
		4.2.1 Heating Methods (Thermal Treatment) for Desoldering	72
	4.3	Semiautomatic PCB Electronic Component-Dismantling Machines	75
		4.3.1 Drum-/Barrel-Type Dismantling Machines	75
		4.3.2 Tunnel-Type Dismantling Machines	76
		4.3.3 Rod- and Brush-Type EC Disassembly Apparatus with IR Heater	77
		4.3.4 Scanning and Laser Desoldering Automated Component-Dismantling Machine	79
	4.4	Sensing Technologies	80
	4.5	Eco-design/Design for Disassembly (DfD) Concept	80
	4.6	Active Disassembly (AD)	80
	References	81
5	**Traditional and Advanced WPCB Recycling**	83
	5.1	Comparison of Traditional and Advanced WPCB Recycling Processes	83

5.2	Traditional Processes for Mixed WPCB Recycling		85
	5.2.1	Uncontrolled Incineration	85
	5.2.2	Mechanical Separation	85
5.3	Advanced Methods for WPCB Recycling		87
	5.3.1	Pyrometallurgy	87
5.4	Microwave Heating		92
5.5	Pyrometallurgical Processes for the Recovery of Metals from E-Waste		93
	5.5.1	Ferrous Scrap	93
	5.5.2	Nonferrous Scrap	93
	5.5.3	Shredder Residues	93
	5.5.4	Metal-Containing Slags/Bottom Ash	93
	5.5.5	Cu Smelters	93
5.6	Limitations of Pyrometallurgical Processes		97
5.7	Emerging Technologies in Pyrometallurgy		97
5.8	Hydrometallurgy		98
	5.8.1	Solvent Leaching	98
	5.8.2	Biometallurgical Leaching	106
5.9	Purification		109
	5.9.1	Chemical Precipitation of Metals	109
	5.9.2	Cementation	109
	5.9.3	Electrowinning (EW)	110
	5.9.4	Solvent Extraction (SX)	115
	5.9.5	Ion Exchange (IX)	118
5.10	Water Treatment		119
References			120
6	**Size Reduction and Classification of WPCBs**		**123**
6.1	Size Reduction of PCBs		123
6.2	Shredders		125
	6.2.1	Single-Shaft Shredders	127
	6.2.2	Double-Shaft Shredders	128
	6.2.3	Four-Shaft Shredders	129
6.3	Blade/Hammer Mill Pulverizers/Granulators		129
6.4	Fractionator Technology		131
	6.4.1	Physical Principles	131
	6.4.2	Material Preparation	131
	6.4.3	Delamination	132
	6.4.4	Separation into High-Purity Output Fractions	132
	6.4.5	Output Fractions	132
6.5	Electrodynamic Fragmentation (EDF) with High-Voltage Pulses (HVP)		134
	6.5.1	Pulse Power Generation	134
	6.5.2	Material Treated	135
6.6	Classification Screens		137

	6.7	Conveyor Belts	140
	6.8	Filtration Systems	140
		6.8.1 Pulse-Jet Bag Filter Dust Collection	140
	References		142
7	**Sorting and Separation of WPCBs**		**143**
	7.1	Sorting Systems	143
	7.2	Physical Properties of WEEE	144
	7.3	Gravity/Density Separation	147
		7.3.1 Wet Gravity Separation	148
		7.3.2 Dry/Air Gravity Separation	155
	7.4	Electrostatic Separation (ES)	159
		7.4.1 Conductive Materials	161
		7.4.2 Insulators/Nonconductors	161
		7.4.3 Types of Electrostatic Separators	162
	7.5	Magnetic Separation (MS)	170
	7.6	Froth Flotation	172
	7.7	Pyrometallurgy+Supergravity+Hydrometallurgy and/or Electrolysis Separation	173
	References		174
8	**Industrial-Scale E-Waste/WPCB Recycling Lines**		**177**
	8.1	Unpopulated/Populated WPCB Recycling Lines	177
	8.2	Umicore's Integrated Smelters-Refineries (ISRs), Hoboken, Antwerp	182
		8.2.1 Off-Gas Treatment	184
		8.2.2 Resource Efficiency	185
	8.3	Austrian Müller-Guttenbrunn Group (MGG)	185
	8.4	Eldan Recycling, Spain	186
	8.5	Daimler Benz in Ulm, Germany	187
	8.6	NEC Group in Japan	188
	8.7	DOWA Group in Japan	189
	8.8	PCB Manufacturing Waste Recycling in Taiwan	190
		8.8.1 Recovery of Cu from Edge Trim of PCBs	191
		8.8.2 Recovery of Sn Metal from Sn/Pb Solder Dross	192
		8.8.3 Recovery of Cu Oxide from Wastewater Sludge	192
		8.8.4 Recovery of Cu from Spent Basic Etching Solution	193
		8.8.5 Recovery of $Cu(OH)_2$ from $CuSO_4$ Solution in PTH Process	193
		8.8.6 Recovery of Cu from the Rack Stripping Process	193
		8.8.7 Recovery of Cu from Spent Sn/Pb Stripping Solution in the Solder Stripping Process	194

Contents   xix

8.9	Sepro Urban Metal Process in Canada	195
	8.9.1 WEEE/E-Waste Pyrolysis Process of Sepro	197
8.10	Shanghai Xinjinqiao Environmental Co., Ltd., and Yangzhou Ningda Precious Metal Co., Ltd., in China	197
8.11	SwissRTec AG	198
8.12	WEEE Metallica, France	200
8.13	Hellatron Recycling, Italy	201
8.14	Aurubis Recycling Center in Lünen (Germany)	203
8.15	Attero Recycling, Roorkee, India	206
8.16	Noranda Smelter in Quebec, Canada	207
8.17	Rönnskar Smelter in Sweden	207
8.18	Comparison of Academic Research and Industrial-Scale E-Waste Recycling Practices	208
References		209

9   Recycling of NMF from WPCBs   211
    9.1   Direct and Chemical Recycling of NMF from WPCBs   211
    9.2   Direct Recycling of NMF from WPCBs   212
    9.3   Chemical Recycling of NMF from WPCBs by Degradation/Modification/Depolymerization of Thermoset Organic Polymers by Solvents   213
    9.4   Pyrolysis   214
    References   218

10   Hydrometallurgical/Aqueous Recovery of Metals   221
    10.1   Solder Stripping Leach   222
    10.2   Base Metal Leaching   228
    10.3   Extraction of Precious Metal by Leaching   233
         10.3.1   Au Leaching   241
    10.4   Full PM Recovery   256
    10.5   Brominated Epoxy Resin (BER) Leaching   256
    10.6   Purification Technologies for Precious Metals from Leachates   258
         10.6.1   Gold Recovery Using Nylon-12 3D-Printed Scavenger   263
         10.6.2   Solvent Extraction (SX)   264
    10.7   Industrial-Scale E-Waste Precious Metal Separation and Refining Solutions   266
    10.8   Occupational, Health, and Safety Hazardous Characteristic Determination Tests for PCBs   267
    References   268

11   Hydrometallurgical Recovery of Critical REEs and Special Metals from WEEE   277
    11.1   Rare-Earth Elements and Special Metals   277
    11.2   Recovery of Magnet Scraps   279

	11.3	Recovery of Lamp and CRT Phosphors	281
	11.4	Indium (In) Recovery from LCDs	283
	11.5	Photovoltaic (PV) Cell/CIGS Solar Panels Recovery	284
	11.6	Cathode Ray Tube (CRT) Recycling at Attero Plant	285
	11.7	Flat Panel Display Unit Recycling	286
	11.8	Cobalt (Co) Recovery from Lion Batteries	287
	References	288
12	**Perspectives of WPCB Recycling**	289
	12.1	Economy Perspective of Recycling	290
	12.2	Environmental Perspectives of Recycling	292
	12.3	Marketability Perspective	293
	12.4	Market for Recycling Technologies	294
	12.5	Eco-design Guidelines for Manufacturing	295
	12.6	Limitations of Current WPCB Recycling Technologies	296
	12.7	Future Development Perspectives	297
	12.8	Toward Sustainability and Zero-Waste Scheme	298
	References	299
13	**Conclusions**	301
	13.1	Summary and Comparison of WPCB Recycling Technologies	301
	13.2	Primary Production Versus Secondary Production	302
	13.3	Manual Dismantling Versus Automated Dismantling	303
	13.4	Conventional Versus Novel WPCB Recycling Technologies	304
	13.5	Recycling Solution	312
	References	313

Index 315

Abbreviations/Acronyms

A	Ampere
A/O	Aqueous/Organic
ABS	Acrylonitrile Butadiene Styrene
AD	Active Disassembly
ADSM	Active Disassembly Using Smart Material
AEC	Aluminum Electrolytic Capacitor
Ag	Silver
AGP	Accelerated Graphics Port
AHP	Analytic Hierarchy Process
Al	Aluminum
Am	Americium
AR	Aqua regia
As	Arsenic
ASTM	American Society for Testing and Materials
ATMI	eVOLV Process Owner Company
ATO	Automatic Throw-Over
Au	Gold
BAM	Barium Aluminate
BAT	Best Available Technique
Be	Beryllium
BER	Brominated Epoxy Resin
BFR	Brominated Flame Retardant
BGA	Ball Grid Array
Bi	Bismuth
BM	Base Metal
BMO	Base Metal Operation
Br	Bromine
C	Carbon
CAS	Cyclone Air Separator
CES	Corona Electrostatic Separation
CAT	Calcium Tungstate

CC	Concentration Criterion
Cd	Cadmium
Ce	Cerium
CFC	Chlorofluorocarbon
HCFC	Hydrochlorofluorocarbon
CO	Carbon Monoxide
Co	Cobalt
COD	Chemical Oxygen Demand
CPU	Central Processing Unit
Cr	Chromium
CRT	Cathode Ray Tubes
CSE	Centre for Science and Environment
Cu	Copper
CuS	Copper Monosulfide
dB	Decibel
DC	Direct Current
DEPHA	Di-(2-ethylhexyl)phosphoric Acid
DfD	Design for Disassembly
DfS	Design for Sustainability
DFSM	Dry Film Photoimageable Solder Mask
DINP	Diisononyl Phthalate
DMA	Dimethylacetamide
DMF	Dimethylformamide
DMSO	Dimethyl Sulfoxide
DNA	Deoxyribonucleic acid
DOP	Outotec SX Unit
DR	Disassembling Rate
DRAM	Dynamic random-access memory
DVD	Digital Versatile Disc
Dy	Dysprosium
EC	Electronic Component
ECS	Eddy Current Separator
EDF	Electrodynamic Fragmentation
EDTA	Ethylenediaminetetraacetic Acid
EEE	Electrical and Electronic Equipment
EHEHPA	2-Ethylhexyl 2-ethylhexyl-Phosphonic Acid
ELV	End-of-Life Vehicle
EMF	Electromotive Force
ENIG	Electroless Nickel Immersion Gold
EoL	End of Life
EPA	Environmental Protection Agency
EPROM	Erasable Programmable Read-Only Memory
ER	Epoxy Resin
Er	Erbium

ES	Electrostatic Separation
E-Scrap	Electronic Scrap
ESD	Electrostatic Discharge
E-Sorting	Electrostatic Separation
Eu	Europium
EW	Electrowinning
E-waste	Electronic Waste
F	Fahrenheit
F	Fluorine
FBS	Fluidized Bed Separator
g	Gram
Ga	Gallium
GC/MS	Gas Chromatography/Mass Spectrometry
Gd	Gadolinium
GEI	General Environmental Index
h	Hour
H	Hydrogen
HASL	Hot Air Solder Levelling (Pb-Free Surface Finish)
HCFC	Hydrochlorofluorocarbon
HClO	Hypochlorous Acid
HCN	Hydrogen Cyanide
HD	Hard Disk
HD	High Definition
HDD	Hard Disk Drive
HDPE	High-Density Polyethylene
HDT	Heat Deflection Temperature
Hg	Mercury
HIMS	High-Intensity Magnetic Separator
HIPS	High-Impact Polystyrene
HM	Heavy Metal
Ho	Holmium
HP	Horsepower
HPOL	High-Pressure Oxidative Leaching
HSE	Health, Safety, and Environment
HVP	High-Voltage Pulse
IC	Integrated Circuit
ICP - AES	Inductively Coupled Plasma-Atomic Emission Spectroscopy
ICT	Information and Communication Technology
ID	Identification Data
IDE	Integrated Drive Electronics
IMC	Intermetallic Compound
In	Indium
IR	Infrared

ISR	Integrated Smelter and Refinery
IT	Information Technology
ITO	Indium Tin Oxide
IX	Ion Exchange
JEDEC	Joint Electron Device Engineering Council
K	Kilo
K	Potassium
kPa	Kilopascal
KRS	Kayser Recycling System
KW	Kilowatt
La	Lanthanum
LAP	Lanthanum Phosphate
LCA	Life Cycle Assessment
LCD	Liquid Crystal Display
LCM	Life Cycle Management
LDA&SDA	Large Domestic Appliances and Small Domestic Appliance
LED	Light-Emitting Diode
LIBS	Induced Breakdown Spectroscopy
LIMS	Low-Intensity Magnetic Separator
LIX84I	2-Hydroxy-5-nonylacetophenone oxime
LIX860IC	5-Dodecylsalicylaldoxime
LIX984	5-Dodecylsalicylaldoxime and 2-hydroxy-5-nonylacetophenone oxime
Li	Lithium
Lion	Lithium Ion Battery
LLE	Liquid-Liquid Extraction
LOI	Loss on Ignition
LDPE	Low-Density Polyethylene
LPSM	Liquid Photoimageable Solder Mask
LPT	Parallel Port
Lu	Lutetium
M	Molarity
MB	Motherboard
Me	Metal
MEK	Methyl Ethyl Ketone
MF	Metal Fraction
mg	Milligram
MGG	Müller-Guttenbrunn Group
MIBC	Methyl Isobutyl Carbinol
MIBK	Methyl Ketone
Min	Minute
MLCC	Multilayer Ceramic Capacitor
mm	Millimeter
Mn	Manganese

Mo	Molybdenum
Mpa	Megapascal
MS	Magnetic Separation
MSA	Methanesulfonic Acid
mV	Millivolt
N	Normality
N12	Nylon-12
Nb	Niobium
Nd	Neodymium
NEC	Nippon Enterprise Company
NFM	Nonferrous Metals
NHE	Normal Hydrogen Electrode
Ni	Nickel
NMF	Nonmetallic Fraction
NMP	N-methyl-2-pyrrolidone
NSES	National Strategy for Electronics Stewardship
O	Oxygen
°C	Celsius
ODS	Ozone Depleting Substance
OECD	Organization for Economic Co-operation and Development
OEM	Original Equipment Manufacturer
ORP	Oxidation and Reduction Potential
OSHS	Occupational Safety and Health Standard
OSP	Organic Solderability Preservative
PCB	Polychlorinated Biphenyl
PBDD	Polybrominated Dibenzo-p-Dioxin
PBDE	Polybrominated Diphenyl Ether
PBDF	Polybrominated Dibenzo-Furan
PC	Personal Computer
PC	Polycarbonate
PcB	Polychlorinated biphenyl
PCB	Printed Circuit Board
PCDD	Polychlorinated Dibenzo-p-Dioxins
PCDD/F	Polychlorinated Dibenzo-p-Dioxins and Dibenzofuran
PCI	Peripheral Component Interconnect
PE	Polyethylene
PEM	Proton Exchange Membrane
PET	Polyethylene Terephthalate
PGM	Platinum Group Metal
PLC	Programmable Logic Control
PLS	Pregnant Leach Solution
PM	Precious Metal
Pm	Promethium
PMC	Phenolic Molding Compound

PMO	Precious Metal Operation
PO	Pyrolytic Oil
PP	Polypropylene
Ppm	Part per Million
PPO	Polyphenylene Oxide Blend
Pr	Praseodymium
PS	Polystyrene
Pt	Platinum
PTH	Plated Through Holes
PUR	Polyurethane
PV	Photovoltaic
PVC	Polyvinylchloride
PWB	Printed Wiring Board
RAM	Random Access Memory
REE	Rare-Earth Element
REM	Rare-Earth Metal
RFID	Radio-Frequency Identification
RoHS	Restriction of Hazardous Substance
ROM	Read-Only Memory
rpm	Revolutions per Minute
RSS	Ramp-Soak-Spike
RTS	Ramp to Spike
Ru	Ruthenium
SAT	System Engineering and Automation
Sb	Antimony
SBH	Sodium Borohydride
SCMO	Supercritical Methanol
SCW	Supercritical Water
SCWD	Supercritical Water Depolymerization
SCWO	Supercritical Water Oxidation
SE	Scare Element
Se	Selenium
SEM	Scanning Electron Microscope
SHE	Standard hydrogen electrode
Sm	Samarium
SM	Special/Scarce/Critical Metal
SMA	Shape Memory Alloy
SMC	Surface-Mount Component
SMD	Surface-Mount Device
SMP	Shape Memory Polymer
SMT	Surface-Mount Technology
Sn	Tin
SPIROK	Outotec SX Unit Mixer
SPLP	Synthetic Precipitation Leaching Procedure

SRT 2	Delamination and Separation Plant
SRT1	Shredding and Separation Plant
SS	Stainless Steel
StEP	Solving the E-Waste Problem
STP	Standard Temperature and Pressure
SX	Solvent Extraction
Ta	Tantalum
TAL	Time Above Liquidus
Tb	Terbium
TBE	Tetrabromoethane
TBP	Tributyl Phosphate
TBRC	Top Blown Rotary Converter
TCLP	Toxicity Characteristic Leaching Procedure
Te	Tellurium
TEHA	Tris(2-Ethylhexyl)Amine
TES	Triboelectric Separation
TFT	Thin-Film Transistor
TGA	Thermogravimetric Analysis
THC	Through-Hole Component
THD	Through-Hole Device
THT	Through-Hole Technology
Ti	Titanium
Tm	Thulium
TOPO	Trioctylphosphine Oxide
tpd	Ton Per Day
TSL	Transport Layer Security
TSS	Tin Stripping Solution
t-T	Time Temperature
TV	Television
UK	United Kingdom
UNDP	United Nations Development Program
UNI-IAS	United Nations University Institute for the Advances Study of Sustainability
US EPA	United States Environmental Protection Agency
USGS	United States Geological Survey
UPS	Uninterruptible Power Supply
V	Vanadium
VGA	Video Graphics Array
VOC	Volatile Organic Compound
VSF	Vertical Smooth Flow
W	Tungsten
W	Watt
WEEE	Waste Electrical and Electronic Equipment
WPCB	Waste Printed Circuit Board

WPWB	Waste Printed Wire Board
Wt	Weight
XRF	X-Ray Fluorescence
XRD	X-Ray Diffraction
Y	Yttrium
YAG	Yttrium Aluminum Garnet
Yb	Ytterbium
YBCO	Yttrium Barium Copper Oxide
YOX	Yttrium Oxide
YVO$_4$	Yttrium Vanadate
YZS	Yttria-Stabilized Zirconia
Z	Atomic Number
Zn	Zinc
Zr	Zirconium
ΔG	Difference in Gibbs Free Energy
μm	Micrometer
μg	Microgram

About the Author

Muammer Kaya was born in Eskisehir, Turkey, in 1960. He obtained his B.Sc. in Mining Engineering from Eskisehir Osmangazi University (ESOGU) in 1981. He received his M.Sc. and Ph.D. from the Metallurgical Engineering Department of McGill University in Canada in 1985 and 1989, respectively. He has more than 175 national and international publications/presentations and is interested in mineral processing, flotation, leaching, solvent extraction, ion exchange, coal preparation, deinking, recycling, and environmental protection. He has been working as a Full Professor at ESOGU since 1999. He has worked as the Director of ESOGU Research Center (TEKAM) for 15 years and also is Founder and First Manager of ESOGU Vocational School (ESOGU-EMYO) in Eskisehir, Turkey.

He is a Member of Minerals, Metals & Materials Society (TMS), Canadian Institute of Mining, Metallurgy and Petroleum (CIM), and American Institute of Mining, Metallurgical, and Petroleum Engineers (AIME).

Chapter 1
E-Waste and E-Waste Recycling

> *"We do not inherit the Earth from our ancestors; we borrow it from our children."*
> –Native American Proverb

Abstract This chapter introduces the definition, classification, and fundamentals of e-waste, which is the fastest growing waste stream in the world and grows three times faster than the municipal waste. Significance and characteristics of waste electrical and electronic equipment (WEEE) and e-waste, which contain both valuable inorganic/organic materials and hazardous substances, are clarified. Harmful effects of toxic material and possible adverse health effects are covered. History and cornerstones of e-waste problem and recycling are reviewed. Representative sample preparation, sampling, and analysis for e-waste are described. Lifecycle of electronics and e-waste management hierarchy are given. Life cycle assessment and life cycle management principles, steps, and tools are defined. Objectives of waste printed circuit board (WPCB) and e-waste recycling opportunities are presented. General driving forces, concerns, and challenges for e-waste processing are expressed. E-waste recycling protects environment, saves energy, and conserves resources.

Keywords E-waste · WEEE · EEE · WPCBs · Recycle · Reduce · Reuse

1.1 Definition and Classification of E-Waste

There is no standard definition for WEEE or e-waste. They comprise various forms of EEE that have no value to their owners. Any household or business item with circuitry or electrical/electronic components with power or battery supply is defined EEE according to Solving the E-Waste Problem (StEP) Initiative White Paper. The reported definitions of e-waste in literature are described below:

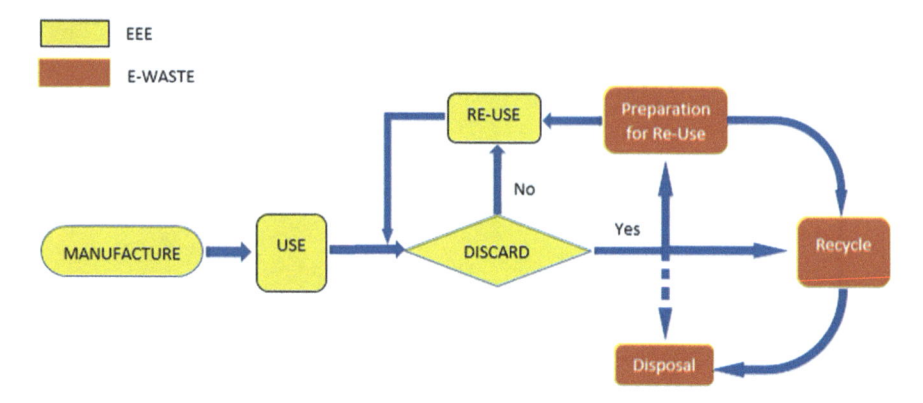

Fig. 1.1 EEE to e-waste with reuse flow

European WEEE Directive: "Electrical or electronic equipment which is waste... including all components, sub-assemblies and consumables, which are part of the product at the time of discarding." WEEE or e-waste is divided into ten categories based on European WEEE directives [1, 2].

Basel Action Network: "E-waste encompasses a broad and growing range of electronic devices ranging from large household devices such as refrigerators, air conditioners, cell phones, personal stereos, and consumer electronics to computers which have been discarded by their users." [3]

StEP: "E-waste is a term used to cover items of all types of EEE and its parts that have been discarded by the owner as waste without the intention of reuse." [4]

Figure 1.1 shows the EEE to e-waste with reuse flow. Definitions of the terms in Fig. 1.1 are given in Table 1.1.

1.2 Fundamentals of E-Waste and E-Waste Recycling

It is essential to understand the fundamental issues underlying e-waste and its recycling. These are independent of the recycled material, the device, and recycling location or region and address the:

- Significance of e-waste for resource management and toxic control
- General structure, main steps, and interfaces of recycling chain
- Objectives to achieve
- General frame conditions, which impact process selection

Table 1.1 Definition of the terms in Fig. 1.1

Manufacture	The phase of the EEE product lifecycle where it is manufactured
Discard	The decision by the owner to discard or not discard product. Discarding indicates it becomes e-waste, whereas not discarding and routing to reuse indicate it is not waste
Reuse	Reuse of EEE or its components is to continue the use of it, for the same purpose for which it was conceived, beyond the point at which its specifications fail to meet the requirements of the current owner and the owner has ceased use of the product. Products could be donated as charity or treated before or in this phase
Preparation for reuse by repair/refurbishing	Preparation for reuse comprises any operation performed to bring used EEE or its components into a condition to meet the requirements of the next potential owner
Recycle	The phase of the product lifecycle where due to lack of functionality, cosmetic condition, or age the product is broken down into component materials and recycled into raw materials for use in the manufacture of new EEE or other products. Waste-to-material and waste-to-product are kinds of recycling
Disposal	Material that cannot be recycled into raw materials for use in manufacture of new EEE or other products would need to be disposed of using other methods, such as energy recovery (waste-to-energy) or landfill. Items that are disposed of in household bin may move directly to this phase avoiding any opportunity of reuse or recycling

1.2.1 Significance and Characteristics of WEEE and E-Waste Recycling (Amounts and Compositions)

E-waste is one of the fastest growing waste streams in the world, and it has been estimated that these items already constitute about 8% of municipal waste [5]. Therefore, e-waste represents a rapidly growing disposal problem worldwide. According to the United Nations University Institute for the Advances Study of Sustainability (UNI-IAS), the Global E-Waste Monitor Report, 41.8 Mt e-waste was produced, and around 6.5 Mt of e-waste was reported as formally treated by national take-back systems in the world in 2014 [6]. It is expected that the amount of e-waste will reach 49.8 Mt in 2018. In 2014, average e-waste generated per inhabitant was 5.9 kg in the world. The rate of e-waste generation is increasing by 10% every year [7]. Most of the e-waste was generated in Asia: 16 Mt in 2014. This was 3.7 kg for each inhabitant. The highest per inhabitant e-waste quantity (15.6 kg inh.$^{-1}$) was generated in Europe. The whole region (including Russia) generated 11.6 Mt. The lowest quantity of e-waste was generated in Oceania and was 0.6 Mt. However, the per inhabitant amount was nearly as high as Europe's (15.2 kg inh.$^{-1}$). The lowest amount of e-waste per inhabitant was generated in Africa, where only 1.7 kg inh.$^{-1}$ was generated in 2014. The whole continent generated 1.9 Mt of e-waste. The Americas generated 11.7 Mt of e-waste, which represented 12.2 kg inh.$^{-1}$ [6]. In

Europe, e-waste is generated with a resultant increase by 16–28% every 5 years. E-waste amount is 5–30 kg per person per year and grows at three times faster than the municipal waste [8–10]. In the world, most of WEEEs are landfilled, incinerated, or recovered without pretreatment [11]. Due to the globally increased sales, the manufacturing of EEE is a major demand sector for PMs (gold (Au), silver (Ag), and platinum group (PGM) and special/scarce/critical metals (SMs) (selenium (Se), tellurium (Te), bismuth (Bi), antimony (Sb), and indium (In)) with a strong further growth potential [12]. Actually, after the use phase, the WEEE could be utilized as an important source to recover these "trace elements." [13]

Metal resources according to minimum availability can be roughly divided into three categories: very scarce, scarce, and less scarce. If the minimum availability of a metal (i.e., life expectancy) is less than 20 years, it is very scarce; between 20 and 40 years, it is scarce; and more than 40 years, it is less scarce. Ag, Au, Sb, Zn, As, Sn, Pb, Zr, and Cd are very scarce metals; Hg, Cu, Mn, Ni, Mo, Bi, Nb, and Y are scarce metals; and Ta, Co, Ti, Be, Li, PGM, V, and Cr are less scarce metals [14]. Some metals are critical and scarce in most of the countries. Metal criticality is not in line with the ranking of metal scarcity. It indicates that criticality is a very regional indicator and it is at the same time highly related to the economy status, structure, and government policies [14]. Metal scarcity is a reality requiring long-term attentions. With this concern, recycling of metals should be broadened, and the strategy of WEEE management or treatment needs to be emphasized. WEEE and WPBCs contain lots of valuable resources together with plenty of heavy metals and hazardous materials, which are considered both an attractive polymetallic secondary source, referred to as "urban mines," and an environmental contaminant. Therefore, recycle of valuable metallic and/or nonmetallic materials from them is necessary and compulsory in many developed/developing countries. Metal recycling business is very lucrative. In the 1980s, 11 elements and in the 1990s 15 elements were used in ICT circuitry. After the 2000s, EEEs include around 60 valuable elements found in the periodic table (Fig. 1.2) [10].

Table 1.2 shows some of the nonhazardous and hazardous (toxic/poisonous/harmful) substance in WEEE. Some of them are harmless, and the others have possible adverse effects to the human health. Common pollutants generated from WPCB treatment or recycling include heavy metals, secondary particulates/emissions, and plastic additives, which are considered as obstacles to WPCB recycling. For useful metals the occurrence of certain impurities can be detrimental and reduce recycling values. Certain concentrations of impurities can be tolerated. Electronic products use a staggering 320 t of Au (i.e., about 11% of world Au production) and more than 7.500 t of Ag annually worldwide [16]. A quarter of the annual global Cu production is destined for digital devices – 0.96–1.12% in PCBs. Hg is used in relays, switches, batteries, liquid crystal displays (LCDs), and gas discharge lamps (i.e., fluorescent tubes in scanners and photocopiers). Yearly, about 22% of the Hg produced in the world is used in electronic industry; but, its use in new EEEs is restricted by the Restriction of Hazardous Substances (RoHS) in the EU. Rechargeable batteries contain Pb, Cd, Li, and Ni. Old TVs, personal computers (PCs), and cathode ray tubes (CRTs) contain Pb in cone glass, Ba in electron gun

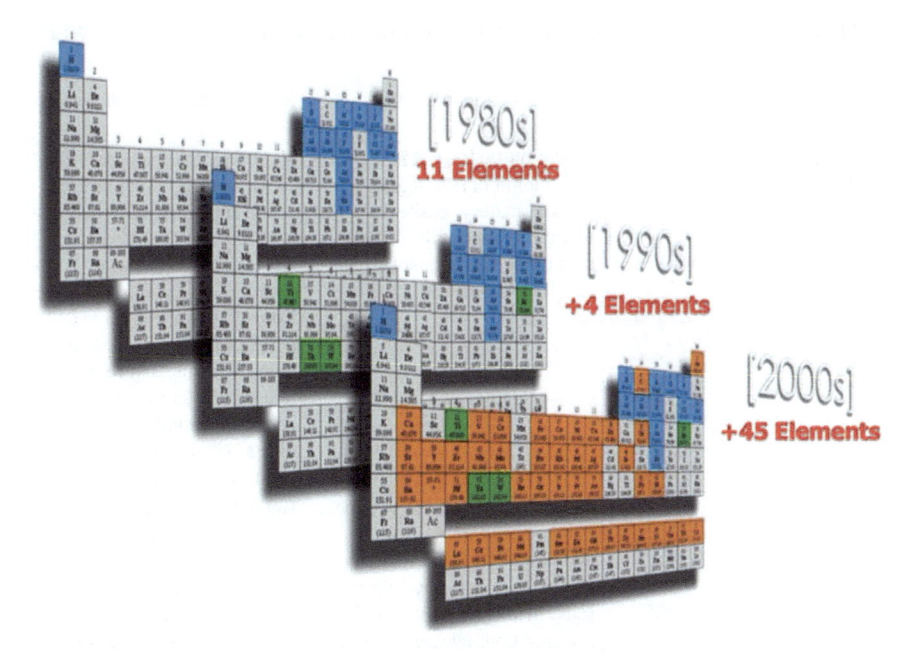

1.2 Fundamentals of E-Waste and E-Waste Recycling

Fig. 1.2 Growth of technology metals used in ICT circuitry [15]

getter, and Cd in phosphors. PCBs have Pb, Sn, and Sb in solder; and Cd and Be are found in switches. Polyvinylchloride (PVC) and brominated frame retardants (BFRs) are main components of plastics. Cr^{+6} are found in data types and floppy discs. Condensers and transformers contain polychlorinated biphenyls (PBBs). Chlorofluorocarbon (CFC) can be found in cooling units and insulation foams. Americium (Am) can be found in smoke detectors. LCDs include liquid crystals embedded between thin layers of glass and electrical control elements. Liquid crystals are mixture of 10–20 substances which belong to the groups of substituted phenyl cyclohexanes, alkylbenzens, and cyclohexylbenzenes. These substances contain O, F, H, and C and are suspected to be hazardous. Due to risk of ignition during soldering of ECs on PCB or impact with electric current, the PCB matrix is often a bromine (Br)-containing, fire retarded matrix likely to contain 15% of Br. Fire retardance can be attained either using additive or reactive fire retardance. BFRs generate polybrominated dibenzo-p-dioxins (PBDD) and polybrominated dibenzofurans (PBDF) at a moderate heating. Now, in new EEEs, PVC, BFR, and Cr^{6+} are limited by RoHS, and there are many substitutes for hazardous substances. E-waste recycling is more than just reducing the waste mountain and prevents the contamination of environment.

Table 1.2 Some of the hazardous and useful elements/chemical substances contained in WEEE/EEE

Hazardous material (harmful/toxic)	Materials and components usage in EEE	Possible adverse health effects
Lead (Pb)/PbO PbS	Solder on PCBs, CRT TVs/monitors (Pb in glass), Pb-acid battery, gasket in monitors, light bulbs, fluorescent tubes, cabling, solar batteries, and photocells	Vomiting, diarrhea, convulsions, anemia, coma or even death, appetite loss, abdominal pain, constipation, fatigue, sleeplessness, irritability, skin damage, and headache. Damage to the central and peripheral nervous systems, blood systems, and reproductive system and kidney damage. Gastric and duodenal ulcers. Affects brain development of children
Mercury (Hg)	Fluorescent tubes, tilt switches, flat screen switches, some alkaline batteries, switches, PCBs, sensors, relays, thermostats, medical equipment, data transmission, telecommunication, mobile phones, flat screens, thermostat, LCD, CRT	Brain, kidney, and central nervous system effects and liver chronic damage. Causes anemia. Respiratory and skin disorders due to bioaccumulation in fishes
Chromium VI (Cr^{6+}) (Corrosion protection of untreated and galvanized steel plates or hardener for steel housing)	Data tapes, floppy discs, metal housings, oxidative stress	Highly toxic. Human carcinogens, impacts on neonates, reproductive and endocrine functions. Irritating to the eyes, skin, and mucous membranes. Asthmatic bronchitis. DNA damage. Defect in neurodevelopment. Multiple organ failure
Barium (Ba) and strontium (Sr)	Getters in CRT, funnel glass	Brain swelling, muscle weakness, damage to the heart, liver, and spleen
Nickel (Ni) and cadmium (Cd)	NiCd rechargeable batteries, contacts and switches, fluorescent layer (CRT screens), printer ink and toners, SMD chip resistors, IR detectors, semiconductors, pins, PCBs	Highly biotoxic and bioaccumulation in the environment. Symptoms of poisoning (weakness, fever, headache, chills, sweating, and muscle pain), lung cancer, and kidney damage. Increased risk of cardiovascular disease, neurological deficits, and developmental deficits in childhood and high blood pressure. Respiratory system, bone problem, and defects in neurodevelopment of fetus. Toxic on human health. Accumulates in the kidney and liver. Causes neural damage. Teratogenic

(continued)

1.2 Fundamentals of E-Waste and E-Waste Recycling

Table 1.2 (continued)

Arsenic (As) and gallium (Ga)	Gallium arsenide in light-emitting diodes (LEDs), laser, photo/tunnel diodes	Skin diseases, decrease nerve conduction velocity, lung and bladder cancer. Affect breathing, cardiovascular diseases. Liver and renal disease. Gastrointestinal disturbances. Carcinogenic
Copper (Cu) and CuO	Wires, cables PCBs, coils, diodes	Copperiedus
Americium (Am) (Radioactive source)	Smoke alarm detectors	Radioactive element
Antimony (Sb)	Flame retardants in plastics, CRT glass, alloying element, photocells (ATO)	Carcinogenic potential
Beryllium (Be), Cu-Be alloys, BeO ceramics	Motherboards, connectors, heat sinks, power supply boxes	Affects organs such as the liver, kidneys, heart, nervous system, and lymphatic system. Carcinogenic (lung cancer). Inhalation of fumes and dust causes chronic berylliosis. Skin disease such as warts
Indium (In)	LCD display units (indium tin oxide (ITO) (90% In_2O_3+ 10% SnO_2)), chips, white LEDs, thin film PV cells, InGaN in LEDs Diodes, transistors, semiconductors, PVs (In_2O_3, Cu(In, Ga)Se_2, ICs (InP)	Taking in by mouth might results in damage to the kidney, heart, liver, and other organs. Breathing in indium might irritate the lungs. Applying In to the skin might cause skin irritation
Selenium (Se)	Diodes, electro-optic, photocopy machines, solar cells, semiconductors	Gastrointestinal disorder, hair loss, sloughing of nail, fatigue, irritability, and neurological damage
Bromide (Br) (HBr)	Flame retardant in plastics (BFR)	Disrupts thyroid function, increases risk of preterm birth and birth defects, slow neural and cognitive development, DNA damages, carcinogenic potential, toxic to kidney, hearing loss, skin disorder
Barium (Ba)	Sparkplugs, fluorescent lamps, and CRT gutters in vacuum tubes (2–9% Ba)	Causes brain swelling, muscle weakness/twitching, damage to the heart, liver, and spleen though short-term exposure. Low blood potassium, respiratory failure, dysfunction paralysis, high blood pressure

(continued)

Table 1.2 (continued)

Lithium (Li) and cobalt (Co)	LiON (23–25% Li, 2–3% Co), NiCd, NiMH rechargeable batteries (SmCo, NdFeB), LiPF$_6$ electrolyte, LiCoO$_2$, CRTs, and PCBs	Nausea, diarrhea, dizziness, muscle weakness, fatigue, and a dazed feeling. Fine tremor, frequent urination, and thirst. Affect gastrointestinal and neurological system. Co is absorbed by the skin and the respiratory tract. Co itself may cause allergic dermatitis, rhinitis, and asthma
Zinc (Zn)	CRTs (30–35%), metal coatings, batteries. Plating/galvanizing for steel parts	Increased risk of Cu deficiency (anemia, neurological abnormalities)
Phosphors (P)	CRT	White phosphorus is widely poisonous, and exposure to it will be fatal. Too much phosphorous can cause kidney damage and osteoporosis. Phosphors can cause skin burns
Chlorofluorocarbon (CFC/HCFCs)	Cooling units, insulation foams	Deleterious effect on the ozone layer, increased incidence of skin cancer, and/or genetic damages
Polychlorinated biphenyls (PCB)	Condensers, transformers, older capacitors, heat transfer fluids	Cancer, effects on the immune systems, reproductive system, nervous system, endocrine system, and other health effects. Liver damage
PBDEs, PBBs; PBDD/Fs; PCDD/Fs	Flame retardants in plastics, PCBs, plastic housings, keyboards, and cable insulation	Hormonal effects, under thermal treatment possible formation of dioxins and furans
Polyvinylchloride (PVC)	Insulation for cables, plastic parts, computer housings and monitors	Carcinogenic furan and dioxins when burned which are highly persistent in the environment and toxic even in very low concentrations. Reproductive and developmental problems. Immune system damage. Interference with regulatory hormones. HCl can cause respiratory problems.
1,2-Ethanediol	Al electrolytic capacitor (AEC) electrolyte primary solvent or additive	Harms the liver, kidney, and central nervous system, causes kidney failure and brain death

(continued)

1.2 Fundamentals of E-Waste and E-Waste Recycling

Table 1.2 (continued)

Methyl cellosolve	Al electrolytic capacitor (AEC) electrolyte primary/secondary solvent	Damages the nervous system, blood, bone marrow, kidney, and liver as well as could lead to a reproductive and developmental toxicity
Useful material (Harmless)	Material and components usage in EEE	Elements detrimental to recycling (reduce the recycling value)
Tin (Sn)	Solder, Pb-free solders, ITO in LCDs, PV cells, miniaturized capacitors	
Copper (Cu) and CuO	Wires, cables PCBs, coils, diodes	Hg, Be, _(As, Sb, Ni, Bi, Al)_
Aluminum (Al)	Heat sinks, capacitors, cables, casings	Cu, Fe, _(Si)_
Iron (Fe), ferrous metal	Steel chassis, casings, fixings, magnets, magnet coils	_(Cu, Sn, Zn)_
Magnesium (Mg)	Casings, body of cameras	
Silicon (Si)	Glass, transistors, ICs, PCBs	
Gold (Au)	Connector plating, pins, contacts, EPROM, ROM, RAM, CPU, IC, transistors, diodes, switches	
Silver (Ag)	Batteries, solders, switches, relays, USPs, stabilizers, contacts, capacitors, RFID chips, PV cells	
PGM-palladium (Pd)	Multilayer capacitors, connectors, transistors, diodes, Ag-Cu-Pd solder	
Ruthenium (Ru)	HD, plasma displays	
PGM-platinum (Pt)	HD, thermocouples, fuel cells, sensors, switching contacts	
Germanium (Ge)	The 1950s and 1960s transistors, semiconductors, diodes, PV cells, glass fiber	
Tantalum (Ta)	Capacitors	
Tellurium (Te)	Thin film PV cells, photoreceptors, photoelectric devices	
Gallium (Ga)	ICs (GaAs), PVs (GaN, InGaN, GaAs, Cu(In, Ga)Se$_2$), LED (GaN, InGaN, GaAsP, GaAs, GaP, AlGaInP), LED (AlGaInP), photodetectors, PV cells, integrated switches	

(continued)

Table 1.2 (continued)

REE-neodymium (Nd)	HDD drive magnet, microphones, speakers
REE-cerium oxide (CeO$_2$/Ce$_2$O$_3$)	CRT faceplate glass, fluorescent lighting, LEDs
REE-yttrium (Y), terbium (Tb), europium (Eu)	Used as phosphors in display units (PV panels, fluorescent lamps (5–7% Y), CRTs (15–20% Y) (Y$_2$O$_3$, Eu$_2$O$_3$, Tb$_2$O$_3$), LEDs)
REE-lanthanum (La)	NiMH batteries (La$_2$O$_3$)
REE-gadolinium (Gd) and yttrium (Y)	White LEDs
Sulfur (S)	Pb-acid battery
Carbon (C)	Steel, plastics, resistors. In almost all electronic equipment
Bismuth (Bi)	Solders, capacitors, heat sinks
Niobium (Nb), nickel (Ni)	Gold coatings, super alloy magnets, capacitors, Nb steel alloys
Tungsten (W)	Tungsten carbide, electrodes, cables, CRT tubes

1.3 History and Cornerstones of E-Waste Problem and E-Waste Recycling

A number of efforts have been launched to solve the global e-waste problem. The efficiency of e-waste recycling is subject to variable national legislation, technical capacity, consumer participation, and even detoxification. E-waste management activities result in procedural irregularities and risk disparities across national boundaries. The best option in a waste treatment is recycling. WPCBs are key components of the WEEE recycling process which is becoming one of the most attractive and profitable market. The E-scrap is growing every year following the compulsory practices required by the current environmental protection rules and laws. In this context, versatile and flexible technologies should be used in order to develop e-waste recycling machines. The best e-waste recycle plants recover the most of metals and PMs by separating from nonmetals. The careful design of the machines allows to reduce drastically the loss of PMs making the business profitable and attractive. Equipment should be compact and with reduced initial investment. The layout of the WEEE recycling machinery is custom designed for getting the capacity required by the end user in accordance with type of WEEE to be processed.

In the past few decades, recycling of WEEE by environment-unfriendly, hazardous, and crude primitive technologies in China, India, and some African countries has increased. Massive amount of dumping or open burning of WEEE took place. Crude primitive recycling (open burning for Cu and Fe metal recovery and acid

1.3 History and Cornerstones of E-Waste Problem and E-Waste Recycling

Fig. 1.3 WPCB recycling flow sheet

washing for PM recovery) is common in China, India, and Nigeria. Figure 1.3 shows crude WPCB recycling flow sheet. Generally, poor women, children, and prisoners work at primitive crude e-waste recycling jobs. Guangdong Province of South China and Zhejiang Province of East China are becoming two regions with the most intensive unregulated e-waste recycling activities [1, 2]. The researchers found that PBDE concentration was very high in Guiyu; meanwhile, the level of PCDD/Fs (65–2760 pg/m^3) was also reported in Guiyu. On account of all these adverse consequences, primitive crude WPCB recycling operations are regarded as illegal

Fig. 1.4 Illegal primitive WPCB recycling operation photographs in China and India

and therefore supposedly prohibited by domestic laws and regulations [3]. Thereby, it is necessary and urgent to explore and develop a proper metal and nonmetal recovering technology from WPCBs without adverse effects to the environment and human health.

Environmental pollution from primitive crude recycling methods of WPCBs: (a) recovering PMs using acid washing; (b) WPCBs dismantling without any protection; (c) open burning of WPCBs [4]. Figure 1.4 shows primitive, illegal, and environment-unfriendly WPCB recycling photographs from China and India.

Resource reutilization and safe disposal of metals from WEEE have a great environmental protection. In the initial stage of WEEE recycling, simple and rough recovery processes were widely adopted. Manual dismantling, wet gravity separation (i.e., shaking tables), strong acid leaching, etc. were the first crude WEEE recycling lines. These methods both have low recovery rates and damage the human

1.3 History and Cornerstones of E-Waste Problem and E-Waste Recycling

Table 1.3 International legislation and initiative programs combating e-waste problem

Participating countries	Governing body	Initiative	Date adopted
167 countries of the UN (excluding Afghanistan, Haiti, and the USA)	UN Environmental Programme	Basel Convention (No transboundary movement of hazardous waste)	1989
The EU	EUREKA Project (EU1140)	A comprehensive approach for the recycling of electronics (CARE) "Vision 2000"	1994
Japan	Ministry of the Environment, Government of Japan	Home Appliance Recycling Law (HARL) (Home appliance manufacturers must take back and recycle end user products)	2001
The USA	Silicon Valley Toxics Coalition		2002
International	Basel Convention	Mobile Phone Partnership Initiative (MPPI)	2002
The EU	Recycling WEEE	2002/96/EC	2002
The EU	European Parliament	RoHS Directives (Restriction of Hazardous Substances). This Directive aims to limit the use of six hazardous substances (Pb, Hg, Cd, Cr^{6+}, polybrominated biphenyls, polybrominated diphenyl ethers) and applies to all new products put on European market, whether imported or manufactured in the EU. Lead-free solder use	2003
The EU	European Parliament	WEEE directives (Reuse/recycle electronic parts: manufacturers internalize take-back/recycling costs) (Recycling rate 25–40%)	2003–2007
International	G8 (UNCRD)	3Rs Initiative	2005
Intended for OECD countries	Organization for Economic Cooperation and Development (OECD)	Environmentally Sound Management of Water: Reclaim E-Waste	2003
International	UNU-IAS SCYCLE	Solving the E-Waste Problem (StEP)	2007
International	Basel Convention	Partnership for Action on Computer Equipment (PACE)	2008
The EU	European Parliament	HydroWEEE-231962, (Innovative Hydrometallurgical Processes to recover metals from WEEE including lamps, CRTs, PCBs, and batteries)	2009
The EU	ErP Directive		2009

(continued)

Table 1.3 (continued)

Participating countries	Governing body	Initiative	Date adopted
		Energy-related products (2009/125/EC)	
China	Ministry of Industry and Information Technology (MIIT)	China RoHS 1 and RoHS 2 Reduction of Pb, Hg, Cd, Cr, and BFR	2006–2016
China	Ministry of Commerce	Circular economy promotion law	2008–2009
China (Guiyu city)	Ministry of Environmental Protection	A pilot program of an integrated e-waste recycling industrial park (2500 acres)	2010–2015
The USA	National Strategy for Electronic Stewardship (NSES)	Improve design of electronic products	2011
India	Ministry of Environment and Forestry	E-waste management and handling, similar to EU WEEE and EPR directives	2011–2012
The EU	European Parliament	Better regulate WEEE generation and disposal. Encourage the reuse and recycle metals and resins (2012/19/EU)	2012
The EU	European Parliament	RECLAIM (Reclamation of Gallium, Indium, and Rare-Earth Elements from photovoltaics, Solid-State Lighting, and Electronics Waste)	2013
The EU	European Parliament	SCARE (Strategic Comprehensive Approach for Electronics Recycling and RE-use)	2013
The EU	European Parliament	TV Target (Eco-efficient Treatment of TV Sets and Monitors)	2013
The EU	European Parliament	CONCEERN - CONex Central European Electr(on)ics Recycling Network	2013
The EU	European Parliament	ReLCD (Liquid Crystal Display Re-use and Recycling)	2013
The EU	European Parliament	MobileRec (Collection, Disassembly, and Recycling of Mobile Telecommunication Equipment)	2013
The EU	European Parliament	Demonstration Plant for the Economic Disassembly of Printed Circuit Boards	2013
The EU	European Parliament	AREP (Advanced Recycling, Recovery, and Reuse)	2013
The EU	European Parliament	European Innovation Partnership (EIP) on Raw Materials	2013

(continued)

1.3 History and Cornerstones of E-Waste Problem and E-Waste Recycling

Table 1.3 (continued)

Participating countries	Governing body	Initiative	Date adopted
The EU	EU Member States	ERA-MIN roadmap coordination of RM-related research funding between EU Member States	2013
The UN	The UNEP	StEP E-Waste Academy (Solving the E-Waste Problem)	2013
UNEP	UNEP Resource Panel	www.unep.org/resourcepanel with new report "Metal Recycling: Opportunities, Limits, Infrastructure"	2013
The UK	WRAP-ESAP	EEE sustainability action plan	2014
The UK	The UK	CLEVER (Closed Loop Emotionally Valuable E-Waste Recovery)	2013–2016
African countries	Nigeria, Kenya, S. Africa, etc.	Poor regulatory frameworks and poor policing of the industry. Illegal e-waste trade	

health and environment. At the beginning of the 1970s, physico-mechanical separation methods began to be utilized to treat the WEEE [5]. Table 1.3 summarizes international legislation and initiative programs combating e-waste problems. Transboundary movement of hazardous waste has been banned by Basel Convention since 1989. The Basel Convention entered into force in 1992. Japan, the EU, and OECD countries adopted laws/directives related to e-waste between 2001 and 2016. Key features of the Basel Convention were the reduction of hazardous waste (excluding radioactive waste), the generation and promotion of sustainable handling techniques at the place of disposal, the restriction of transboundary movements of hazardous wastes, and the formation of legislation frameworks where transboundary movements are permissible. The EU's WEEE directive promotes reuse compared to recycling. Key features were free take-back schemes for consumers and to reduce WEEE to landfills through collection and recovery targets, WEEE to be collected separately from other wastes, and to stimulate "eco-design" – product design that reduces WEEE and increases its ease of recovery and producer responsibility for end-of-life treatment of their products. Key features of StEP initiative were research and piloting on e-waste treatment, strategy and goal setting to eliminate the e-waste problem, training and development on e-waste issues, and establishment of communication and awareness among members and throughout the industry. G8 countries agreed to establish the 3R (reduce, reuse, and recycle) initiative in Japan in 2005. The key features were promoting 3Rs and building a "sound-material-cycle society." In 2011 the USA implemented the "National Strategy for Electronic Stewardship (NSES)" as the bases for improving the design of electronic equipment and enhancing the management of used or discarded electronic items.

Fig. 1.5 Error pyramid for sample analysis

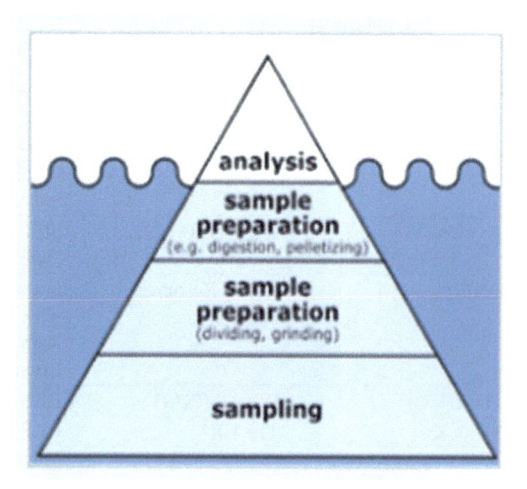

As can be seen from Table 1.3, there is a considerable divide between developed and developing countries in international legislation and initiative programs combating e-waste management.

1.4 Representative Sample Preparation and Analysis

E-waste and WPCBs are very complex and heterogeneous materials. Representative analysis results require adequate sample preparation. Figure 1.5 shows the error pyramid for sample analysis. In an analog to an iceberg, where the greatest part is under water, only a small part of the sum of errors is perceived, whereas the major part of potential errors is not taken into account. Sampling and sample preparation are done in a traditional way which has become a routine over the years and is no longer regarded as having a critical influence on the subsequent analyses. Moreover, the errors of each step add up, i.e., the error increases during the process (error propagation). The question is now how these errors occur and what can be done to minimize them.

Sampling is difficult for e-waste composition analysis due to the inhomogeneous and composite nature of the materials. Large numbers and kinds of small components are attached to the PCBs. Generally, PCBs are crushed into small sizes (less than 1–2 mm). After shredding and fine grinding, fine powder can be obtained (Fig. 1.6). Using representative samples taken after shredding and pulverizing, chemical analysis can be conducted using acid/aqua regia (AR) leaching, loss on ignition (LOI), and inductively coupled plasma-atomic emission spectroscopy (ICP-AES).

Figure 1.7 shows the main materials found in WEEE. Of course, chemical compositions of different WPCBs are quite different. Metals, especially Fe, Cu, and Al, form most of the material weight of today's e-waste. Ferrous (48%) and

1.4 Representative Sample Preparation and Analysis

Fig. 1.6 WPCB before grinding and after shredding and sieving from 4 mm, fine grinding and sieving from 0.5 mm, and fine grinding and sieving from 0.25 mm

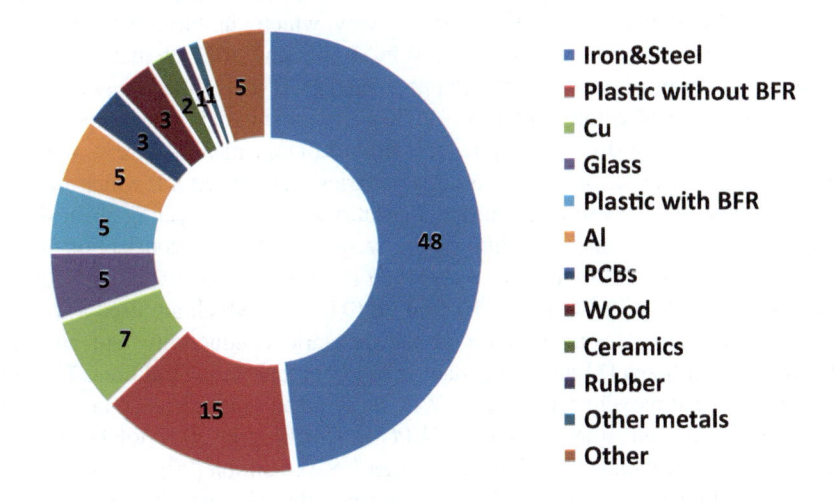

Fig. 1.7 Percent main materials found in electrical and electronic equipment (EEE)

nonferrous (12%) metallic (Cu, Al) substances are about 60%; organic materials (plastics 20%, rubber 1%, and wood 3%) are 24%; glass is 5%; PCBs are 3%; etc [16]. Generally, metals in e-waste can be grouped into BMs (Cu, Al, Ni, Sn, Zn, and Fe), PMs (Au, Ag), PGMs (Pd, Pt, Rh, Ir, and Ru), metals of concern/hazardous (MCs) (Hg, Be, In, Pb, Cd, As, and Sb), and special metals/scarce elements (SEs) (Te, Ga, Se, Ta, and Ge) (Table 1.4). Acrylonitrile butadiene styrene (ABS),

Table 1.4 Classification of metals in WEEE

Precious metals (PMs)	Au, Ag
Platinum group metals (PGMs)	Pd, Pt, Rh, Ir, and Ru
Base metals (BMs)	Cu, Al, Ni, Sn, Zn, and Fe
Concern/hazardous metals (MCs)	Hg, Be, In, Pb, Cd, As, and Sb
Special/scarce metals (SEs)	Te, Ga, Se, Ta, and Ge

polycarbonate (PC), high-impact polystyrene (HIPS), and polyphenylene oxide blends (PPO) are major different polymers found in WEEE [17]. Most of the plastics contain BFRs and make recycling difficult. Electric cables typically consist of Cu covered with polyvinyl chloride (PVC). PVC contains about 57% Cl and additives such as plasticizers, stabilizers, flame retardants, and fillers. Thermal treatments of PVC suffer from the following drawbacks: extra energy for heating is necessary, and acidic HCl gas is emitted, which has to be captured by absorbents such as Ca(OH)$_2$ to prevent damage to the industrial equipment. If the temperature is low, there is the additional risk of producing chlorinated organic such as dioxins [18].

WPCBs, which are resource-rich, are generally referred to as "urban mines." This term is increasingly used by authorities and scholars to describe the recovery/reclaim of resources/raw materials from the waste streams/spent products generated within urban environments. The major difference between urban mining and conventional mining is the source of materials for recovery which, in turn, determines the appropriate mining process. Typically, WPCB contains 40% of metals, 30% of organics, and 30% of ceramics. Bare/unpopulated PCB assemblies represent about 23% of the weight of whole WPCBs [19]. There is a great variance in composition of WPCB coming from different appliances, from different manufacturers, and from different ages. After removing hazardous batteries and capacitors, which must be recycled separately according to current legislations, WPCB recycling chain should start. At a rough estimate, one-third of the weight of WPCBs consists of metals, mainly Cu (~16%), Sn (~4%), Fe (~3%), Ni (~2%), and Zn (~1%). In addition, PMs, like Au (0.039%), Ag (0.156%), and Pd (0.009%), which are used as contact materials or plating layers because of their electric conductivity and chemical stability are ten times [8] more abundant in WPCBs than in natural ores. Thereby, it is obvious that recycling PMs from WPCBs is greatly significant. For example, recycling of Au and other PMs from WPCBs in China in 2007 totaled US$2.6 billion, with other metals contributing a further US$0.4 billion [20]. However, these recycling activities are dominantly occurring in informal recycling sector in the world without pollution controls. From the sustainable viewpoint, the recycling of WPCBs becomes more and more attractive because of the low ore concentration, difficult mining conditions, and other factors [20]. E-waste contains valuable resources which offer opportunities for urban mining and new job creation. In the world, the Environmental Protection Agency (EPA) and UN estimate that only 15–20% of e-waste is recycled and the rest of these consumer electronics go directly to landfills and incineration [21]. The vast majority of WPCBs (about 85%) are consigned to landfill either directly or within the original equipment in 2014. This represents a non-sustainable loss of finite material sources and is placing a dramatically increasing burden on landfill [17]. According to Assocham report in India, only 1.5% of total e-waste is recycled by formal recycling sector in an environment-friendly way. The rest is recycled by informal recycling sector in India [22].

Recycling PCBs and electronic components (ECs) salvage expensive PMs and scarce rare earths, which are critical to the EC industry. PMs have recycling rates above 50% due to intrinsic values. But, extraction of rare-earth and other materials from certain devices is not only uneconomic; it is also exceptionally difficult. White LEDs, for example, contain In, Ga, Ce, Eu, Gd, and Y in miniscule amounts – all of which are exceedingly difficult to extract from such components. However, given that more than 25 billion white LEDs were produced in 2011 alone (according to the National Academy of Sciences assessment on advanced solid-state lighting), these tiny amounts quickly add up to very large amounts of rare elements that are largely unrecovered and unrecoverable. In other words, many tons of rare-earth materials simply end up in landfill every year. This is not to say that research isn't being conducted by forward-thinking organizations [23].

As a matter of fact, the e-waste recycling and management are not simple and straightforward at present [24]. The "mineralogy" of WPCBs is much different compared to the natural ores for metal refining: [20] first, up to 60 different elements are closely interlinked with complex assemblies and subassemblies [24, 25], such as Cu, Fe, Al, Pb, Sn, etc., as well as PMs, whose physical and chemical properties are much different; second, the metals contained in WPCBs usually cross-link to organic plastics which are usually toxic and have the potential to bioaccumulate, such as BFRs, polychlorinated dibenzo-p-dioxins and dibenzofurans (PCDD/Fs), polybrominated diphenyl ethers (PBDEs), chlorinated dioxin, and polycyclic [26] During unregulated disposal activities, these hazardous substances are usually released, resulting in an extreme damage to the environment and human health [9, 27]. Hence, the composition of WPCBs is much more complex than natural ores that make processes for recycling valuable metals more complex, resulting in efficient and environmentally sound processing of the WPCBs requiring special attention [20].

1.5 Life Cycle of Electronics and E-Waste

The short useful life expectancy of electronic products, driven by rapid innovation, miniaturization, and affordability, has led to a major increase in the accumulation of toxic e-waste. Old/used electronics do not die; they pile up in attics/backyards at homes or e-waste graveyards in China, India, Brazil, Nigeria, or many other countries all over the world (Fig. 1.8).

Figure 1.9 shows the life cycle of electronics from postproduction to disposal. WEEEs are produced by manufacturers, sold by retailers, used by consumers, and collected by recyclers. Proper e-waste management for all countries is necessary, because e-waste pollutes the ground water, acidifies the soil, generates toxic fume and gas after burning, accumulates fastest in municipal disposal areas, and releases

Fig. 1.8 E-waste mountains in graveyards [28]

carcinogenic substances into the air, if it is not properly managed. If WEEE can be leached to the environment, due to moisture, oxidation or raining can generate serious environmental problems. For a proper e-waste management, waste prevention/reduction lowers the amount of waste and conserves scarce resources; minimization reduces material usage, and reuse uses materials/parts repeatedly. They are the most favored options and are on top of the e-waste hierarchy pyramid. Burning (i.e., incineration or pyrolysis) for energy recovery prior to disposal and safe disposal by landfilling are the least favored options in the e-waste management pyramid. Disposal does not conserve any resources. Recycling e-waste is an intermediate polymetallic secondary resource recovery option. Principles and practices of waste management hierarchy are given in Fig. 1.10. Open dumping is the most common form of e-waste disposal in most underdeveloped/developing countries. Burial or landfill disposal allows heavy metals to be leached into the groundwater or methane off gassing. Combustion of organic substances in waste by incineration makes hazardous material airborne and generates ashes and heat. Leaching of the ashes

1.5 Life Cycle of Electronics and E-Waste

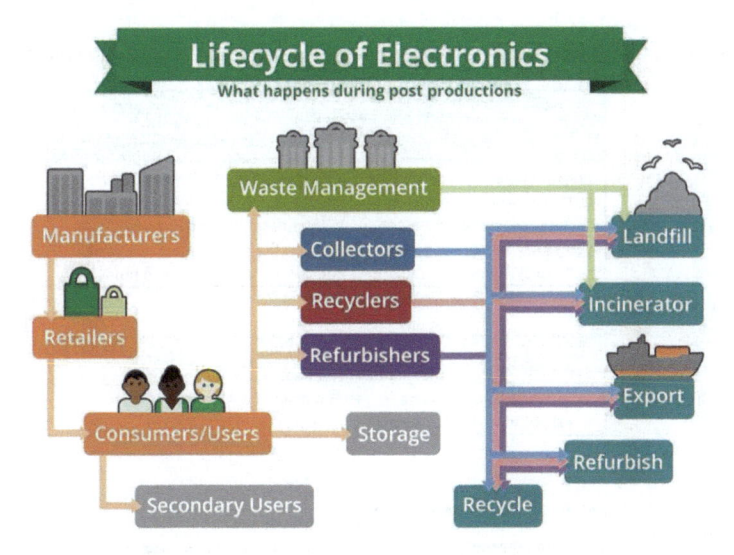

Fig. 1.9 Life cycle of electronics from postproduction to disposal [16]

may cause water and soil contamination. E-waste constitutes about 40% of Pb and 70% of heavy metals in landfills [7, 29].

In landfill, e-waste is placed in a hole, compacted, and covered/buried with soil. It reduces the amount, lessens the danger of fire, and decreases the bad odor. A double liner system (i.e., compacted plastic clay and plastic geomembrane liner) at the bottom prevents liquid waste from seeping into the groundwater and collects leachate to seep through the solid waste. Improper treatment of e-waste generates serious soil, air, and water pollution problems. Globally, exporting any e-waste to underdeveloped and developing countries is prohibited by the Basel Convention in 1989 [30] Landfill is a WPCB treatment method with a long history and wide application worldwide because of simplicity in operation. However, the land for WPCB disposal is thereby regarded as wasteland and normally incapable of being exploited again in predictable decades, which is not suitable for countries/communities where there is a lack of land space. In addition, the environmental issues aroused by WPCB landfilling such as the leachate formation in landfill sites and evaporation of hazardous substances raise safety concerns. It is reported that almost 70% of the HM (i.e., Hg, Cd, Pb, etc.) in landfill sites come directly from WEEE, which indicates that a considerable part of heavy metals actually come from WPCB. Concurrently, the co-landfilling of WPCB with various kinds of municipal solid waste makes the formation of leachate containing brominated toxic compounds and HMs possible due to the reactions during the landfilling process. The synergic pollution will contaminate the atmosphere, soil, and groundwater, and this has been proven by several research groups.

Current improper e-waste handling includes:

- Open burning of WPCBs and cables for metals
- Acid/cyanide stripping of valuable metals
- CRT cracking and dumping

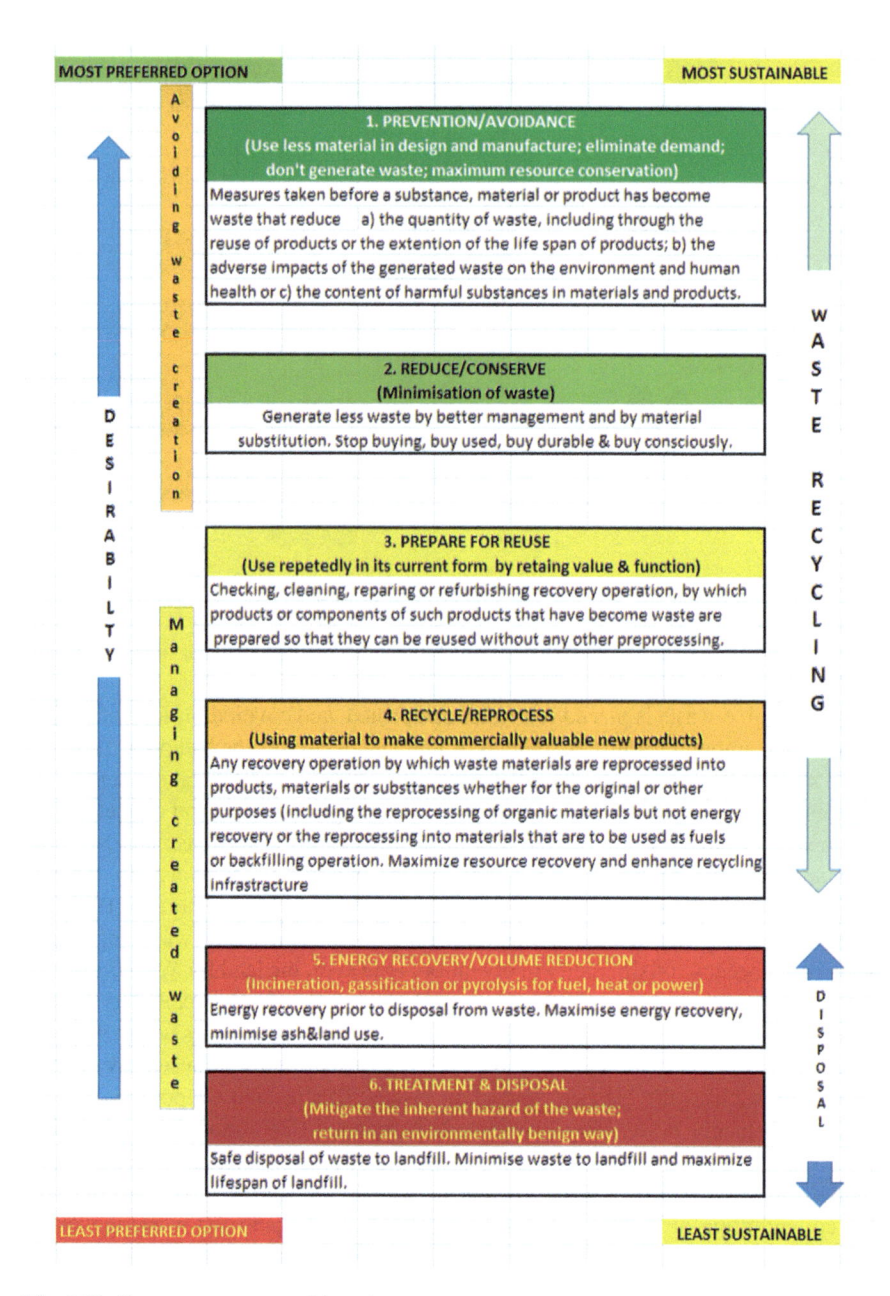

Fig. 1.10 E-waste management hierarchy

1.6 Life Cycle Assessment (LCA) and Life Cycle Management (LCM)

The challenge of sustainable development at the beginning of the twenty-first century has become a systematic one with environmental, social, and economic dimensions on an equal footing. Promotion of resource efficiency improved material recycling and life cycle thinking. This part provides unrivaled science to inform policymakers about how the recycling of metals can be optimized on an economic and technological basis along product life cycles in the move toward sustainable metal management.

The focus needs to be on optimizing the recycling of entire products at their EoL instead of focusing on the individual materials contained in them. The global mainstreaming of a product-centric view on recycling will be a remarkable step toward efficient recycling systems, resource efficiency, and a green economy in the context of sustainable development and poverty eradication. BATs should be used for the separation and recovery of different components of a product.

LCA is a tool/approach that facilitates understanding and quantification of the ecological/environmental and human health impacts of a product, service, or system over its complete life cycle. It must be linked to rigorous simulation tools to quantify resource efficiency. There are currently two standards that specify LCA:

- ISO 14040: It defines the principles and framework of LCA.
- ISO 14044: This standard further clarifies the requirements for LCA and guidelines for conducting a LCA.

Life cycle management (LCM) is a business approach that can help companies in achieving sustainable development. It helps reducing, for instance, a products' carbon, material, and water footprints, as well as improving its social and economic performance. LCM is about making life cycle thinking and product sustainability operational for businesses that aim for continuous improvement.

1.6.1 LCA Principles

An Approach Based on the Life Cycle The LCA is based on several principles. Firstly, it is an approach based on the life cycle. It is necessary to integrate all stages of product life. We generally distinguish the following phases (Fig. 1.11):

- The extraction of raw materials
- The manufacture of the product
- The distribution
- Use of the product
- The EoL: recycling, incineration, landfill, etc.

Fig. 1.11 Life cycle analysis phases

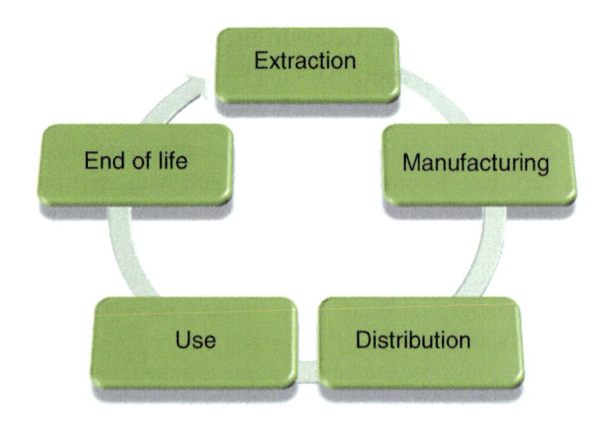

There are also all transport stages that occur during the life cycle of the product. This life cycle-oriented approach allows a systematic approach and avoids any transfer of impact of a life cycle phase to another [31].

An Environmental Approach The second principle is that LCA is an environmental approach. This approach focuses on the environmental impacts of a product; therefore, the social or economic impacts are not considered in this approach.

A Relative Approach It is also important to know that LCA is a relative approach. Indeed, first of all it is necessary to establish the functional unit to be used for LCA. This functional unit quantifies the service provided by a product. An example of functional unit of a mobile phone is "using a mobile phone for 11 minutes per day over a period of 2 years." The notion of functional unit makes the results of a LCA relative, there are no absolute results, and comparison of results of two LCAs can only be done on the basis of an identical functional unit.

A Transparent Approach Transparency is also one of the essential principles when conducting a LCA, to ensure proper use of data and a correct interpretation of results.

A Scientific Approach Conducting a LCA is primarily based on a scientific approach. Finally, the analysis of life cycle is based on the principle of completeness, that is to say that this is a transversal approach taking into account the largest possible number of environmental aspects (impact on air, water, soil, etc.)

1.6.2 The Several Steps of LCA

A life cycle analysis has four stages:

- The definition of the goal and scope of the study: This is during the phase that is defined the purpose and scope of the study, the functional unit, if it is a comparative study or the recipients of the study (industrial, general public, other).
- The inventory phase of life cycle: This is the most important phase of the implementation of a LCA. This is to collect all the information on inflows and outflows on the life cycle of the product. Inflows are raw materials, energy use, etc. Outflows related emissions to air, water, soil, or the production of recycled materials.
- Impact assessment: Once the inventory is completed, the next step is to assess the impacts for all environmental impacts selected for the study associated with each flow listed during the inventory. The assessment of these impacts can be done through particular characterization models.
- Result interpretation: Once the impact assessment is carried out, the last phase of LCA is the interpretation of results. This interpretation is to present the results of the LCA in accordance with the objectives of the study. The interpretation may result in recommendations for the use of the product or ways for redesign to reduce potential environmental impacts of the product studied. During interpretation, it is sometimes possible to distinguish the phase of the life cycle that has the greatest impact or determine the origin of the most significant environmental impacts.

1.6.3 LCA Tool

There are currently a very large number of life cycle analysis tools. Some of these softwares include databases for performing the life cycle inventory. These programs also offer assessment methods to classify and characterize the impacts on a number of indicators. These programs include the following:

- SimaPro – Edited by the PRe Consultants http://www.pre.nl/
- GaBi – Edited by the PE International www.gabi-software.com
- Umberto – Edited by IUF Hamburg http://www.umberto.de/en/
- EIME – Edited by the Bureau Veritas CODDE http://www.codde.fr/
- TEAM – Edited by the ECOBILAN http://ecobilan.pwc.fr/en/boite-a-outils/team.jhtml
- SIEC – http://www.acv-siec.fr/siec/index.php
- ecoinvent, www.ecoinvent.ch

Some software is dedicated to specific sectors such as construction (Eco-Bat, Equer, etc.). There are several regulatory texts on EEE. These texts come from the European law. Among these, some relate specifically to WEEE and waste management (collection targets, recycling). Others don't affect only EEE or WEEE but control the use of certain substances or even set up eco-design rules for EEE. All these texts are aimed at preventing and reducing risks to human health and environment.

1.7 Objectives of WPCB/E-Waste Recycling Opportunities

Recycling can greatly increase the amount of available metals and nonmetals for society, provided the potential sources and recycling technologies exist. One of the most promising recycling sources is WEEE, containing many of the high-tech and high-value metals with rising demand. Recycling WPCBs are salvaging valuable metals. The main objectives to achieve are:

- Take care of harmful/toxic substances contained in e-waste in an environmentally sound manner while preventing secondary and tertiary emissions.
- Recover valuable raw materials (metals and nonmetals) as effective as possible.
- Create economically and environmentally sustainable businesses (optimize eco-efficiency).
- Consider the social implications and local context of operations (e.g., new employment opportunities, available skills and education, etc.).

The benefits of recycling are potentially very significant: it reduces the future scarcity of some high-demand elements, creates economic value, reduces greenhouse gas emissions, and limits other environmental harms. It forms a vital part of the transition to a green economy, where societal progress is decoupled from unsustainable natural resource depletion, and it provides a source of metals in high demand for sustainability-enabling technology, reducing the risk of metal scarcity, particularly of metals needed for transition to a green economy [32].

1.8 General Driving Force for E-Waste Processing

In order to achieve the call "from waste to resources," integrated waste policy and management, which address environmental impacts along the whole life cycle of products, materials, and processes, are crucial. Primary production (conventional mining) plays the most important role in the supply of metals for EEE applications since secondary metals (recycle) are only recovered in limited quantities so far. Environmental impact/footprint of primary metal production is significant, especially for PMs and SMs which are low in concentration in natural ores. Considerable amounts of land are used in mining, wastewater and $SO_{2(g)}$ are created, and the energy consumption and CO_2 emissions are large. According to 3R principle (reduce, reuse, and recycle), recycling reduces waste going to final disposal, decreases consumption of natural resources, and improves energy efficiency. The treatment of e-waste may be considered as an integration of a generic waste treatment hierarchy within which economic, legislative, and technology drivers determine the structure and methodology of approach. The deployed in order of increasing environmental impact may be considered as reduction, reuse, recovery, recycle, and disposal [17]. There are three general reasons for e-waste processing: environmental concerns, energy savings, and resource (material and water)

1.8 General Driving Force for E-Waste Processing

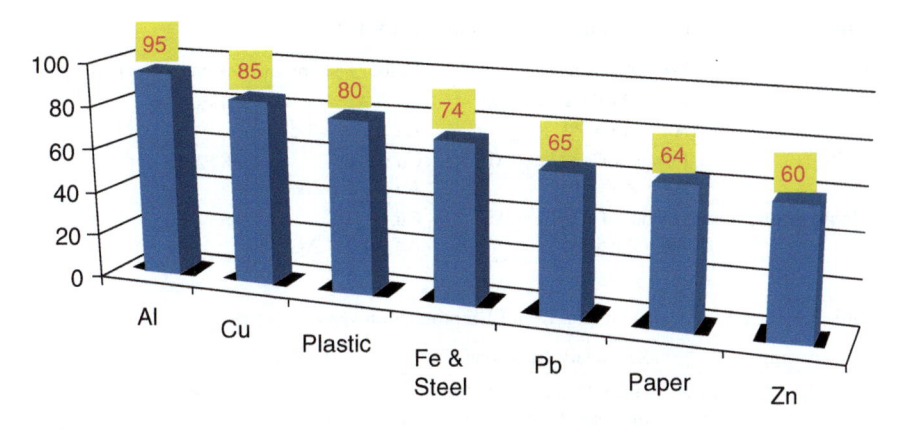

Fig. 1.12 Recycled material energy savings over virgin materials [34]

conservation efficiency. Altogether, metal production today represents about 8% of the total global energy consumption and a similar percentage of fossil fuel-related CO_2 emissions [32].

Improved recycling processes can be much cheaper than primary production, primarily because they can use much less energy in production of the metal. The energy savings for a number of common metals and materials are summarized in Fig. 1.12. About 60–95% energy savings are possible with WEEE recycling. Energy savings are more than 80% for Al, Cu, and plastics. Recycling often only involves the remelting of metals. Benefits of scrap Fe and steel usage are savings in energy (74%), savings in raw materials use (90%), reduction in air pollution (86%), reduction in water use (40%), reduction in water pollution (76%), reduction in mine wastes (97%), and reduction in consumer wastes generated (100%). Without any loss of performance, Cu is 100% recyclable. The recycling of Cu requires up to 85% of less energy than primary production. The recycling of old and new scrap produces 9 million tons of Cu per year. Around the world, Cu recycling only saves 40 million tons of CO_2 [33].

Moreover, processing of e-waste will reduce burden on mining ores for primary metals. Therefore, scarce resources especially for PMs could be conserved, e.g., metals that exist at low concentrations in primary ores and consume significant energy during extraction. Factually, e-waste is a rich source of PMs compared to their primary ores. The amount of Au recovered from 1 ton of e-waste from PCs is more than that recovered from 17 tons of Au ore. Generally, 3 g of Au can be extracted from 1 ton of average natural Au ore; but 300 g Au can be extracted from 1 ton of mobile phone recycling [35]. The processes for recovering PMs from e-waste, in limited cases, are easier than their primary ores. If PMs and SEs are unrecovered, it will be a significant loss of precious scarce resource. The recoveries of PMs and BMs are important for e-waste management, recycling, sustainability, and resource conservation. Weight distribution share of plastics, Cu, Fe, and Al are high in WEEE. Value share changes Au > Ag > Pd. The value distribution of PMs in

Table 1.5 Weight and value distribution of some WEEE [34]

WEEE	Weight % share	Value % share	Sum of precious metals
TV boards	Fe = plastic > Cu = Al Ag > Au > Pd	Cu > Al > Fe Au > Ag = Pd	43%
PCBs	Plastic > Cu > Fe > Al Ag > Au > Pd	Cu > Al Au > Pd > Ag	85%
Mobile phone	Plastic > Cu > Fe > Al Ag > Au > Pd	Cu Au > Pd > Ag	93%

Table 1.6 Levels of toxic emission from EEE production and e-waste recycling

Primary emissions	Secondary emissions	Tertiary emissions
Hazardous substances that are contained in e-waste	Hazardous reaction products for improper treatment (improper incineration and smelting)	Hazardous substances/reagents used during recycling (leaching and amalgamation)
Pb, Hg, As, Cr^{6+}, PCBs, CFCs, etc.	Dioxins, furans, BFRs	Inorganic acids, cyanide, Hg
EU RoHS Directive fully restricts these elements/substances	EU RoHS Directive partly restricts these substances	There are no necessary legislative approaches even in cleanest/greenest inappropriate recycling

mobile phones is more than 85%. After PMs, Cu is the next highest value metal to be extracted from e-waste. It is worth noting that sustainable resource management demands the isolation of hazardous metals from e-waste and also maximizes the recovery of PMs. The loss of PMs during the recycling chain will adversely affect the process economy. The extraction of PMs and BMs from e-waste is a major economic drive due to their associated value, as summarized in Table 1.5.

Calculations of $CO_{2(g)}$ emissions associated with primary metal production (for Cu, Co, Sn, In, Ag, Au, Pd, Pt, and Ru) used for EEE manufacture in 2007 showed total 23.4 and in 2010 33.6 Mt/a [26]. Between 2007 and 2010, 65–73% of this amount comes from Cu, 17–22% from Au, 6.2% from Sn, and 2.1–3.7% from Ag metal production [32, 36]. Cooling and freezing equipment employ ozone-depleting substances (ODS) in the refrigeration systems. These substances, such as CFCs and HCFCs, have a huge global warming potential. It must be mentioned that particularly the older devices contain ODS with a high global warming impact (i.e., CFC 11/12/113/114/115 and HCFC 22/141b); the newer devices use alternative substances (i.e., Halon 1211/2402) [37]. CO_2 emissions reduced 13.2 kg CO_2/kg recovered metal from powders/residues from fluorescent lamps, 19.2 kg CO_2/kg from CRTs, 27.0 kg CO_2/kg from Li-ion batteries, and 25.9 kg CO_2/kg from PCB recycling [38]. On a local-level point of view, uncontrolled discarding or inappropriate waste management/recycling generates significant hazardous emissions, with severe impacts on health and environment. In this context, there are three levels of toxic emissions (Table 1.6).

Some of the collected WEEE can be refurbished or reusable. Recycling can generate some functional ECs for repairment of reusable WEEE. In 2015,

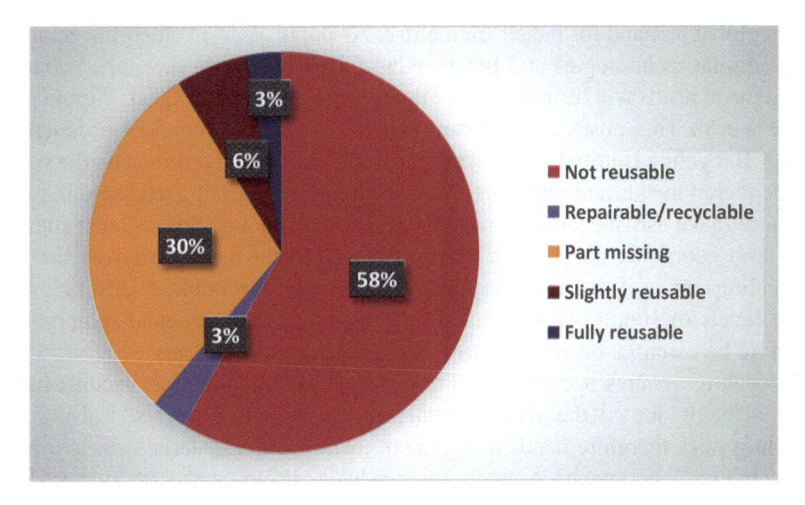

Fig. 1.13 Functionality of collected WEEE

El-Kretsen in Sweden determined the degree of functionality of collected WEEE in 1265 cages. The breakdown is given in Fig. 1.13. About 58% of the collected WEEE is not reusable, 3% repairable, 30% part missing, 6% slightly reusable, and 3% fully reusable [39].

WEEEs usually contain up to 61% different BMs and PMs and up to 21% polymers which renders them attractive as secondary resources to the primary ores [40]. There are 15 different types of plastics are found in e-waste.

1.9 E-Waste Concerns and Challenges

Primary raw ores and secondary metal source of e-waste have different properties for recycling. Natural ores contain oxidized or sulfidized minerals, while e-wastes contain pure metals or metal alloys in their structure. Ores are fragile or brittle and broken easily. But e-waste is electric/ductile material. Metal contents of natural ores (Cu ore, 0.5–1.0%; Au ore, 1–10 ppm) are very low as compared to e-waste (Cu, 20%, and Au, 250 ppm). Primary ores are homogeneous, while e-wastes are heterogeneous and very complex. Ores liberate at fine particle size; but e-waste liberates at relatively coarse particle size. Ores have uniform particle shapes, while e-wastes have rod, plate, or various particle shapes. Natural ores can easily be dissolved by acids or bases; however, e-waste requires oxidizing conditions for dissolution.

The increasing demand for metals in the course of the last century, putting permanent pressure on natural resources, has revealed that metals are a priority area for decoupling economic growth from resource use and environmental degradation. The imperative of decoupling will become even more pressing in the future with a global demand for metals on the rise. As populations in emerging economies adopt similar technologies and lifestyles as currently used in developed countries, global metal needs will be three to nine times larger than all the metals currently used in the world. This poses a significant call for increased secondary production of metals. Recycling is primarily an economic industrial activity, driven by the value of the recovered metals and materials. An infrastructure for optimized recycling would therefore make use of economic incentives. Those economic drivers must align with long-term economic goals, such as conserving critical metal resources for future applications, even if their recovery may be currently uneconomic. Ensuring appropriate levels of supply while reducing the negative environmental footprints will therefore be essential on our way toward a global green economy.

Recycling systems need to adjust to the fact that recycling has become increasingly difficult due to the rising complexity of WPCB products. Raising metal recycling rates therefore needs realignment away from a material-centric toward a product-centric approach. A focus on products discloses the various trade-offs between, for example, achieving weight-based policy targets and the excessive energy consumed in efforts to meet these targets.

The following are the challenges for e-waste recycling:

- There are no accurate figures/estimates of the rapidly increasing e-waste generation, disposal, and imports in the world.
- Low/little level of awareness among manufacturers, consumers, and e-waste workers on the hazards of incorrect e-waste disposal/recycle.
- Major portion of e-waste in the world is processed by the informal/unorganized backyard scrap dealers using rudimentary acid leaching and open-air burning techniques which results in severe environmental damages and health hazards.
- Informal recyclers use vulnerable social groups like women, children, immigrants, prisoners, and immigrant workers for high-risk backyard recycling operations.
- Informal recycling only recovers Au, Ag, Pt, Cu, etc. with substantial losses of material value and resources.

The present book responds to the pressing need to optimize current recycling schemes with the help of a better understanding of the limits imposed by physics, chemistry, thermodynamics, and kinetics, as well as by the technological, economic, and social barriers and inefficiencies encountered. Much is at stake when thinking about how to improve recycling systems: closing loops, reducing related environmental impacts, safeguarding the availability of metals, minimizing metal prices, and promoting meaningful and safe jobs for poor people in developing countries.

References

1. European Parliament (2003) Directive 2002/96/EC of the European Parliament and of the Council of 27 January 2003 on waste electrical and electronic equipment (WEEE). Off J Eur Union L37:24–38

References

2. European Parliament (2012) Directive 2012/19/EU of the European Parliament and of the Council of 4 July 2012 on waste electrical and electronic equipment (WEEE). Off J Eur Union L197:38–71
3. Puckett J, Byster L, Westervelt S, Gutierrez R, Davis S, Hussain A, Dutta M (2002) Exporting harm—the high-tech trashing of Asia; The Basel Action Network (BAN) Silicon Valley Toxics Coalition (SVTC): Seattle, WA, USA
4. http://www.stepinitiative.org/files/step/_documents/StEP_WP_One%20Global%20Definition%20of%20E-waste_20140603_amended.pdf
5. Widmer R, Oswald KH, Sinha-Kheeetriwal D, Schnelmann M, Boni H (2005) Global perspectives on e-waste. Environ Impact Assess Rev 25:436–458. https://doi.org/10.1016/j.eiar.2005.04.001
6. https://i.unu.edu/media/unu.edu/news/52624/UNU-1stGlobal-E-Waste-Monitor-2014-small.pdf
7. Sakunda P (2013) Strategy of e-waste management. http://www.slideshare.net/ketanwadodkar/e-waste-tce-r2?related=2
8. Zhou Y, Qiu K (2010) A new technology for recycling materials from waste printed circuit boards. J Hazard Mater 175(1–3):823–828. https://doi.org/10.1016/j.jhazmat.2009.10.083
9. Kaya M (2016) Recovery of metals from electronic waste by physical and chemical recycling processes. Int J Chem Nucl Mater Metall Eng 10. scholar.waset.org/1307-6892/10003863
10. Hagelüken C (2006) Recycling of electronic scrap at Umicore's integrated metals smelter and refinery. World Metals-Erzmetall 59(3):152161
11. Chancerel P, Meskers CEM, Hagelüken C, Rotter VS (2009) Assessment of precious metal flows during preprocessing of waste electrical and electronic equipment. J Ind Ecol 13:791–810. https://doi.org/10.1111/j.1530-9290.2009.00171.x
12. Chancerel P (2010) Substance flow analysis of the recycling of small waste electrical and electronic equipment—an assessment of the recovery of gold and palladium. Technische Universität, Berlin
13. Wang F, Zhao Y, Zhang T, Duan C, Wang L (2015) Mineralogical analysis of dust collected from a typical recycling line of WPCBs. Waste Manag 43:434–441. https://doi.org/10.1016/j.wasman.2015.06.021
14. Sun Z, Cao H, Xiao Y, Sietsma J, Jin W, Agherhuis H, Yang Y (2017) Towards sustainability for recovery of critical metals from electronic waste: the hydrochemistry processes. ACS Sustain Chem 5:21–401. https://doi.org/10.1021/acssuschemeng.6b00841
15. http://globalsmt.net/articles_&_papers/inemi-report-state-metals-recycling/
16. Kaya M (2018) Current WEEE recycling solutions, Chap. 3. In: Veglio F, Birloaga I (eds) Waste electrical and electronic equipment recycling, aqueous recovery methods, pp 33–93. https://doi.org/10.1016/B978-0-08-102057-9.00003-2
17. Kellner D (2009) Recycling and recovery. In: Hester RE, Harrison RM (eds) Electronic waste management, design, analysis and application. RSC Publishing, pp 91–110
18. Xu J, Tazawa N, Kumagai S, Kameda T, Saito Y, Yoshioka T (2018) Simultaneous recovery of high-purity copper and PVC from electrical cables by plasticizers extraction and ball milling. RSC Adv 8:6893–6903. https://doi.org/10.1039/c8ra00301g
19. Duan H, Hou K, Li JH, Zhu XD (2011) Examining the technology acceptance for dismantling of waste printed circuit boards in light of recycling and environmental concerns. J Environ Manag 92:392–399. https://doi.org/10.1016/j.jenvman.2010.10.057
20. Hagelüken C, Corti CW (2010) Recycling of gold from electronics: cost effective use through "design for recycling". Gold Bull 43:209. https://doi.org/10.1007/BF03214988
21. https://i.unu.edu/media/unu.edu/news/52624/UNU-1stGlobal-E-Waste-Monitor-2014-small.pdf
22. http://www.assocham.org/newsdetail.php?id=5725
23. https://www.eeweb.com/profile/eeweb/articles/recycling-pcbs-part-2-salvaging-precious-metals
24. Kaya M (2016c) Recovery of metals from electronic waste by physical and chemical recycling processes. In: Proceeding Part VII, 18th international conference on waste management, recycling and environment (ICWMRE 2016), Barcelona, Spain, pp 939–950

25. Mesker CEM, Hagelüken C, Van Damme G (2009) TMS 2009 Annual meeting & exhibition, San Francisco, California, USA, EPD Congress 2009 proceedings Ed. by. SH Howard, P Anyalebechi, L Zhang, pp 1131–1136, ISBN No: 978-0-87339-732-2
26. Jonhson J, Harper EM, Lifset R, Gradel TE (2007) Dining at the periodic table: metals concentration as they relate to be recycling. Environ Sci Technol 41(5):1759–1765. https://doi.org/10.1021/es060736h
27. Wang JB, Xu ZM (2015) Disposing and recycling waste printed circuit boards: disconnecting, resource recovery and pollution control. Environ Sci Technol 49:721–733. https://doi.org/10.1021/es504833y
28. https://www.pcworld.co.nz/slideshow/524767/pictures-old-electronics-don-t-die-they-pile-up/
29. Kaya M (2016) Recovery of metals and nonmetals from electronic waste by physical and chemical recycling processes. Waste Manag 57:64–90. https://doi.org/10.1016/j.wasman.2016.08.004
30. http://www.basel.int/Portals/4/Basel%20Convention/docs/text/BaselConventionText-e.pdf
31. http://eco3e.eu/en/base/lca/
32. http://wedocs.unep.org/bitstream/handle/20.500.11822/8423/-Metal%20Recycling%20Opportunities%2c%20Limits%2c%20Infrastructure-2013Metal_recycling.pdf?sequence=3&isAllowed=y
33. Ning C, Lin CSK, Hui DCW (2017) Waste printed circuit board (PCB) recycling techniques. Top Curr Chem (Cham) 375:43. https://doi.org/10.1007/s41061-017-0118-7
34. Khaliq A, Rhamdhani MA, Brooks G, Masood S (2014) Metal extraction processes for electronic waste and existing industrial routes: a review and Australian perspective. Resources 3(1):152–179. https://doi.org/10.3390/resources3010152
35. http://www.recupel.be/en/whats-new/where-does-the-gold-in-my-wedding-ring-come-from/
36. http://www.ewasteguide.info/files/UNEP_2009_eW2R.PDF
37. www.igsd.org/docs/Ozone_and_Climate_FINAL.pp
38. Rocchetti L, Veglio F, Kopacek B, Beolchini F (2013) Environmental impact assessment of hydrometallurgical processes for metal recovery from WEEE residues using a portable prototype plant. Environ Sci Technol 47:1581–1588. dx. https://doi.org/10.1021/es302192t
39. http://www.elkretsen.se/sites/elkretsen_eng/files/media/Dokument/ELK_Functionality_test_2015.pdf?370
40. Tuncuk A, Stazi V, Akcil A, Yazici EY, Deveci H (2012) Aqueous metal recovery techniques from E-scrap: hydrometallurgy in recycling. Miner Eng 25:28. https://doi.org/10.1016/j.mineng.2011.09.019

Chapter 2
Printed Circuit Boards (PCBs)

> *"The greatest threat to our planet is the belief that someone else will save it."*
>
> –Robert Swan, Author

Abstract Generally, e-waste contains 3–5% PCBs, which are the most valuable part of WEEE. Structure, material composition, sources, and value of WPCBs are covered in detail in this chapter. Bare and populated WPCB and electronic component chemical compositions are compared. WPCB grades and prices are given along with valuable metal contents. Value chain and economic value of WPCB recycling are explained graphically. PCB types and assembly structure, methods of fastening electronic components on PCBs, and soldering methods and desoldering for electronic component are described elaborately. The effects of e-waste recycling on metal resources are explained. Characterization and amount of wastes from PCB manufacturing processes are clarified.

Keywords PCB · Value chain · Electronic components · Soldering · Desoldering

2.1 Printed Circuit Boards (PCBs)

Printed circuit boards, the base of electronics, are essential and common elements of almost all of the electronic systems. PCB is a thin board made of epoxy resin and glass fiber, which is coated with layers of thin Cu film. PCBs are used to mechanically support and electrically connect ECs using conductive pathways, tracks, or signal traces etched from Cu sheets laminated onto a nonconductive substrate. PCBs, which provide interconnection between software and hardware, are found in all EEEs. PCBs contain many ECs such as resistors, relays, capacitors, transistors, heat sinks and integrated circuits/chips (ICs), switches, processors, etc. PCB boards can be classified as a bare/unpopulated and populated PCB with ECs (Fig. 2.1). WPCBs may be new/unused PCB trimmings/scraps from new PCB production and used (EoL) PCBs.

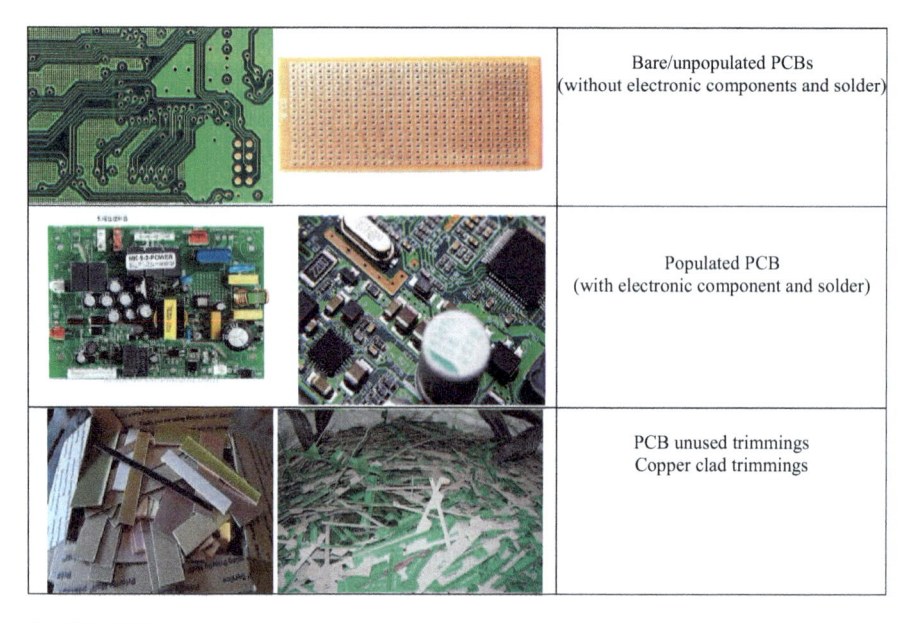

Fig. 2.1 PCB types

2.2 Structure and Contents of PCBs

PCB material composition consists of substrate, Cu clad, solder mask, and silkscreen layers. Generally, PCBs have a similar sandwich structure and are composed of three main layers: substrate (several layers of woven fiberglass and two Cu foils), upper, and lower compound units (which can be defined as an assembly of isolating fiberglass layers, solder joints, conductive tracks, contacts, and solder mask; all these elements are adhered by static friction and epoxy resin) (Fig. 2.2). PCBs comprise multilayers of fiberglass (representing PCB substrate), two Cu foil layers (upper and lower), Cu tracks, through-hole pads, and two solder mask layers (upper and lower). The outer layer material consists of sheets of fiberglass, pre-impregnated with epoxy resin. The shorthand for this is called prepreg. Cu content/pureness varies between 86% and 93%. Bromine content changes from 9% to 29% [1].

All PCB assemblies contain a green/yellow board and ECs attached to it and essentially consist of three basic parts:

- A nonconducting substrate or laminate
- Conducting Cu substrate printed on or inside the laminate
- The electronic components attached to the substrate (chips (Ga, In, Ti, Si, Ge, As, Sb, Se, and Te), connectors and contactors (Au, Ag), multilayered ceramic capacitors (MLCCs) (Ta), aluminum capacitors (Al), ICs, etc.))

PCBs are a mixture of phenolic/cellulose paper (yellow and low grade) or epoxy (green and high grade) resins, woven glass fiber, and multiple kinds of metals (Cu,

2.2 Structure and Contents of PCBs

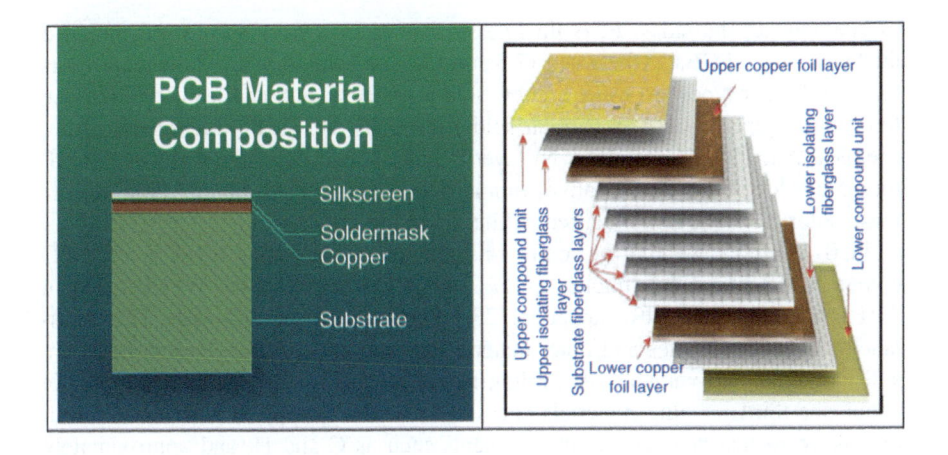

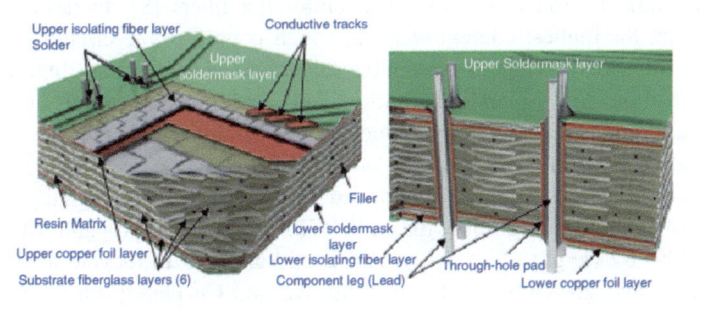

Fig. 2.2 Layers and components of WPCBs [1, 2]

Sn, Pb, etc.). The basic structure of the PCBs is the Cu clad laminate consisting of glass fiber-reinforced epoxy resin and a number of metallic materials including PMs. The concentration of PMs especially Au, Ag, Pd, and Pt is much higher than their respective primary resources, making waste PCBs an economically attractive urban mining source for recycling. Additionally, PCBs also may/could contain different hazardous elements including heavy metals (Cr, Hg, Cd, etc.), rare elements (Ta, Ga, etc.), and flame retardants (Br and Cl) that pose grave danger to the ecosystem during conventional waste treatment of landfilling and incineration [3]. Many research works have revealed that the composition of metals, ceramics, and plastics in PCBs with ECs could reach 40%, 30%, and 30%, respectively [4–6]. WPCBs have been paid much more attention from researchers and enterprises, not only due to their rich resource content but also due to their potential risk for the environment and human health with informal recycling. Therefore, factors affecting the extraction of metals are economic feasibility, recovery efficiency, and environmental impact.

Depending on the structure and alignment, PCBs can be classified as single-sided, double-sided, or multilayered. Single- and double-sided PCBs have the conducting layer on one or both sides of the laminates and with or without plated through-holes to interconnect the sides. PCB thicknesses can change from 0.2 to 7.0 mm. Cu thickness on the PCBs are between 17.5 and 175 µm. Minimum drill hole diameter is 0.2 mm. Solder types are water-soluble solder paste, leaded, and lead-free. Surface finishing is generally HASL (hot air solder leveling), HASL Pb-free, chemical Sn, chemical Au, ENIG (electroless Ni layer is coated with thin immersion Au), OSP (organic solderability preservative), immersion Au/Ag, or Au plating type. Solder mask color can be green, blue, red, white, or yellow.

PCB assembly weights of some consumer electronics are CRT TVs 7%, PC computers 18.8%, mobile phones 21.3%, and LCD screens 11.9%. WPCBs contain 33% semiconductors, 24% capacitors, 23% unpopulated circuit boards, 12% resistances, and 8% switches and other materials by weight [7]. Many components are still functional and valuable. According to Takanori et al. (2009), WPCBs contain about one-third metallic materials such as Cu and Fe, approximately one-fourth organic resin materials containing elements such as C and H, and approximately one-third glass materials used as resin reinforcing fibers [8]. In terms of metal composition, the highest content was Cu, which is used in the circuitry, followed by Fe, Al, Sn, and Pb, which is used in the soldering and lead frames. In terms of the PM composition, Au, Ag, and Pd are found in ICs as contact materials or as plating layers due to their high conductivity and chemical stability [8, 9]. Table 2.1 shows representative material compositions of WPCBs by wt % used in 31 previous studies. The materials can be categorized into three groups: organic materials, metals, and ceramics. Bare/unpopulated WPCBs without any ECs and solder represent 65–70% by weight of an average PCB and are easy to recycle to obtain Cu, epoxy resin, and glass fiber with some Au, Ni, and Cu plating on their surfaces. Typically, a WPCB from PC can contain 7% Fe, 5% Al, 20% Cu, 1.5% Pb, 1% Ni, 3% Sn, and 25% organic compounds (determined by LOI), together with 250 ppm (g/t) Au, 1000 ppm Ag, 110 ppm Pd, and trace amount of As, Sb, Ba, Br, and Bi. In common computer PCB, Cu and Au contents are, respectively, 20–40 times (Cu) and 25–250 times (Au) higher [10].

Gu et al. (2017) reviewed 18 articles for mobile phone WPCB material composition determination and found that the average Au content of 0.12%, Ag 0.31%, Pd 0.08%, Cu 38.15%, Ni 1.56%, Pb 1.89%, Sn 1.91%, Fe 2.63%, and Zn 0.85%. WPCBs from PCs and mobile phones have the highest PM contents [14]. Chancerel et al. (2009) analyzed 1 ton of mixed e-waste for PM distribution. PCB amounts by weight in different EEEs changed by weight in the following order [12]: mobile phone and phone (22%) > laptop (15%) > PC (13%) > video recorder and DVD player (10%) > mouse, printer-fax, and Hi-Fi (8%) > LCD (4%) > keyboard and loudspeaker (2%). Average WPCB weight value was 9%. Au content was maximum for mobile phones (980 g t^{-1}), LCD monitors (490 g t^{-1}), and PCs and laptops (250 g t^{-1}). Ag content was maximum for mobile phones (5440 g t^{-1}), telephones (2244 g t^{-1}), LCD monitors (1300 g t^{-1}), and laptops and PCs (1000 g t^{-1}). Pd is maximum for mobile phones (285 g t^{-1}), telephones (241 g t^{-1}), and PCs and laptops (100 g t^{-1}). For 1 ton of mixed e-waste, average Ag was 67.6 g, Au 11.2 g, and Pd 4.4 g.

2.3 Sources and Value of WPCBs

Table 2.1 PCB material composition (compiled from [11–13])

Metals up to 40%	Percentage (%)	Ceramics up to 30%	Percentage (%)	Plastics up to 30%	Percentage (%)
Cu	6–31	SiO_2	15–30	PE	10–16
Fe	0.7–15.2	Al_2O_3	6.0–9.4	PP	4.8
Al	1.3–11.7	Alkali-earth oxides	6.0	PS	4.8
Sn	0.7–7.4	(BeO)	3.0	Epoxy	4.8
Pb	0.8–6.7	Titanates-micas		PVC	2.4
Ni	0.2–5.4			PTPE	2.4
Zn	0.2–2.2			Nylon	0.9
Sb	0.2–0.4				
Au (ppm)	9.0–2050				
Ag (ppm)	110–5700				
Pd (ppm)	3.0–4000				
Pt (ppm)	5–40				
Co (ppm)	1–4000				

Table 2.2 Unpopulated PCB and EC compositions

Composition (%)	Bare PCB	ECs	Total (populated PCB)
Sn	**10.12**	**3.20**	6.0
Pb	3.20	0.68	1.7
Cu	**21.62**	**13.80**	19.9
Fe	0.21	**19.49**	11.8
Al	1.36	6.91	4.7
Zn	0.056	5.66	3.4
Ni	0.036	0.65	0.4
Cr	0.027	0.53	0.3
Cd (g t^{-1})	0.53	14.45	8.9
Au (g t^{-1})		**40.76**	24.4
Ag (g t^{-1})	194.91	112.68	**145.7**

Yang et al. (2017) determined WPCBs from desktop and laptop PCs after CPU and RAM had been removed. Unpopulated PCB is rich in Cu, Sn, and Ag, and ECs are rich in Cu, Fe, Sn, Ag, and Au (Table 2.2). Recycling of these metals is becoming more and more important, in order to counteract the depletion of mineral resources, especially as the demand for these metals continues to increase [15].

2.3 Sources and Value of WPCBs

The primary sources of WPCBs are from original equipment manufacturers (OEMs), PCB manufacturers, end users (corporate or individual), and dismantlers. WPCBs may be scraps, faulty, malfunctioning, redundant, and overcapacity products. A mobile phone can contain more than 40 elements including BMs (Cu, Sn), SMs

(Co, In, Sb), and PMs and PGMs. But, not all WPCBs have a high value. If WPCBs have sufficient economic value, they will be recycled when appropriate technological infrastructure exists for recovering their contained element. There are three grades of WPCBs according to values: low, medium, and high grades. Hagelüken (2006) uses Au content of e-waste for grouping [16]. If Au content is less than 100 ppm, it is low value; if Au content is between 100 and 400 ppm, it is medium value; and if Au content is more than 400 ppm, it is high value. TV boards (without CRTs), audio scraps, calculators, power supply units, alkaline batteries, lead batteries, phone batteries, Ni-Cd batteries, laptop batteries, laminate offcuts, transformers, and heat sinks are low-grade WEEEs. Their prices change less than 1 Euro kg^{-1}. Pins, keyboards, PWBs, car electronics, lamps, phosphors, CRTs, ICT wastes, PC boards, laptops, handheld computers, WEEE fines, and edge connectors are medium-grade WPCBs, and prices change from 1 to 8 Euros kg^{-1}; Au ICs, optoelectronic devices, PCB scraps, mobile phones, NdFeB magnets, ICs, multilayer ceramic capacitors (MLCCs), some boards, and main frames are high-grade WPCBs, and prices change from 8 to 25 Euros kg^{-1} (Table 2.3). High-grade and high-value WPCBs are estimated about 15% of total PCBs [17]. The value of WPCBs changes from 1000 to 25000 Euros per ton. Table 2.4 shows average metal contents of some PCBs and ECs from WEEE along with geochemical earth's crust abundances, adequacy of world reserves (i.e., life expectancy in years), and approximate minimum grade of ore resources to be mined [18, 19]. PMs are rare and expensive, have high melting point, and are more ductile than other metals. Their contents are the highest for metallic pin connectors (ports, sockets, slots) and mobile phones. Figure 2.3 shows the metals found in WPCBs. In ECs, Cu content changes from 61% to 88%, Zn 0.07 to 26%, Sn 1.0 to 11.7%, Pb 0.4 to 4.1%, Ni 0.84 to 2.74%, Au 50 to 1273 ppm, Ag 8 to 579 ppm, and Pd 93 to 134 ppm. Valuable metal content of WPCBs from mobile phones and PCs is higher than LCD and CRT monitors. Therefore, WEEEs are called "urban mining" resources. Crustal abundances of metals show that all of the metals are very rare in the earth crust. Thus, e-waste recycling is a significant urgent issue and necessary for a scarce natural

Table 2.3 WPCB grades and values

Low Grade	Medium Grade	High Grade
< 100 ppm Au	100-400 ppm Au	> 400 ppm Au
TV boards	Pins	Au ICs
Power supply units	Edge connectors	Opto electric devices
Laminate off cuts		High precious metal content boards
Heavy transformers		Au pin boards
Al heat sinks		Pd pin boards
		Main frames
< 1000 Euros/t	1000-8000 Euros/t	8000-25000 Euros/t

2.3 Sources and Value of WPCBs

Table 2.4 Average metal content of some PCBs and ECs from WEEE along with value share of PMs and useful life

PCB	Au (ppm)	PMs Ag (ppm)	Pd (ppm)	BM Cu (%)	Value Share of Total PMs (%)	Useful Life (years)
PC board	200-250	900-1000	80-110	15-25	86	5
CRT monitor	15	280	10	10	47	
LCD monitor	123	1100	14	20		5
Mobile phone	340-980	1380-5540	120-285	13-25	93	1-2
TV boards	20	280	10	10		
Portable audio	10	150	4	21	21	
DVD	15	115	4	5	52	
PCB	110	280	-	10		

Component	Mass (g)	(%) Cu	(%) Zn	(%) Sn	(%) Pb	(%) Ni	Au (ppm)	Ag (ppm)	Pd (ppm)
Ethernet ports	4.2	87	0.07	4.1	0.4	1.4	1273	170	134
CPU sockets	28.7	87	0.16	11.7	4.1	2.74	938	579	98
USB ports	6.5	76	0.57	9.7	1.3	0.86	344	48	157
IDE connectors	61.5	62	25	1.0	0.7	0.77	219	8	95
Head strips	50.8	61	25	3.1	1.3	0.70	216	27	93
DRAM memory slot	85.4	72	14	4.9	1.1	1.0	198	18	113
AGP slot	49.4	72	18	2.6	0.5	0.9	187	14	109
Keyboard/mouse ports	14.0	65	21	2.6	0.7	1.2	175	16	101
Audio ports	33.4	88	12	3.5	0.4	1.1	178	26	117
Parallel/VGA ports	84.8	69	22	2.4	0.7	1.1	93	16	100
PCI slot	195.3	62	26	1.9	1.0	0.84	58	19	99
Geochemical earth's crustal abundances (ppm)		55	70	2.0	12.5	75	0.004	0.07	0.01
Life expectancy (years) according to USGS		25-50	10-25	25-50	10-25	50-100	10-25	10-25	+100
Minimum grade of ore to be mined (%/ppm)		0.5-3.0	2.5	0.5	3.0	1.0	2-10	50	2.0
World mine production t/a (2011-USGS)		16200000		261000			2500	22200	229
EEE demand (%)		7174000		129708			327	7554	44
EEE demand/mine production (%)		44		50			13	34	19

Grade higher than minimum cut-off grade to be economically and technically mined ore resources.

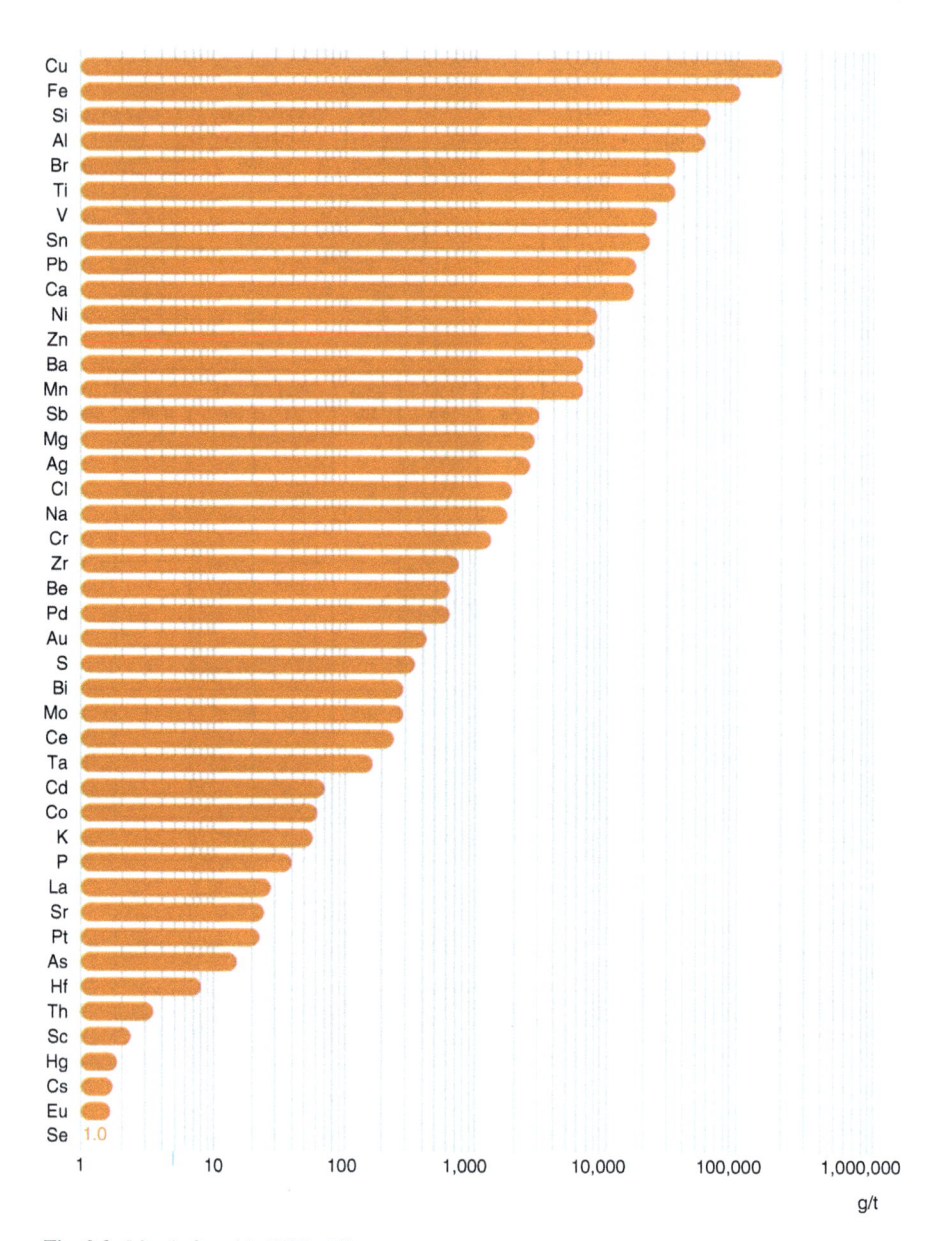

Fig. 2.3 Metals found in PCBs [4]

resource conservation point of view. The adequacy of world mineral reserves is very low (10–25 years) for Au, Ag, Zn, and Pb; medium (25–50 years) for Cu and Sn; high (50–100 years) for Ni; and very high (+100 years) for Pd. Minimum cutoff grades for metals to be mined show that most of the PCBs and ECs contain more valuable metals than raw ore deposits [18, 19].

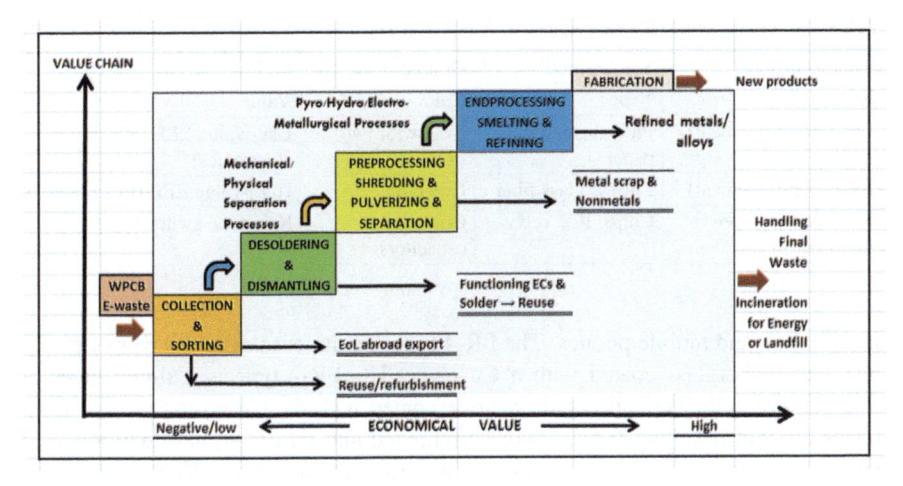

Fig. 2.4 Graphically value chain and economic value of WPCB recycling

According to the above classification, mobile phones contain high-grade WPCBs, PCs and LCD monitors medium-grade WPCBs, and CRT monitors low-grade WPCBs. Some 50,000 t y^{-1} of WPCBs is generated within the UK. Eighty percent of these are populated boards, and the remaining is unpopulated boards such as laminate offcuts. About 15% of these total amount of WPCBs are subject to any form of recycling, with the balance being consigned to landfilling and offshore. High-grade WPCBs are treated pyrolytically within a smelter. More than 90% of the intrinsic material value of boards, which may be classified as medium-grade WPCBs, is in the Au and Pd content. Figure 2.4 graphically shows value chain and economic value of e-waste recycling. Economic value increases with physico-mechanical and metallurgical processing [20].

2.4 Characterization of WPCBs

WPCBs are the most complex, hazardous, and valuable component among e-waste. As a consequence of continuous modifications of function and design of EEE, WPCBs are a highly heterogeneous mix of materials. Due to the diverse and complex nature and multilayered feature of WPCBs, characterization in terms of types, structure, components, and composition is very difficult and important to establish the route and process for recycling. Typically, metals are embedded or entrapped in laminated structure of polymers which will hinder recycling. Liberation of metals is considered to be one of the crucial steps during recycling (such as wet chemical WEEE treatment). Thin films of Sn or Ag are used in the PCBs to protect against oxidation [21]. The BMs mainly found in PCBs are used because of their conductive properties. There are two types of PCBs (i.e., FR-4 and FR-2) normally

Table 2.5 Populated PCB types, contents, and properties

Board/substrate	Single sided	Double sided	Multiple layered
Resin	Type	Color	Value
FR-2 (reinforcement)	Phenolic/cellulose paper	Yellow/brown	Low-value EEE (TV, home electronics)
FR-4 (reinforcement)	Epoxy, glass fiber	Green/blue	High-value EEE (PC, phones)
Electronic components (ECs)	Chips, ICs, relays	Connectors, capacitors	Resistors, switches

used in PCs and mobile phones. The FR-4 type is composed of multilayer of epoxy resins and fiberglass coated with a Cu layer. The FR-2 type is a single layer of fiberglass or cellulose paper and phenolic coated with the Cu layer [22, 23]. Both resins are thermosetting (i.e., cannot be remelted and reformed) and thermoplastic such as bromine-based resins which are used as flame retardants. The FR-4 type is used in small devices such as mobile phones, and the FR-2 type is used in TVs and household appliances such as PCs [24]. The FR-4 epoxy resins are green/blue in color and have high value, while the FR-2 phenolic resins are yellow/brown in color and have low value. Table 2.5 shows the types, contents, and properties of PCBs with ECs. Bare PCBs without ECs contain about 30% metals and 70% nonmetals. The NMFs consist of cured thermosetting resins, glass fiber, ceramics, BFRs, residual metals (Cu and solder), and other additives. NMF composition contains 65% glass fiber, 32% epoxy resin, and impurities (Cu, < 3%; solder, < 0.1%) by weight [25]. Resins are organic plastic polymers and have high-cost and low-quality products. If resins are landfilled or incinerated like in the past, they create potential environmental problems. Glass fibers, which are about 50–70% of PCBs, are reinforcing material in PCBs.

Material content of the nonmetallic recycled from WPCBs by air classification showed that residual Cu and glass fiber were in fine-size class and resins were in coarse-size class [6, 26–29]. The thermal stability of NMF is very important both for physical and chemical recycling methods; but, the demands for the thermal stability in two methods are just the opposite. In physical recycling, NMFs have to be thermally stable in the injection or compression molding. For chemical recycling, the energy cost will be economical if the degradation temperature of the cured thermosetting resins in the NMF is too high [30]. Br in epoxy resin (ER) starts to evaporate intensively, and weight loss increases dramatically in the temperature range of 310–360 °C (65–70% mass loss). ER decomposition range is 260–400 °C and weight loss 75% [1]. NMF thermogravimetric analysis (TGA) studies showed that degradation was greatest at 343 °C and the onset temperature was 323 °C. At 471 °C about 60% and at 800 °C about 33% residual weight remained, and most of the resins were decomposed. Because the upper temperature in most molding process is below 323 °C, the thermal stability of the NMF is good enough for physical recycling methods for the NMFs. As for chemical recycling methods, the degradation temperature below 471 °C is reasonable [30].

2.5 Impact on Metal Resources

E-waste can contain up to 60 different elements (Fig. 2.5). Most of the elements are transition metals, and the rest is nonmetals and precious and heavy metals. Generally, the concentrations of PMs, which are much lower than those of BMs, are very high than natural ores. WEEE stream generated in the EU is predicted to be over 12 million tons by 2020. An effective recovery method for Au is needed. A mobile phone can contain over 40 elements from periodic table. Metals represent 23% of the weight of a phone; majority is Cu, while the remainder is plastic and ceramic material. Figure 2.6 shows the PMs in notebook PCBs. Au content follows the following order: memory cards > display PCBs > HDD > motherboards > optical drives > small PCBs. Ag content changes in the following order HDD > optical drives > memory cards, and Pd content changes in the following order HDD > memory cards > display PCBs.

Materials and metals are essential and crucial components of today's society. Metals play a key role in enabling sustainability through societies' various high-tech applications. However, the resources of our planet are limited. The maximization of resource efficiency through optimal recycling of metals, materials, and products is essential to this and has been identified as one of the pillars on which to build a resource-efficient world. To systematically fully understand resource efficiency in the context of material use and ensure maximum recovery of elements, metals and compounds from e-waste. Product-centric or material /metal-centric approaches can be used in e-waste recycling. According to the UNDP report, less than one-third of 60 economically important metals are recycled globally at rates greater than 50%. More than half of the metals are recycled at rates less than 1% [32]. Product-centric recycling is the application of economically viable technology and methods throughout the recovery chain to extract metals from the complex interlinkages within designed "minerals," i.e., products, gleaning from the deep know-how of recovering metals from complex geological minerals. EoL, waste, and residue streams destined for recycling must be processed by the best available technique (BAT), according to specific performance standards and being mindful of environmental and social costs and benefits. BAT processes separate a maximum of valuable metals and create a minimum of secondary residues for optimizing recycling.

Figure 2.7 shows reuse stats postconsumer recycling rates for metals. Recycling of most technology metals still lags way behind. Recycling is very important for a resource-efficient economy, and metal recycling rates must rise within the next decade to conserve and maintain resources. Fe, Pb, Cr, Ni, and Mn have recycling rates that are higher than 50%. In contrast, recycling rates fall below 1% for Li, used as rechargeable batteries and Ce in catalysts, In used in semiconductors, and LCDs and LEDs in low concentrations. WEEE PM recycling rates below 15%. When the metal prices are low, there is less incentive to recycle low-concentration metals. As the prices go up, recycling becomes more cost-effective.

One ton of mobile phone handset (without battery) includes 3.5 kg Ag, 340 g Au, 140 g Pd, as well as 130 kg Cu. For a single phone unit, the PM content is 250 mg

Periodic Table of Elements

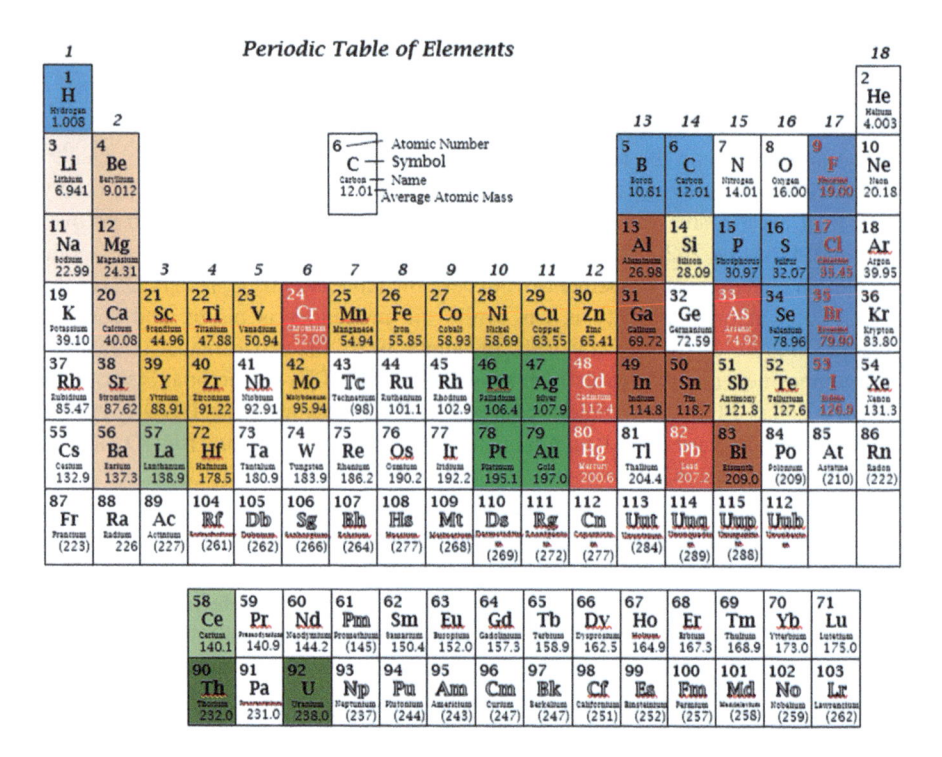

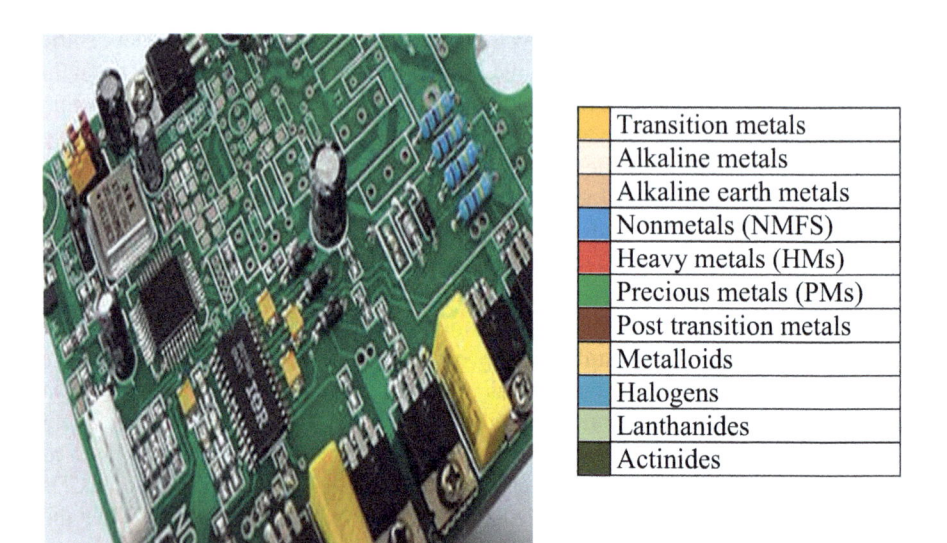

1																	18
1 H Hydrogen 1.008	2											13	14	15	16	17	2 He Helium 4.003
3 Li Lithium 6.941	4 Be Beryllium 9.012											5 B Boron 10.81	6 C Carbon 12.01	7 N Nitrogen 14.01	8 O Oxygen 16.00	9 F Fluorine 19.00	10 Ne Neon 20.18
11 Na Sodium 22.99	12 Mg Magnesium 24.31	3	4	5	6	7	8	9	10	11	12	13 Al Aluminum 26.98	14 Si Silicon 28.09	15 P Phosphorus 30.97	16 S Sulfur 32.07	17 Cl Chlorine 35.45	18 Ar Argon 39.95
19 K Potassium 39.10	20 Ca Calcium 40.08	21 Sc Scandium 44.96	22 Ti Titanium 47.88	23 V Vanadium 50.94	24 Cr Chromium 52.00	25 Mn Manganese 54.94	26 Fe Iron 55.85	27 Co Cobalt 58.93	28 Ni Nickel 58.69	29 Cu Copper 63.55	30 Zn Zinc 65.41	31 Ga Gallium 69.72	32 Ge Germanium 72.59	33 As Arsenic 74.92	34 Se Selenium 78.96	35 Br Bromine 79.90	36 Kr Krypton 83.80
37 Rb Rubidium 85.47	38 Sr Strontium 87.62	39 Y Yttrium 88.91	40 Zr Zirconium 91.22	41 Nb Niobium 92.91	42 Mo Molybdenum 95.94	43 Tc Technetium (98)	44 Ru Ruthenium 101.1	45 Rh Rhodium 102.9	46 Pd Palladium 106.4	47 Ag Silver 107.9	48 Cd Cadmium 112.4	49 In Indium 114.8	50 Sn Tin 118.7	51 Sb Antimony 121.8	52 Te Tellurium 127.6	53 I Iodine 126.8	54 Xe Xenon 131.3
55 Cs Cesium 132.9	56 Ba Barium 137.3	57 La Lanthanum 138.9	72 Hf Hafnium 178.5	73 Ta Tantalum 180.9	74 W Tungsten 183.9	75 Re Rhenium 186.2	76 Os Osmium 190.2	77 Ir Iridium 192.2	78 Pt Platinum 195.1	79 Au Gold 197.0	80 Hg Mercury 200.6	81 Tl Thallium 204.4	82 Pb Lead 207.2	83 Bi Bismuth 209.0	84 Po Polonium (209)	85 At Astatine (210)	86 Rn Radon (222)
87 Fr Francium (223)	88 Ra Radium 226	89 Ac Actinium (227)	104 Rf (261)	105 Db (262)	106 Sg (266)	107 Bh (264)	108 Hs (277)	109 Mt (268)	110 Ds (269)	111 Rg (272)	112 Cn (277)	113 Uut (284)	114 Uuq (289)	115 Uup (288)	116 Uuh		

58 Ce Cerium 140.1	59 Pr Praseodymium 140.9	60 Nd Neodymium 144.2	61 Pm Promethium (145)	62 Sm Samarium 150.4	63 Eu Europium 152.0	64 Gd Gadolinium 157.3	65 Tb Terbium 158.9	66 Dy Dysprosium 162.5	67 Ho Holmium 164.9	68 Er Erbium 167.3	69 Tm Thulium 168.9	70 Yb Ytterbium 173.0	71 Lu Lutetium 175.0
90 Th Thorium 232.0	91 Pa Protactinium 231.0	92 U Uranium 238.0	93 Np Neptunium (237)	94 Pu Plutonium (244)	95 Am Americium (243)	96 Cm Curium (247)	97 Bk Berkelium (247)	98 Cf Californium (251)	99 Es Einsteinium (252)	100 Fm Fermium (257)	101 Md Mendelevium (258)	102 No Nobelium (259)	103 Lr Lawrencium (262)

	Transition metals
	Alkaline metals
	Alkaline earth metals
	Nonmetals (NMFS)
	Heavy metals (HMs)
	Precious metals (PMs)
	Post transition metals
	Metalloids
	Halogens
	Lanthanides
	Actinides

Fig. 2.5 Elements (metals, nonmetal, and metalloids) used in the PCB manufacture

2.5 Impact on Metal Resources

Fig. 2.6 Precious metals in notebook PCBs [31]

Ag, 24 mg Au, 9 mg Pd, and 9 g Cu on average. Furthermore, the Li-ion battery of a phone contains about 3.5 g Cu. Table 2.6 compares the impact of phones and PCs on metal demand, based on global sales 2007 and 2010. 1.2 billion mobile phones contain 300 t Ag; 29 t Au; 11 t Pd; 11,000 t Cu; and 4500 t Co. 255 million PCs and laptops contain 255 t Ag; 56 t Au; 20 t Pd; 128,000 t Cu; and 6500 t Co. The world's 3% of Ag, 3% of Au, 13% of Pd, 1% of Cu, and 15% of Co mine production may come from e-waste recycling in 2007 [18]. In 2010, the world's 5% of Au, 21% of Pd, and 20% of Co come from e-waste recycling.

Globally, 30% of Ag, 12% of Au, 14% of Pd, 33% of Sn, 30% of Cu, 19% of Co, and 6% of Pt were used in EEE production in 2007. Electronics make up for almost 80% of the world's demand of indium (In) (transparent conductive layers in LCD glass), over 80% of ruthenium (Ru) (magnetic properties in hard discs (HD)), and 50% of antimony (Sb) (flame retardants). Se, Te, and In are used in thin photovoltaic (PV) panels for renewable energy generation; Pt and Ru are proton exchange membrane (PEM) fuel cells. In 2007, the monetary values of the annual use of important EEE metals represent 45.4 billion $ [35]. The recycling of WEEE is mandatory in EU (directive 2002/96/EC), but its weight-based recycling targets do not provide a direct incentive for the recovery of the SMs and PMs. Precious metals

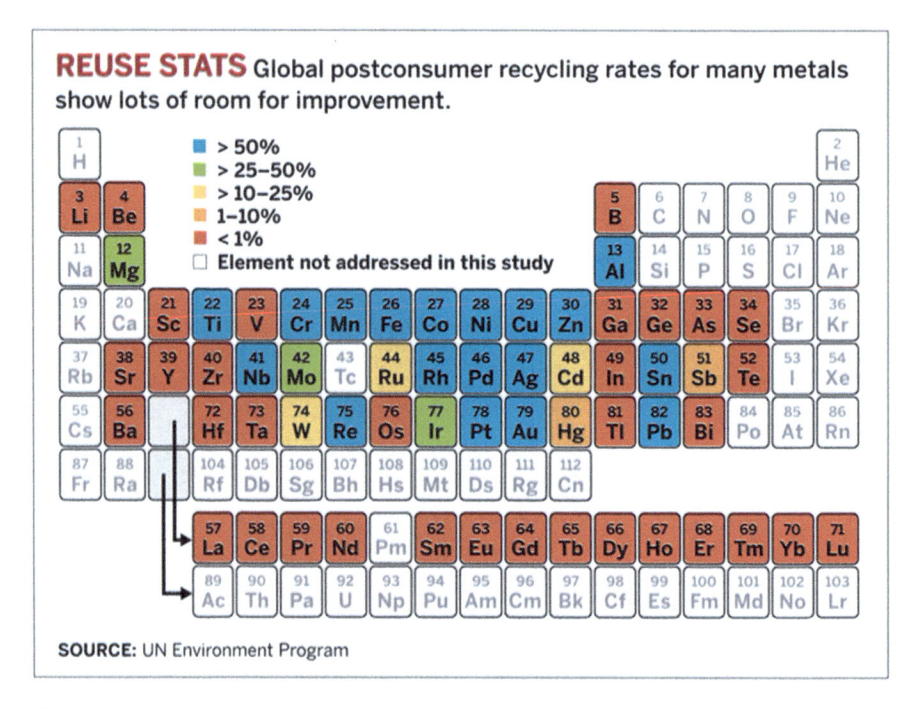

Fig. 2.7 Postconsumer recycling rates for many metals [32]

Table 2.6 Impact of phones and PCs on metal demand, based on global sales 2007 and 2010 [33, 34]

Mobile phones (1)	PCs and laptops (2)	Urban mine (1 + 2)	
2007	2007	*Mine production*	*Share*
1.2 billion unit y^{-1}	255 million unit y^{-1}	2007	(%)
Ag: 250 mg = 300 t	Ag: 1000 mg = 255 t	Ag: 20000 t y^{-1}	3
Au: 24 mg = 29 t	Au: 220 mg = 56 t	Au: 2500 t y^{-1}	3
Pd: 9 mg = 11 t	Pd: 80 mg = 20 t	Pd: 230 t y^{-1}	13
Cu: 9 g = 11000 t	Cu: 500 g = 128000 t	Cu: 16 M t y^{-1}	1
Co: 1200 M∗20 g/battery 3.8 g = 4500 t	Co: 100 M∗65 g = 6500 t	Co: 60000 t y^{-1}	15
2010	2010	2010	(%)
1.6 billion unit y^{-1}	350 million unit y^{-1}	Ag: 22200 t y^{-1}	3
Ag: 250 mg = 400 t	Ag: 1000 mg = 350 t	Au: 2500 t y^{-1}	5
Au: 24 mg = 38 t	Au: 220 mg = 77 t	Pd: 200 t y^{-1}	21
Pd: 9 mg = 14 t	Pd: 80 mg = 28 t	Cu: 16 M t y^{-1}	1
Cu: 9 g = 14000 t	Cu: 500 g = 175000 t	Co: 88000 t y^{-1}	20
Co: 1300 M∗20 g/battery ∗ 3.8 g = 6100 t	Co: 180 M ∗ 65 g = 11700 t		

are recovered for their biggest scrap intrinsic value. In a mobile phone, they account with less than 0.5% of the weight for over 80% of the value. Cu represents 5–15% of the value with 15–20% of the weight. Fe, Al, and plastics, which dominate the weight, have only small value contribution [18]. Recycling of less valuable elements such as Pb, In, Sn, and Ru from EEE is only economically feasible because valuable elements such as Au, Ag, Pd, and Cu are present.

2.6 Methods of Fastening Electronic Components on WPCBs

In many products, different components are joined by welding/soldering, bolting, riveting, gluing, inserting contracted bushings, foaming connections, and chemically connected materials/phases in compounds. Moreover, such components often are nonmetallic, such as rubber, plastics, glass, and ceramics, and the joints have different degrees of difficulty for recycling. The effect of joints on liberation behavior when they pass through the shredder/cutter is very important, and there is a randomness of liberation [4]. Bolting/riveting gives high liberation and high randomness; gluing and foaming give medium liberation and medium randomness; and coating/painting gives low liberation and low randomness of liberation after shredding.

Most of the PMs and SMs in PCBs are mixed with or connected to other metals in contacts, connectors, and solders; connected to ceramics in capacitors and ICs; and connected to resins in the layers of circuit boards. These complex combinations of metals and materials in WEEE require appropriate pretreatment and metal recovery processes that can deal with the mixture of metals and recover them with high efficiency. There are many available choices of fasteners (snap-fit, molded-in hinges, welding, and energy bonding), which enhance disassembly and manufacturing efficiency. Snap-fit fasteners range from cantilevers and annular snaps to traps and darts and easily assemble and disassemble without the use of tolls. There are a number of types of integral and molded-in hinges. Hinges can be attached with ultrasonic energy or plastic rivets. Thermoplastic parts can be ultrasonically welded together. Infrared (IR) welding is used for thin-walled parts. Organic solvent bonding should be avoided for workers' safety. Two push-button fasteners and dart connectors are other types of fasteners. It is recommended to use the least number of different types of fasteners, to use plastic fasteners with the same resin type as the part, and to use fasteners that can be removed without tools [17].

2.7 Methods of Joining Components in PCBs: Soldering

ECs are mounted on PCB assemblies using various types of connections. These connections are typical of the following types: socket pedestal (press-fit), through-hole device (THD) (solder wave type), surface-mounted device (SMD) (solder by reflux), screw joint, and rivet. There are several methods by which these connections can be broken. For example, components with socket pedestal connection can be disassembled directly by nondestructive force; but the method used to disassemble components with SMD or THD connections is always destructive, involving removal of solder or pins [5]. Welding/soldering is a process through which chemically and mechanically two metals are joined at a low melting point. Soldering occurs at a temperature of 40 °C above the melting point of the solder alloy and is valid for any type of solder, including electronic welding. It has a relatively low melting point (183 °C), good wettability, good mechanical and electrical properties, and low cost [36].

2.7.1 Solder

Solder is a fusible metal alloy used to create a permanent bond and electrical connection between electrical wiring and ECs to PCBs. In fact, solder must be melted in order to adhere to and connect the pieces together, so a suitable alloy for use as solder will have a lower melting point than the pieces it is intended to join. The solder should also be resistant to oxidative and corrosive effects that would degrade the joint over time. There are two types of solders. Soft solder typically has a melting point range of 90–450 °C and is commonly used in electronics. Solder alloys that melt above 450 °C is called hard solder, Ag solder, or brazing. Solder has a density between 7.7 and 8.4 g cm^{-3} and Brinell hardness is 15. Surface tension is between 470 and 490 mN m^{-1} at 298–528 °C. Solder starts to pyrolyze after 280 °C and gives toxic fumes [37]. Therefore, removing ECs from WPCBs by desoldering should be performed below this temperature. In electronics generally Sn-Pb, Pb-free solders, or solder pastes are used. Sn-Pb solder alloys commonly used for electrical soldering are 60Sn-40Pb, which melts at 188 °C. 63Sn-37Pb eutectic alloy is used principally in electrical/electronic work and melts at 183 °C.

2.7.2 Pb-Free Solder

After July 1, 2006, EU restricted the use of Pb in most consumer electronics with WEEE and RoHS directives. In the USA, manufacturers may receive tax benefits by reducing the use of Pb-based solder. Legislation promotes new Pb-free welding technologies to reduce the damage to both humans and the environment. The

2.7 Methods of Joining Components in PCBs: Soldering 49

eventual elimination of Pb-based solder has major implications for the processing, assembly, reliability, and electronic packaging cost aspects due to the solder melting temperature, processing temperature, wettability, mechanical and thermomechanical fatigue, and so on [38]. Pb-free solders in commercial use may contain Sn, Cu, Ag, Bi, In, Zn, and Sb and traces of other metals. Most Pb-free replacements for conventional 60Sn-40Pb and 63Sn-37Pb solders have melting points from 5 to 20 °C higher, though there are also solders with much lower melting points. SAC (Sn-Ag-Cu) solders are used by two-thirds of Japanese manufacturers for reflow and wave soldering and by about 75% of companies for hand soldering. The widespread uses of these popular Pb-free solder alloy family are based on the reduced melting point of the Sn-Ag-Cu ternary eutectic behavior (217 °C), which is below the 22Sn-78Ag (wt%) eutectic of 221°C and the 59Sn-41Cu eutectic of 227 °C [39].

2.7.3 Solder Paste

For miniaturized PCB joints with surface-mount components (SMCs), solder paste has largely replaced solid solder. Solder paste is a material used in the manufacture of PCBs to connect SMCs to pads on the board. A solder paste is essentially a mixture of pre-alloyed solder metal powder and flux suspended in a thick medium. Flux is added to act as a temporary adhesive, holding the components until the soldering process melts the solder and makes a stronger physical connection. The paste is gray in color and has a peanut butter-like consistency material. The composition of the solder paste varies, depending upon its intended use. For example, when soldering plastic component packages to a FR-4 glass epoxy circuit board, the solder compositions used are eutectic 63Sn-37Pb or SAC alloys. If one needs high tensile and shear strength, Sn-Sb alloys might be used with such a board. Generally, solder pastes are made of a Sn-Pb alloy, with possibly a third of metal alloyed, although environmental protection legislation is forcing a move to Pb-free solder. According to the Joint Electron Device Engineering Council (JEDEC) standard J-STD-004 "Requirements for Soldering Fluxes," solder pastes are classified into three types based on the flux types: pine tree extract rosin based (need to be cleaned with CFC solvent after soldering), organic materials, and glycol-based water-soluble and resin-based no-clean fluxes.

2.7.4 Solder Flux

The purpose of flux is to facilitate the soldering process. One of the obstacles to a successful solder joint is an impurity at the site of the joint, for example, dirt, oil, or oxidation. At high temperatures, fluxes become reductant and prevent metal oxide occurrences. Flux also acts as a wetting agent to reduce surface tension of molten solder to flow and wet the work pieces more easily during soldering.

Table 2.7 Solder alloys used in PCB packaging [44]

Element	Melting point (°C)	Eutectic composition	Melting point of eutectic solder (°C)
Sn	223.0	67Sn33Cd	176
Cd	320.8	63Sn37Pb	183
Pb	327.5	60Sn40Pb	183–190
Zn	419.4	50Sn50Pb	193–215
Ag	960.8	91Sn9Zn	199
Cu	1083.1	96.5Sn3.5Ag	221
Al	660.1	99.3Sn0.7Cu	227
Mg	651.0	99.5Sn0.5Al	228
		98Sn2Mg	200

Fig. 2.8 Cross-sectional SEM images of eutectic Sn-Pb solder cap on a Cu foil [11]

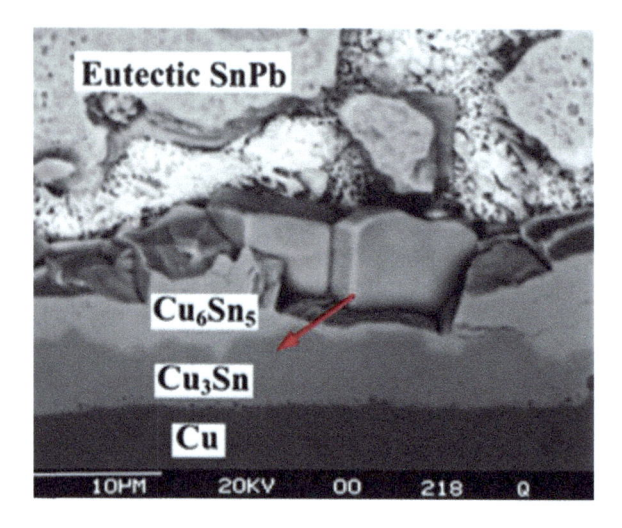

2.7.5 Solder Alloys

The oldest and most common Pb-based solders are 63Sn-37Pb. Table 2.7 shows solder alloys used in PCB packaging. Melting points of solder alloys (for desoldering) change between 176 and 228 °C. Sn-Pb alloys, particularly those near the eutectic composition, are used as solders, while the main substrate or leads are made of Cu. Sn in the solder readily reacts with Cu to form intermetallic compounds (IMCs) as a film at the interface during the solder reflow process. The cross section of the samples in the soldered conditions showed only a single layer of η phase (Cu_6Sn_5) at the interface, and the values are in a range of 1.6–2.3 μm [40]. Cu_6Sn_5 starts to melt at 415 °C, whereas Cu_3Sn starts to melt at 676 °C [41]. Figure 2.8 shows micrographs of the cross section of the joint. The morphology of the phase layer depends on the Pb content of the solder [42]. Formation of IMC film is imperative for good wetting and bonding, but an excessively thick film is harmful because of its brittleness, which makes it prone to mechanical failure under low loads. The mechanical properties for solder joints are sensitive to temperature and strain rate. Melting point of IMC is much higher than soldering Sn. However,

2.8 Soldering Methods of Electronic Products 51

Table 2.8 PCB soldering application types and orders

PCB production type	Common soldering type	Soldered component type	Soldering order
Small scale	Hand (100 W)	Bulkier parts	
	Dip	Surface-mount parts Through-hole parts	3
Mass/bulk	Wave	Through-hole parts	2
	Reflow	Surface-mount parts	1

IMCs are often brittle. Thus, both the assistance of gravity force (or artificial action) and shear stress caused by temperature change are necessary to remove soldering Sn [43].

Pb-Sn solders readily dissolve Au plating and form brittle intermetallics. 60Sn-40Pb solder oxidizes on the surface, forming a complex four-layer structure: Sn (IV) oxide on the surface, below it a layer of Sn(II) oxide with finely dispersed Pb, followed by a layer of Sn(II) oxide with finely dispersed Sn and Pb, and the solder alloy itself underneath.

2.8 Soldering Methods of Electronic Products

Currently, mass-production PCBs are mostly wave soldered or reflow soldered, though hand and dip soldering of production electronics are also still standard practice. Table 2.8 summarizes soldered component types, soldering types, and soldering order according to the PCB production types.

2.8.1 Dip Soldering

Dip soldering is a small-scale soldering process by which ECs are soldered to a PCB. The solder wets to the exposed metallic areas of the board, creating a reliable mechanical and electrical connection. Dip soldering is used for both through-hole PCB assemblies and surface mount. It is one of the cheapest methods to solder and is extensively used in the small-scale industries of developing countries. Dip soldering is the manual equivalent of automated wave soldering. PCB with mounted components is dipped manually into the tank when the molten solder sticks to the exposed metallic areas of the board.

2.8.2 Wave Soldering

Wave soldering is a bulk soldering process used in the manufacture of PCBs. The circuit board is passed over a pan of molten solder in which a pump produces an upwelling of solder that looks like a standing wave. In wave soldering, parts are temporarily kept in place with small dabs of adhesive, and then the assembly is passed over flowing solder in a bulk container. This solder is shaken into waves so the whole PCB is not submerged in solder, but rather touched by these waves. The end result is that the solder stays on pins and pads, but not on the PCB itself. As through-hole components (THCs) have been largely replaced by SMCs, wave soldering has been supplanted by reflow soldering methods in many large-scale electronic applications. However, there is still significant wave soldering where surface-mount technology (SMT) is not suitable (e.g., large power devices and high pin count connectors) or where simple through-hole technology (THT) prevails. Since different components can be best assembled by different techniques, it is common to use two or more processes for a given PCB. For example, SMCs may be reflow-soldered first, with a wave soldering process for the THCs coming next, and bulkier parts hand-soldered last.

2.8.3 Reflow Soldering

Reflow soldering is a process in which a solder paste is used to stick the components to their attachment pads, after which the assembly is heated by an IR lamp, by a hot air pencil, or, more commonly, by passing it through a carefully controlled oven. In reflow soldering, a thermal profile is a complex set of time-temperature (t-T) values for a variety of process dimensions such as slope, soak, time above liquidus (TAL), and peak. Solder paste contains a mixture of metals, flux, and solvents that aid in the phase change of the paste from semisolid to liquid and to vapor and the metal from solid to liquid. For an effective soldering process, soldering must be carried out under carefully calibrated conditions in a reflow oven. There are two main profile types used today in soldering. Thermal temperature-time profiles for the Ramp-Soak-Spike (RSS) and the Ramp to Spike (RTS) are given in Fig. 2.9.

Ramp is defined as the rate of change in T over t, expressed in $°C\ s^{-1}$. The most commonly used process limit is $2\ °C\ s^{-1}$. In the soak segment of the profile, the solder paste approaches a phase change. After the soak segment, the profile enters the ramp-to-peak segment of the profile, which is a given temperature range and time exceeding the melting temperature of the alloy. Successful profiles range in temperature up to 30 °C higher than liquidus, which is approximately 183 °C for eutectic and approximately 217 °C for Pb-free.

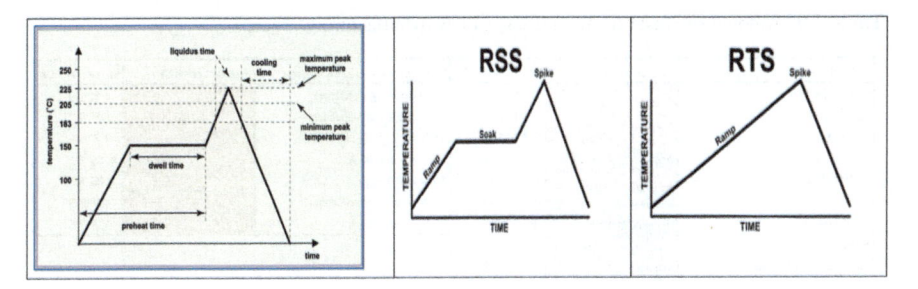

Fig. 2.9 Thermal *t-T* profiles for the Ramp-Soak-Spike (RSS) and the Ramp to Spike (RTS)

2.9 Solder Mask

Solder mask or solder stop mask or solder resist is a thin lacquer-like layer of polymer that is usually applied to the Cu traces of a PCB for protection against oxidation and to prevent solder bridges from forming between closely spaced solder pads. A solder bridge is an unintended electrical connection between two conductors by means of a small blob of solder. PCBs use solder masks to prevent this from happening. Solder mask is not always used for hand-soldered assemblies, but is essential for mass-produced boards that are soldered automatically using reflow or solder bath techniques. Once applied, openings must be made in the solder mask wherever components are soldered, which is accomplished using photolithography. Solder mask is traditionally green but is now available in many colors. Solder mask comes in different media depending upon the demands of the application. The lowest-cost solder mask is epoxy liquid that is silkscreened through the pattern onto the PCB. Other types are the liquid photoimageable solder mask (LPSM) inks and dry film photoimageable solder mask (DFSM). LPSM can be silkscreened or sprayed on the PCB, exposed to the pattern, and developed to provide openings in the pattern for parts to be soldered to the Cu pads. DFSM is vacuum laminated on the PCB, then exposed, and developed. All three processes go through a thermal cure of some type after the pattern is defined.

2.10 Characterization of Wastes from PCB Manufacturing

PCB manufacturing process is very complicated, involving many special chemicals and valuable materials. Some of these materials discharge into the environment in the forms of wastewater, spent solution, and solid waste. The manufacturing process for PCBs is a difficult and complex series of operations. Most of the PCB industries use the subtractive method. In general, this process consists of a sequence of brushing, curing of etching resistor, etching, resistor stripping, black oxide, hole drilling, de-smearing, plating through-hole, curing of plating resistor, circuit plating, solder plating, plating resistor stripping and Cu etching, solder stripping, solder

Table 2.9 Amount of waste from multilayer PCB manufacturing process

PCB Manufacturing Waste	Characterization	kg/m² of PCB	PCB Manufacturing Waste	Characterization	kg/m² of PCB
Waste board	hazardous	0.01~0.3	Acid etching solution	hazardous	1.5~3.5
Edge trim	hazardous	0.01~0.1	Basic etching solution	hazardous	1.8~3.2
Hole drilling dust	hazardous	0.005~0.2	Sn/Pb striping solution	hazardous	0.2~0.5
Cu powder	nonhazardous	0.001~0.01	Sweller solution	hazardous	0.05~0.1
Sn/Pb dross	hazardous	0.01~0.05	Flux solution	hazardous	0.05~0.1
Cu foil	nonhazardous	0.01~0.05	Microetching solution	hazardous	1.0~2.5
Alumina plate	nonhazardous	0.01~0.1	PTH Cu solution	hazardous	0.2~0.5
Film	nonhazardous	0.1~0.4			
Drill backing board	nonhazardous	0.02~0.05			

mask printing, and hot air leveling. Table 2.9 shows the amount of waste generated from a typical multilayer PCB process per square meter board. Solid wastes include edge trim, Cu clad, protection film, drill dust, waste board, and Sn/Pb lead dross. Liquid wastes include high-concentration inorganic/organic spent solutions, low-concentration washing solutions, and resistor and ink solutions. PCB manufacturing process generates about 53% slurry, 11% edge trim, 7% waste board, 8% paper, 5% drill pad, etc. Many spent solutions from PCB manufacturing are strong bases or strong acids. These spent solutions may also have high heavy metal content and high chemical oxygen demand (COD) values. These spent solutions are characterized as hazardous wastes and subjected to tight environmental regulations. Nevertheless, some of the spent solutions contain high concentrations of Cu with high recycling potential. These solutions have been subjected to recycling by several recycling plants with great economic benefit for many years. Recently, several other wastes have also been recycled on a commercial scale. These wastes include PCB edge trim, Sn/Pb solder dross, wastewater treatment sludge containing Cu, CuSO$_4$ PTH (plated through-holes) solution, Cu rack stripping solution, and Sn/Pb spent stripping solution [45].

Annual average world PCB manufacturing rate increases by 8.7% and much higher in China and Southeast Asia [46]. Average PCB amount in e-waste is 3–5% and even more in some EEEs, like TV set 7%, computer 19%, and mobile phone 21% [11]. Most recycling approaches practiced can only recover metal contents of PCB scraps to an extent 30% of the total weight. More than 70% of PCB scraps, which are nonmetallic fraction (NMF), cannot be efficiently recycled and recovered and have to be incinerated or landfilled [3]. After years of research endeavors by academia, research institutes, and the recycling industry, many valuable resources have been identified, and recycling of these resources has been very successful in a commercial scale. Recycling of resourceful wastes generated by the PCB manufacturing industry includes [45]:

- Recovery of Cu metal from edge trim of PCBs
- Recovery of Sn metal from Sn/Pb solder dross in the hot air leveling process
- Recovery of copper oxide from wastewater treatment sludge

- Recovery of Cu from the basic etching solution
- Recovery of copper hydroxide from copper sulfate solution in the plated through-hole (PTH) process
- Recovery of Cu from the rack stripping process
- Recovery of Cu from spent Sn/Pb stripping solution in the solder stripping process

PCB processing, separating, and recycling allows recovering:

- Small ferrous components (stainless steel, Fe)
- Metals (Cu, Al, Sn, etc.)
- Precious metals (Au, Ag, Pd)
- Organic, inert fraction (plastics)
- Glass fibers

References

1. Tatariants M, Yousef S, Sidaraviciute R, Denafas G, Bendikiene R (2017) Characterization of waste printed circuit boards recycled using a dissolution approach and ultrasonic treatment at low temperatures. RSC Adv 7:37729–37738. https://doi.org/10.1039/C7RA07034A
2. Yousef S, Tatariants M, Bendikiene R, Defafas G (2017) Mechanical and thermal characterizations of non-metallic components recycled from waste printed circuit boards. Journal of cleaner production 167, 271–280
3. Li J, Shrivastava P, Gao Z, Zhang HC (2004) Printed circuit board recycling: a state-of-the art survey. IEEE Trans Electron Packag Manuf 27(1):33–42. https://doi.org/10.1109/TEPM.2004.830501
4. http://wedocs.unep.org/bitstream/handle/20.500.11822/8423/-Metal%20Recycling%20Opportunities%2c%20Limits%2c%20Infrastructure-2013Metal_recycling.pdf?sequence=3&isAllowed=y
5. Ghosh B, Ghosh MK, Parhi P, Mukherjee PS, Mishra BK (2015) Waste printed circuit boards recycling: an extensive assessment of current status. J Clean Prod 94:5–19. https://doi.org/10.1016/j.jclepro.2015.02.024
6. Sum EYL (1991) The recovery of metals from electronic scraps. JOM 43(4):53–61
7. Tohka A, Lehto H (2005) Mechanical and thermal recycling of waste from electric and electronic equipment. Helsinki University of Technology, Department of Mechanical Eng. Energy Engineering and Environmental Protection Publications, Espoo
8. Takanori H, Ryuichi A, Youichi M, Minoru N, Yasuhiro T, Takao A (2009) Techniques to separate metal from waste printed circuit boards from discarded personal computers. J Mater Cycles Waste Manage 11:42–54. https://doi.org/10.1007/s10163-008-0218-0
9. Jung LB, Bartel JT (1999) Computer take-back and recycling, an economic analysis for used consumer equipment. J Electron Manuf 9:67–77. https://doi.org/10.1142/S0960313199000295
10. Soare V, Burada M, Dumitrescu DV, Costantian I, Soare V, Popescu ANJ, Carcea Ii Innovation approach for the valorization of useful metals from waste electrical and electronic equipment (WEEE), 2016, IOP conference series. Mater Sci Eng. https://doi.org/10.1088/1757-899X/145 (2/022039. http://researchgate.net/publication/304310103
11. Duan H, Hou K, Li JH, Zhu XD (2011) Examining the technology acceptance for dismantling of waste printed circuit boards in light of recycling and environmental concerns. J Environ Manag 92:392–399. https://doi.org/10.1016/j.jenvman.2010.10.057

12. Chancerel P, Meskers CEM, Hagelüken C, Rotter VS (2009) Assessment of precious metal flows during preprocessing of waste electrical and electronic equipment. J Ind Ecol 13:791–810. https://doi.org/10.1111/j.1530-9290.2009.00171.x
13. Zhang K, Schoor JL, Zeng EY (2012) E-waste recycling: where does it go from here? Environ Sci Technol 46:10861–10867. https://doi.org/10.1021/es303166s
14. Gu F, Summers PA, Widijatmoko SD, Zheng Y, Wu T, Miles NJ, George MW, Hall P (2017) Materials recovery methods for recycling waste mobile phones: a critical review. Waste Manag. (accepted)
15. Yang C, Li J, Tan Q, Liu L, Dong Q (2017) Green process of metal recycling: Coprocessing waste printed circuit boards and spent tin stripping solution. ACS Sustain Chem Eng 5:3524–3535. https://doi.org/10.1021/acssuschemeng.7b00245
16. Hagelüken C (2006) Recycling of electronic scrap at Umicore's integrated metals smelter and refinery. World Metals-Erzmetall 59(3):152161
17. Kellner D (2009) Recycling and recovery. In: Hester RE, Harrison RM (eds) Electronic waste management, design, analysis and application. RSC Publishing, Cambridge, pp 91–110
18. Mesker CEM, Hagelüken C, Van Damme G (2009) TMS 2009 annual meeting & exhibition, San Francisco, California, USA, EPD Congress 2009 Proceedings Ed. by. S.H. Howard, P. Anyalebechi, L. Zhang, pp 1131–1136, ISBN No: 978–0–87339-732-2
19. Mesquita RA, Silva RAF, Majuste D (2018) Chemical mapping and analysis of electronic components from waste PCB with focus on metal recovery, Process Safety and Environmental Protection, 120, 107–117. https://doi.org/10.1016/j.psep.2018.09.002
20. Kaya M (2018) Current WEEE recycling solutions, Chap. 3. In: Veglio F, Birloaga I (eds) Waste electrical and electronic equipment recycling, aqueous recovery methods, pp 33–93. https://doi.org/10.1016/B978-0-08-102057-9.00003-2
21. Veit HM, Diehl TR, Salami AP, Rodrigues JS, Bernardes AM, Tenório JAS (2005) Utilization of magnetic and electrostatic separation in the recycling of printed circuit boards scrap. Waste Manag 25:67–74. https://doi.org/10.1016/j.wasman.2004.09.009
22. William JH, Williams PT (2007) Separation and recovery of materials from scrap printed circuit boards. Resources Conservation and Recycling 51:691–709. https://doi.org/10.1016/j.resconrec.2006.11.010
23. Murugan RV, Bharat S, Deshpande AP, Varughese S, Haridoss P (2008) Milling and separation of the multi-component printed circuit board materials and the analysis of elutriation based on a single particle model. Powder Technol 183:169–176. https://doi.org/10.1016/j.powtec.2007.07.020
24. LaDou J (2006) Printed circuit board industry. Int J Hyg Environ Health 209:211–219. https://doi.org/10.1016/j.ijheh.2006.02.001
25. Yokoyama S, Iji M (1997) Recycling of printed wiring boards with mounted electronic parts. In: Proceedings of the 1997 IEEE Int. Sym, pp 109–114
26. Zhou Y, Qiu K (2010) A new technology for recycling materials from waste printed circuit boards. J Hazard Mater 175(1–3):823–828. https://doi.org/10.1016/j.jhazmat.2009.10.083
27. Koyanaka S, Ohya H, Lee JC, Iwata H, Endoh S (1999) Impact milling of printed circuit board wastes for resources recycling and evaluation of the liberation using heavy medium separation. J Soc Powder Technol Jpn 36:479–483. https://doi.org/10.4164/sptj.36.479
28. Vidyadhar A, Das A (2013) Enrichment implications of froth flotation kinetics in the separation and recovery of metal values from PCBs. Sep Purif Technol 118:305–312. https://doi.org/10.1016/j.seppur.2013.07.027
29. Yamane LH, Moraes VT, Espinosa DCR (2011) Recycling of WEEE: characterization of spent printed circuit boards from mobile phones and computers. Waste Manag 31:2553–2558. https://doi.org/10.1016/j.wasman.2011.07.006
30. Guo J, Guo J, Xu Z (2009) Recycling of non-metallic fractions from waste printed circuit boards: a review. J Hazard Mater 168(2–3):567–590. https://doi.org/10.1016/j.jhazmat.2009.02.104
31. http://www.downtoearth.org.in/coverage/wasted-e-waste-40440

References

32. https://pubs.acs.org/cen/news/89/i22/8922notw4.html
33. Hagelüken C, Corti CW (2010) Recycling of gold from electronics: cost effective use through "design for recycling". Gold Bull 43:209. https://doi.org/10.1007/BF03214988
34. www.umicore.com
35. http://www.ewasteguide.info/files /UNEP_2009_eW2R.PDF
36. Marques AC, Cabrera JM, Malfatt CF (2013) Printed circuit boards: a review on the perspective of sustainability. J Environ Manag 131:298–306. https://doi.org/10.1016/j.jenvman.2013.10.003
37. Khaliq A, Rhamdhani MA, Brooks G, Masood S (2014) Metal extraction processes for electronic waste and existing industrial routes: a review and Australian perspective. Resources 3(1):152–179. https://doi.org/10.3390/resources3010152
38. Guo F (2007) Composite lead-free electronic solders. J Mater Sci Mater Electron 18:129–145. https://doi.org/10.1007/s10854-006-9019-1
39. https://en.wikipedia.org/wiki/Solder
40. Tu K, Zeng K (2001) Tin-lead (SnPb) solder reaction in flip chip technology. Mater Sci Eng 34:1–58. https://doi.org/10.1016/S0927-796X(01)00029-8
41. Fields RJ, Low SR, Lucey GK et al (1991) Physical and mechanical properties of intermetallic compounds commonly found in solder joints. In: Cieslak MJ (ed) The metal science of joining. TMS, Warrendale, pp 165–174
42. Prakash KH, Sritharan T (2004) Tensile fracture of tin-lead solder joints in copper. *Mater Sci Eng A* 379:277–285. https://doi.org/10.1016/j.msea.2004.02.049
43. Flandinet L, Tedjar F, Ghetta V, Fouletier J (2012) Metals recovering from waste printed circuit boards (WPCBs) using molten salts. J Hazard Mater 213–214:485–490. https://doi.org/10.1016/j.jhazmat.2012.02.037
44. Kaya M (2016) Recovery of metals from electronic waste by physical and chemical recycling processes. Int J Chem Nucl Mater Metall Eng 10. scholar.waset.org/1307-6892/10003863
45. https://www.epa.gov/sites/production/files/2014-05/documents/handout-10-circuitboards.pdf
46. Huang K, Guo J, Xu Z (2009) Recycling of waste printed circuit boards: a review of current technologies and treatment status in China. J Hazard Mater 164:399–406. https://doi.org/10.1016/j.jhazmat.2008.08.051

Chapter 3
WPCB Recycling Chain and Treatment Options

"Recycle today for a better tomorrow."

Anonymous

Abstract The recycling chain for e-waste includes collection and sorting; dismantling and size reduction pretreatment; separation, upgrading, and/or extraction preprocessing; and purification and refining endprocessing steps. This chapter briefly covers selective and simultaneous disassembly techniques; size reduction methods for metal-nonmetal fraction liberation and separation; and upgrading and/or extraction of desired valuable materials/metals/electronic components for purification and refining endprocesses. Electronic component recycling for reuse and material recycling can be achieved by mechanical processing, pyrometallurgy, hydrometallurgy, or combination of both pyrometallurgical and hydrometallurgical techniques. Base and precious metals, plastics, batteries, and some valuable electronic components can be recycled, while the remaining nonrecyclable residues should be disposed of properly.

Keywords Recycling chain · Dismantling · Selective/simultaneous disassembly · Size reduction · Preprocessing · Endprocessing

3.1 Recycling Chain

Generally, WPCBs are not bare PCBs, but PCB assemblies. PCBs are always mounted with ECs. PCBs without ECs are called unpopulated/bare PCB or printed wiring board (PWB). The WPCBs are more complicated and diverse in chemical compositions with many kinds of valuable materials. Therefore, the recycling process of WPCBs is much more complicated. Most of the ECs are functional and usable at the time of disposal/dismantling. Therefore, dismantling ECs from WPCBs is a crucial step from both standpoints of materials recovery and EC reuse in the WPCB recycling chain to conserve scarce resources, reuse functioning ECs, and eliminate potential exposure to harmful materials. For example, removal and

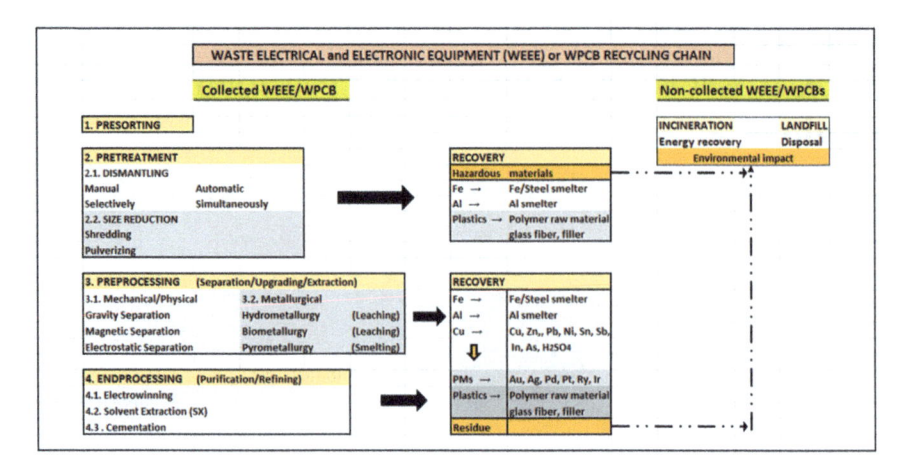

Fig. 3.1 E-waste recycling chain and end-of-life WEEE/WPCB treatment options

disposal of toxic (mainly in electrolytes), explosive, and combustible materials from Al electrolytic capacitors (AEC) are essential and crucial in WPCB recycling. AEC also contains up to 50 wt% Al and Fe and is worth recovering [1]. Current generic material and component recovery methods are typically a combination of the following recycling processes. The recycling chain for e-waste is classified into four main subsequent steps [2]:

- Collection and presorting
- Pretreatment (includes dismantling, size reduction for liberation, and granulometric classification)
- Preprocessing (includes separation, upgrading, and extraction)
- Endprocessing (purification and refining) (Fig. 3.1)

The main collection options for postconsumer goods are collective municipal or commercial collection, individual producer and retailer collection, and collection by the informal sector (waste pickers). The collection of postconsumer waste is very much a logistic challenge. In contrast to Europe, where consumers pay for collection and recycling, in developing countries usually the waste collectors pay consumers for their obsolete appliances and metal scrap. Informal waste sectors are often organized in a network of individuals and small businesses of collectors, traders, and recyclers, each adding value, and creating jobs, at every point in the recycling chain. As many poor people rely on small incomes generated in this chain, this results in impressive collection rates of up to 95% of total generated waste. This shows how strong economic stimulus for collection is a key factor in successful collection, a point that occasionally may be missing in today's formalized takeback schemes.

Many collection programs are in use, but their efficiencies vary from place to place, and improvements of collection rates depend on social and societal factors rather than on collection methods, which is of crucial importance as this determines

3.1 Recycling Chain

the amount of material that is actually available for recycling. About 3% of the people bring the phone to a collection point for recycling and reuse. About half of the phones are hibernating at consumer's attics, and the rest is given away or sold or reused [3]. After sorting, the equipment enters a pretreatment step to separate materials in different streams from which valuable metals can be recovered efficiently. Furthermore, hazardous components such as batteries and capacitors have to be removed prior to mechanical pretreatment. Both disassembly and mechanical processes are used for the liberation and separation of the metallic components from WPCBs in order to expose the metals for subsequent chemical processes. The physico-mechanical process is considered the most environmentally friendly methodology to recover metals. Generally, a physico-mechanical process could contain shredding, grinding, gravity separation, magnetic separation, and electrostatic separation. However, the major challenge for the physical process is poor recovery of BMs and PMs.

During mechanical pretreatment considerable losses of PMs and SMs in WPCB occur. There are two causes for this problem: incomplete liberation of minor and major metals of the complex materials and losses of Pd dusts during shredding of ceramic components [4]. Imperfect liberation causes imperfect separation. Various grades/quantities of recyclates are produced during physico-mechanical preprocessing separation. Extensive pretreatment of EEE is not always necessary; small EEEs such as mobile phones, MP3 players, etc. can after removal of the battery also be treated directly (without size reduction). Mechanical sorting commonly produces imperfect recyclates, and society is probably best suited to discern between the vast combinations of materials, to produce the highest grade of recyclates possible. Typical recyclates created through dismantling are coil (contains Cu, Fe_3O_4, PbO, SiO_2, PVC, and plastics); wires (Cu and PVC); getter (glass, W, BaO, CaO, Al_2O_3, SrO, PbO, SiO_2, and plastics); CRT (PbO, BaO, SrO, Fe_3O_4, Fe, Eu, ZnS, SiO_2, Y_2O_3, glass, and fluorescent powder); PCBs (Au, Ag, Pb, Pd, Sn, Ni, Sb, Al, Fe, Al_2O_3, epoxy, Br, etc.); and housings (wood, plastic, steel, Cl, Br, Sb_2O_3, etc.).

End-of-life WEEEs/PCBs follow four different treatment options: landfill, incineration, shredding, and disassembly (Fig. 3.1). Reuse of some ECs can be achieved with only disassembly. Shredding generates products for recycle, incineration, and landfill. WEEEs/WPCBs can also be disposed by either incineration or landfilling. While collection, dismantling, and preprocessing can differ across different e-waste streams, depending on the constituent components or materials as well as on the technologies available, endprocessing technologies have been developed with the focus on the material streams, regardless of the e-waste device stream they come from. The state-of-the-art WEEE/WPCB reuse and recycling after collection and sorting can be divided into the pretreatment, preprocessing, and endprocessing major steps. Pretreatment is the initial process of WPCB recycling. Separation, metal extraction, refining, and disposal are the following steps.

3.2 Dismantling/Disassembly

For depollution, removal of toxic/environmentally harmful components/substances and useful EC materials (Pb, Hg containing switches/lamps; CRT, P coatings; NiCd/NiMH/LiON and Li polymer batteries; degassing; oil; PcB (polychlorinated biphenyls contained) capacitors; inks; toners; coolants/refrigerants; CFCs/HCFCs in foam, etc.) dismantling is necessary. The aim of dismantling is to liberate the materials and direct them to adequate subsequent pre-/end- treatment processes. Hazardous substances have to be removed and stored or treated safely, while valuable components/materials (PCBs, HDs, etc.) need to be taken out for reuse or to be directed to efficient recovery processes. Dismantling of ECs on WPCBs is the first and the most important step in a recycling chain which can help to conserve scarce resources, establish the reuse of components, and eliminate hazardous materials from the environment. Manuel disassembly is the major cost element and time-consuming operation in recycling technology. Low processing capacities hinder the application of manual disassembly in large-scale industrial processes. The development of automatic disassembly process is vital when considering the amount of WPCBs worldwide. The most valuable part of e-waste (i.e., PCBs) can be removed from the devices by manual dismantling, mechanical treatment, or a combination of both. Manual removal of PCBs prior to e-waste shredding will prevent losses of PMs and SMs. Manual dismantling holds a dominant position in pretreatment of WPCBs; has low investment cost, utilizing simple electrical/pneumatic tools; and can be done by people with little or low education after appropriate training. Manual or automated disassembly can be used selectively or simultaneously.

3.2.1 Selective Disassembly

Targeting hazardous or valuable components for special treatment. Dismantling is generally performed manually and takes a long time.

3.2.2 Simultaneous Disassembly

The entire ECs are evacuated from WPCBs together. Mixed EC disassembly is good for time-saving. Dismantling is performed semiautomatically/automatically. Automated disassembly provides a cost-effective recycle at minimum labor force on a large scale. Radio-frequency identification (RFID) tags will be very helpful for recognition of ECs on WPCBs [5]. Separation and dismantling criteria for e-waste are summarized in Table 3.1 [6].

During dismantling and downward recycling streams, critical issue with regard to occupational health should be applied, and safety and massive dust exposure should

3.3 Size Reduction

Table 3.1 Separation and dismantling treatments for e-waste

	Desired treatment/action
1. Separate before treatment	
(a) Toxic/hazardous materials	
Cooling fluids and foam	Controlled removal and disposal
Mercury backlights	Controlled depot
PCB capacitors	Controlled depot
Batteries	Sort and process in specialized plants
(b) High value materials	
Reusable components	Refurbish and sell
Circuit boards (high and medium grade)	Process in integrated nonferrous/copper smelters
Circuit boards (low grade)	Upgrade (manually) and process in integrated smelters
2. Dismantle, liberate, sort	
Clean plastics	Process further with appropriate technologies
(CRT) glass	Process further with appropriate technologies; glass to glass producer, CRT glass to CRT glass producer or lead smelter
Ferrous metals	To integrated steelmaking facility or to steel scrap remelter (electric arc furnace)
Nonferrous metals Al, Mg	To secondary aluminum or magnesium remelter or other appropriate technology[a]
Nonferrous metals Cu, Pb, Sn, Ni, PM	Process further with appropriate technologies
Others	Process further with appropriate technologies

[a]Low-quality scrap can also be used in steelmaking as a reducing agent (feedstock recycling)

be avoided as much as possible. For hard disc recycle, confidential data should be destroyed completely. Disassembly may be considered to have an impact upon overall future recycling strategies. Limitations of purely mechanical process routes are effectively concerned with PM loss from component structures on populated boards due to the nature of the metal-nonmetal interface, and an effective automated-disassembly methodology could well expand the potential for mechanical turnkey approaches for all grades of WPCBs.

3.3 Size Reduction

E-waste recycling and management is not a simple and straightforward process. The mineralogy of WPCBs is much different compared to the natural ores for metal refining. Many elements are closely interlinked with complex assemblies and sub-assemblies, and secondly metals are cross-linked to toxic organics. In addition to the aims of liberation, the particle size must suit the chosen processing method. The

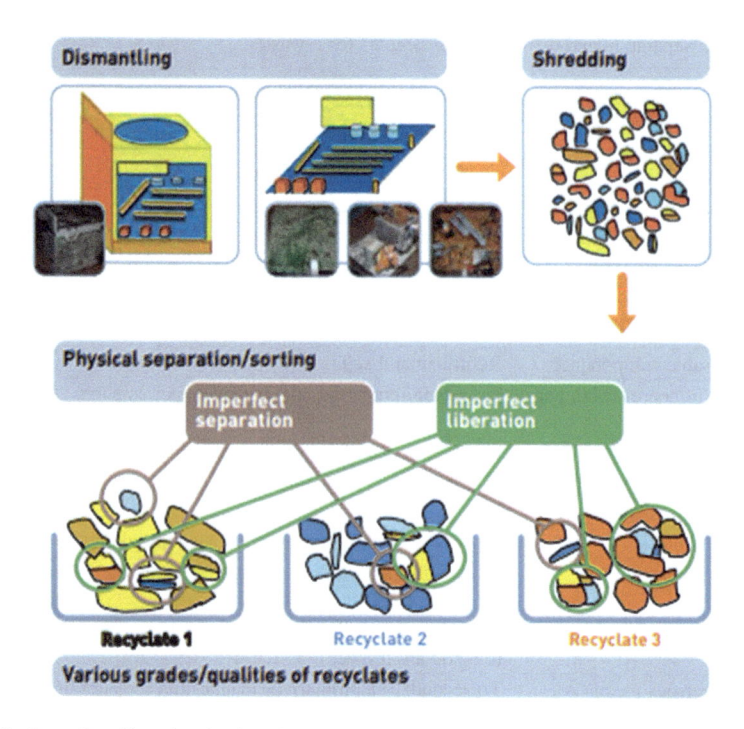

Fig. 3.2 Imperfect liberation leads to imperfect separation

different material behaviors will dictate the method of size reduction. Size reduction is performed to liberate metals from organic nonmetals and from other metals by shredding, pulverizing, and sequential sorting/classification using mechanical processing methods. Incomplete material liberation from WPCBs is the key reason of resource loss. Imperfectly liberated particles land in the wrong recyclate streams due to the randomness and to their physical properties but also due to the connected materials that affect the physical properties and therefore separation (Fig. 3.2).

Size reduction involves crushing, grinding, or shearing, generically called "shredding," though this rarely leads to complete liberation. Imperfect liberation leads to imperfect separation. Particle size significantly affects both the separation and extraction subsequent processes. Mechanical separation of the metallic components from WPCBs is easier to accomplish with a smaller particle sizes and a narrow size distribution [7, 8]. During leaching, dissolution of metals at the smallest particle size fractions due to the high surface area is highest. Previous studies used particle sizes between 37 and 4750 μm [9]. Industrial-scale plants use much coarser particle sizes (in mm size, i.e., 5–50 mm) than laboratory-scale studies (in μm size, i.e., 53–150 μm). Calcination after size reduction is an environmentally unacceptable pretreatment in PCB recycling [10]. Chancerel et al. (2009) found that shredding increases the loss of PMs and has a negative impact on PM recovery [4]. Some of the PMs are lost in plastics and ferrous metals. More shredding results in the decrease of the concentration of PMs in PCBs. Pd is found in ceramics that are broken down to dust during shedding and ends up in filter dust. To reduce the losses of PMs during shredding and sorting, the first and most straightforward approach is to reduce the quantity of PMs entering the shredder by manually sorting step at the beginning of the process to remove most PM-rich materials. For PM-rich WEEEs, shredding should be avoided and directly fed into appropriate metallurgical recovery processes for PMs and SMs. During shredding and pulverizing, noise and powder which may contain Cr, Cu, Cd, and Pb on the air should be minimized. According to the Occupational Safety and Health Standards (OSHS), the maximum permissible limit equivalent continuous sound level is 90 dB for workers for 8 h d^{-1} in the workshop. Particles below 100 μm diameter should be less than 500 μg m^{-3}, and particles below 10 μm diameter should be less than 250 μg m^{-3}. Sound insulation and personal protective devices such as masks must be used.

3.4 Separation/Upgrading/Extraction

Separation preprocessing upgrades/concentrates and extracts desired materials/ metals and prepares materials for purification and refining endprocesses. Separation exploits physical characteristic/property differences (such as size, density, magnetic and electrostatic properties) of the materials. Material extraction depends on chemical properties and includes pyro- and hydrometallurgical processing techniques. Physics primarily determine the potential and limits of metal recycling, the thermodynamic properties of each metal being particularly important. When metals have similar thermodynamic properties, heat-based processing technology cannot fully separate them, resulting in impure, mixed alloys that may have limited or no value. In such cases, hydrometallurgy is required, for example, as used during the separation of chemically similar REEs that are separated by several solvent extraction steps. These physical realities have consequences for all the links in the chain of activities that support recycling. Separation processes can be dry or wet operation. Dry processes can produce dust containing both metals and other possible hazardous components. Wet process needs water and creates sludge. Dry separation methods include manual and ballistic sorting and also (i) magnetic separation, (ii) eddy current separation, (iii) air separation/zigzag wind sifter, (iv) screening, (v) fluidized-bed separation, (vi) sensor-based sorting (by image analysis, color, X-ray, spectroscopy), (vii) Cu wire sorting, and (viii) electrostatic sorting. Wet separation methods include (i) sink-float, (ii) heavy-medium cyclones, (iii) jigging, (iv) shaking tables, and (v) flotation [11]. Solvolysis and reprocessing are other material extraction processes which will not be covered here. Separation methods will be covered in detail in the following chapters.

Smelting is pyrometallurgical and leaching is hydrometallurgical processing. Pyrometallurgy uses high temperatures (sometimes above 2000 °C) to drive chemical reactions as well as melting (just remelting metals) or smelting (chemical conversion from compound to metal) processes. The processes vary depending on the metal and materials being smelted. The advantage of smelting is that it concentrates metals into metal alloys or similar, therefore decreasing their entropy. Hydrometallurgical leaching puts metals into a low-temperature aqueous solution, reaching 90 °C under atmospheric conditions and 200 °C when under pressure for separating different elements and compounds. Again, various processes exist, based on either acid (low pH) or basic (high pH) aqueous solutions. Hydrometallurgy generally first increases the entropy of metals, before recovering the ions in solution with a significant input of energy for lowering the entropy again.

Pyro- and hydrometallurgy are commonly used in tandem for obtaining valuable metals, particularly noncarrier metals. Optimizing the entropy thus maximizes the resource efficiency. For many economically viable metallurgical plants, depending on the metal, pyrometallurgy does the first rough separation (concentration into speiss, metal alloy, matte, flue dust, etc.), and hydrometallurgy produces the final high-quality metals and other types of valuable products. Alloying usually happens during a second stage, for example, to produce specific alloy types in alloying and ladle furnaces. Contrary to such pragmatic pyro- and hydrometallurgical solutions, there have been various attempts to push pure hydrometallurgical recycling solutions. These often overlook that, when bringing elements into solution from complex waste such as WEEE, complex electrolyte solutions are created that have to be purified before metals can be produced from them. The purification inevitably creates complex sludge and residues that must be dealt with in an environmentally sound manner, which is often economically impossible and creates dumping/containment costs.

3.5 Purification and Refining Endprocesses

Purification and refining are metallurgical endprocessing for pure material production. Hydro- and electrometallurgical refining involves cementation, precipitation, solvent (liquid/liquid) extraction, or electrowinning/electrolysis/electrodeposition processes. In precipitation, Cu can be precipitated with Fe, Zn, or Al and Ag with Zn dust or NaCl at certain pH values. Cementation is the precipitation of a metal by other metals, which are more electropositive. An important aspect of cementation process is the use of metals which are already present in solutions, avoiding contamination with other ions. The general reaction of the cementation process of the cementation process is

$$Cu^{+2} + Me = Cu + Me^{+2} \qquad (3.1)$$

where Me is the cementing metal.

For Cu cementation Fe is the best precipitant [12].

In electrowinning and electrorefining of Cu, Zn, Au, Ag, etc., direct electrical current in electrolyte solution is used. More than 99% metal pureness can be obtained with electrowinning. Ferrous fraction is directed to steel plants for recovery

3.5 Purification and Refining Endprocesses

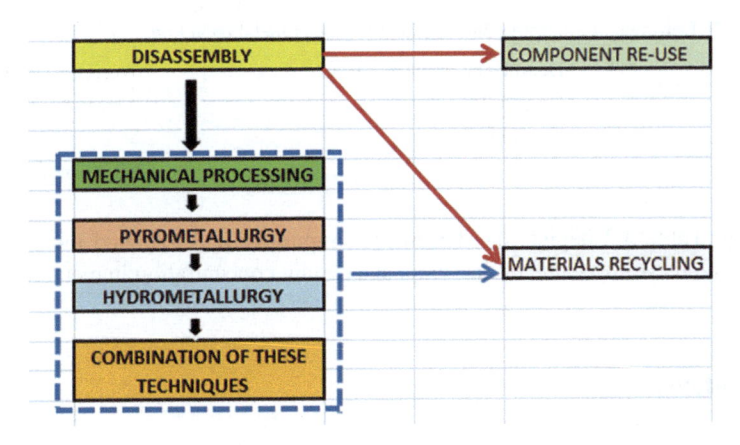

Fig. 3.3 E-waste disassembly and upgrading methods for recovery of components and recycling of materials

of Fe and Al fractions are going to Al smelters. At the same time, Cu/Pb fractions, PCBs, and other PMs-containing fractions are going to integrated metal smelters, which recover PMs, Cu, and other nonferrous metals while isolating the hazardous substances. The followings are also in the final processing steps:

- Base metal refinery
- Precious metal refinery
- Plastics recycling
- Batteries recycling
- Other component treatment
- Disposal of nonrecyclable residues

It is necessary to preprocess big electrical and electronic devices (i.e., fridges, TVs, washing machines, etc.) during recycling. It has to be noted that preprocessing of e-waste is not always necessary. Small and highly complex electronic devices such as mobile phones, MP3 players, etc. (after battery removal) can also be treated directly by an endprocessor to recover metals. Residuals will go to landfill or incineration. In terms of obtained products from WPCBs, there are two recycling categories, component recycling and materials recycling, whereas in terms of recycling techniques, five categories have been noted (Fig. 3.3) [13, 14]. Essentially, disassembly of WEEE necessitates the need to remove materials and ECs for reuse or remanufacture; the need to remove materials having negative environmental impact (Restriction of Hazardous Substances (RoHS) Directive bans placed on the EU market of new EEE containing more than agreed levels of Pb, Cd, Hg, Cr^{+6}, PBB, PBDE flame retardants); and the need to effectively segregate material streams to enhance yield in subsequent recycling processes.

It is a commercial prerequisite that the labor costs for dismantling and separation must be lower than the gained increase in the scrap value. Some impurities have a detrimental impact upon metal recycling and reduce the value of the recovered

Table 3.2 Detrimental impurities and tolerated metals in metal recycling

Metal	Detrimental impurity metals	Tolerated metals
Cu	Hg and Be	As, Sb, Ni, Bi, and Al
Al	Cu and Fe	Si
Fe		Cu, Sn, and Zn

fraction. Table 3.2 depicts the detrimental impurities and tolerated metals, which reduce the recycling value of the scrap, at a certain concentration in metal recycling.

References

1. Wang JB, Xu ZM (2015) Disposing and recycling waste printed circuit boards: disconnecting, resource recovery and pollution control. Environ Sci Technol 49:721–733. https://doi.org/10.1021/es504833y
2. Kaya M (2018) Waste printed circuit board (WPCB) recycling technology: disassembly and desoldering approach. In: Reference module in materials science and materials engineering/Encyclopedia of renewable and sustainable materials volume. Elsevier. https://doi.org/10.1016/B978-0-12-803581-8.11246-9
3. Mesker CEM, Hagelüken C, Van Damme G (2009) TMS 2009 Annual Meeting & Exhibition, San Francisco, California, USA, EPD Congress 2009 Proceedings Ed. by. SH Howard, P Anyalebechi, L Zhang, pp 1131–1136, ISBN No: 978-0-87339-732-2
4. Chancerel P, Meskers CEM, Hagelüken C, Rotter VS (2009) Assessment of precious metal flows during preprocessing of waste electrical and electronic equipment. J Ind Ecol 13:791–810
5. Kellner D (2009) Recycling and recovery. In: Hester RE, Harrison RM (eds) Electronic waste management, design, analysis and application. RSC Publishing, Cambridge, pp 91–110
6. http://www.ewasteguide.info/files/UNEP_2009_eW2R.PDF
7. Cui J, Forssberg E (2003) Mechanical recycling of waste electric and electronic equipment: a review. J Hazard Mater 99(3):243–263
8. Ogunniyi IO, Vermaak MKG, Groot DR (2009) Chemical composition and liberation characterization of printed circuit board comminution fines for beneficiation investigations. Waste Manag 29:2140–2146. https://doi.org/10.1016/j.wasman.2009.03.004
9. Yang C, Li J, Tan Q, Liu L, Dong Q (2017) Green process of metal recycling: Coprocessing waste printed circuit boards and spent tin stripping solution. ACS Sustain Chem Eng 5:3524–3535. https://doi.org/10.1021/acssuschemeng.7b00245
10. Cao R, Xiao S (2005) Recovery of gold, palladium and silver from waste mobile phone (in Chinese), rec. Metal 26(2):13–15
11. http://wedocs.unep.org/bitstream/handle/20.500.11822/8423/-Metal%20Recycling%20Opportunities%2c %20Limits%2c%20Infrastructure-2013Metal_recycling.pdf?sequence=3&isAllowed=y
12. Zhang K, Schoor JL, Zeng EY (2012) E-waste recycling: where does it go from here? Environ Sci Technol 46:10861–10867. https://doi.org/10.1021/es303166s
13. Kaya M (2018) Current WEEE recycling solutions, Chap. 3. In: Veglio F, Birloaga I (eds) Waste electrical and electronic equipment recycling, aqueous recovery methods, pp 33–93. https://doi.org/10.1016/B978-0-08-102057-9.00003-2
14. Kaya M (2018) Waste printed circuit board (WPCB) recycling: conventional and emerging technology approach. In: Reference module in materials science and materials engineering/Encyclopedia of renewable and sustainable materials volume. Elsevier. https://doi.org/10.1016/B978-0-12-803581-8.11246-9

Chapter 4
Dismantling and Desoldering

"Recycle, reduce, reuse (3Rs)… close the loop!"

Anonymous

Abstract This chapter emphasizes dismantling and desoldering techniques for electronic component and solder removal from the e-waste/WPCBs. Details and development in thermal and chemical desoldering for disassembly are covered extensively. Melting stoves, IR heating, hot fluid heating, and industrial waste heat use are introduced in thermal treatment process. Drum; tunnel; rod-brush and scanning-laser type desoldering automatic/semi-automatic WPCB dismantling machines are broadly explained. Industrial-scale equipment details and applications are presented. Heating with heat transfer liquids and chemical dissolving treatment, which damages ECs, are also mentioned. Sensing technologies, eco-design, design for disassembly, and active disassembly concepts are introduced.

Keywords Dismantling · Desoldering · Disassembly · Thermal treatment · Dismantling machine

4.1 Dismantling/Disassembly Process/System

Although WPCBs are designed for durability of 500,000 h, the average EoL for ECs is 20,000 h or less than 5% of its designed life span. Hence, many ECs are still functioning and usable at the time of disposal as e-waste. Therefore, disassembling ECs from WPCBs is a first and crucial step in WPCB recycling chain to conserve scarce resources and eliminate potential exposure to hazardous materials [1]. ECs are usually divided into two parts: (i) functional and usable ECs which could be reused for new products and repair and (ii) damaged ECs which could be disposed for metal recovery. Disassembly (demanufacturing/inverse manufacturing) process employs to segregate ECs and/or materials that are reusable, identifiable, or hazardous in such a manner as to maximize economic return and to minimize environmental pollution,

Table 4.1 Characteristics of connection types related to their specific degree and nonrandomness of liberation behavior after shredding with different connection complexities

Connection type	Liberation after shredding
Bolting/riveting	High liberation, high randomness
Gluing	Medium liberation, medium randomness
Coating/painting	Low liberation, low randomness
Foaming	Medium liberation, medium randomness
Connected materials/components of different levels of complexity	High/medium liberation from structure, high/medium randomness
Heterogeneous/high number of connections per surface area	Low liberation on component, low randomness
Material properties, low complexity, low number of joints	High liberation, high/medium randomness

enabling subsequent processes to be performed more efficiently. Disassembly is the systematic removal of ECs, parts, a group of parts, or a subassembly from e-waste.

ECs are connected to the PCBs mainly by soldering, bolting/riveting, gluing, coating/ painting, and foaming. Table 4.1 shows the characteristics of connection types related to their specific degree and nonrandomness of liberation behavior after shredding with different connection complexities. For ECs' disassembly, fracturing, drilling, ungluing, heating, and lubricating operations can be used. For disassembly, special tools, simple tools, or by hand can be used.

In electronics, desoldering is the removal of solder and ECs from a circuit board for troubleshooting, repair, replacement, and salvage (recycle/reuse). Large numbers of researches focusing on EC dismantling from WPCBs are conducted, which are mainly processed in two steps (Fig. 4.1) [2].

- Remove solder between baseboards and ECs, including wearing down the solder joints on the backside of PCBs by grinders [3]; dissolving the solder by chemical reagents [4]; melting the solder by infrared heaters and [5] electronic heating tubes [6, 7]; hot air [8, 9]; and special hot liquids [10–13] like molten salts (LiCl-KCl), melted solder, diesel, kerosene, paraffinic oil and methyl phenyl silicone oil. Special liquors can be used as the medium to transmit heat to melt solder from the PCB assemblies in desoldering pretreatment process [14–16].
- Dislodge ECs from WPCBs, including mechanical sweep [17], gas jet [8, 9], and centrifugation [18]. However, almost all of these technologies could not be applied for industrialization attributed to their various limits, such as low efficiency and high cost.

Current WPCB dismantling procedures have restrictions on e-waste recycling due to low efficiency and negative impact on environment and human health [20]. Therefore, there is a need to seek an environmental-friendly dismantling process. Manual disassembly and sorting are major cost element and high labor burden within any WEEE recycling methodology. Disassembly can be performed either partially or completely. The components reuse oriented selective disassembly technology for WPCBs with wet chemical selective desoldering which may prevent

4.1 Dismantling/Disassembly Process/System

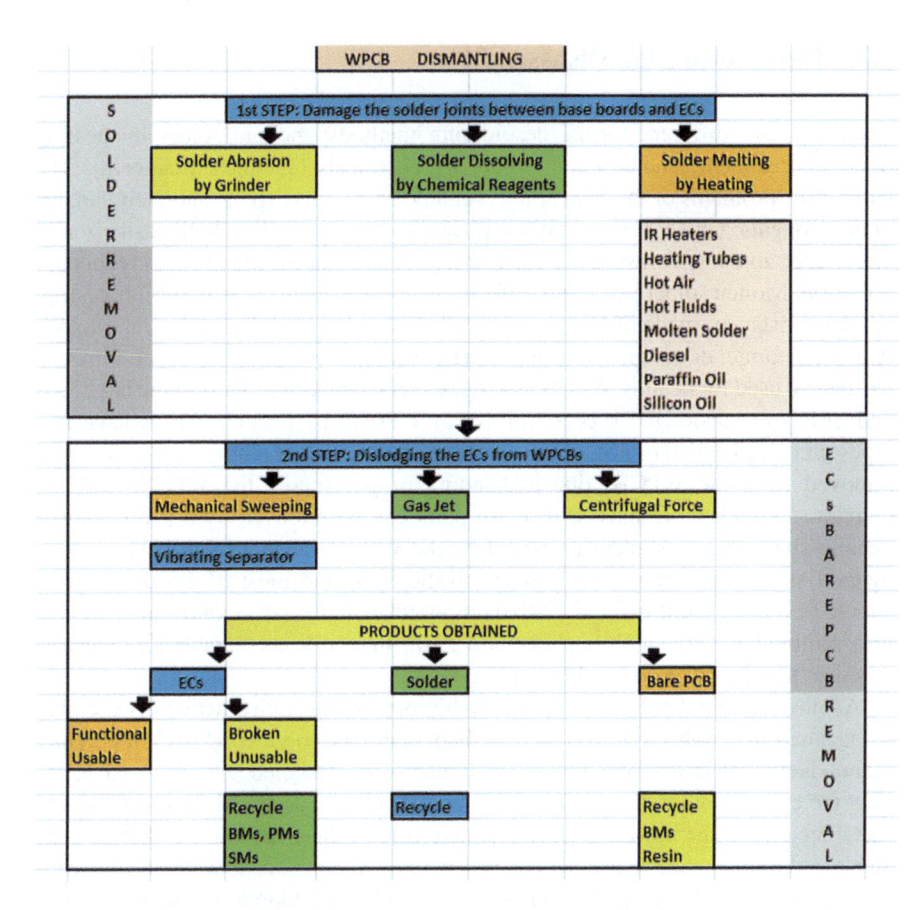

Fig. 4.1 WPCB dismantling steps [19]

also environmental pollution and recover expensive solder/tin. Desoldering is necessary for dismantling ECs from WPCB assemblies. The current informal manual selective dismantling uses chisels, hammers, and cutting torches to open solder connections and separate various types of metals and components. Cooking on a coal-/electric-heated plate and melt solder in order to sell the chips and other recovered components to acid strippers for further processing is another manual process. Manual informal dismantling has a bad smell, black fumes, and banned process disadvantages.

4.2 Desoldering for Disassembly

There are two different formal desoldering methods: melting solder by heating (thermal treatment) and leaching solder by chemicals (chemical treatment). The latter faces problems of the selection of suitable chemical reagents and their damage to components, which prevents this application in practice. Leaching easily generates a large amount of waste acid, alkaline liquid, and sludge, which cause secondary pollution. Molten solder baths and different heating elements (i.e., wire/ribbon/strip/tabular resistances made of Nichrome (80/20 NiCr), Kanthal (FeCrAl), Cupronickel (CuNi), ceramic, or composite), infrared (IR) sources, hot air guns, hot air blowers, etc. can be used for heating WPCB assemblies for desoldering. Most of the large ECs and SMCs are dislodged successfully from WPCBs by heating. THCs have two types: pins are bended or not. THCs cannot be disassembled although all solder was removed from the WPCBs due to bending of pins under the WPCBs and big components with more pins. But, these parts can be removed by a plyer without heating since there no solder remained on the WPCBs after heating [21]. Heating method should give minimum damage to the ECs, and most of the components should be reused. Dodbiba et al. (2016) proposed an underwater explosion method to disassemble the whole mobile phone units; but, since it is not suitable for industrial practice, it will not be covered here [22].

Another potential solution to disassembly problem is called active disassembly using smart materials (ADSM) (such as shape-memory polymers (SMP) of polyurethane). New materials – based on bio-polymers – are designed and produced, to see how the component parts can be quickly and efficiently accessed for recycling and metal recovery. The most important characteristic of these materials is that they are stable and robust while in use but can be triggered to decompose when the device is to be taken apart for recycling. Prototype modularized mobile phone based on a "skeleton" made of plastic/cellulose composite which can be dissolved into sugars in the presence of engineered bacteria, while components such as battery, screen, motherboard and memory attached to the skeletons as organs.

4.2.1 Heating Methods (Thermal Treatment) for Desoldering

Working temperature, which is around 225–265 °C, is above the melting temperature of the solder. Critical temperature to generate toxic fumes of PCBs is 270–280 °C [23]. During heating certain amount of toxic gases including acetaldehyde, benzene, xylene, styrene, and lead fume can be released and cause a secondary pollution. Heating of WPCBs is performed in a tin melting stove and in a rotating drum dismantling machines, horizontal tunnel furnaces, or IR-heated rods/lamps (250 W) and brushes. Heating can be performed by resistances, which are 6–8 cm away from WPCB or IR lamps in tunnels and hot air guns or air blowers in rotating drums. ECs are removed from heated boards manually after tin melting stoves and

4.2 Desoldering for Disassembly

Fig. 4.2 Tin melting stove layout and operation for EC dismantling from WPCBs

semiautomatically after rotating drum and tunnel-type dismantling machines. Heating methods have very low efficiencies in melting solders, and they frequently result in damaged components because of the differences in heat absorption rates and temperature gradients. Explosion (electrolytic capacitors may explode at 140 °C) and burning of ECs (some plug-ins may burn) should be prevented during heating period. Disassembling rate (DR) percent can be calculated from

$$\text{DR }(\%) = \frac{\text{Number of ECs released from WPCBs}}{\text{Total number of components on WPCBs}} * 100 \qquad (4.1)$$

4.2.1.1 Tin Melting Stoves

ECs are connected to WPCBs by solder. In order to disassemble ECs, firstly solder must be melted and removed (Fig. 4.2). In tin melting stoves, electricity is used in heating. When the solder melts, ECs are removed by hitting and shaking the board. Most of the ECs fall from the board. Industrial MX-300 melting stoves have a power of 4.0 KW, a capacity of 0.4–0.8 t/h/worker/table, and a dimension of 720*500*390 mm [24]. One worker works at one tin melting stove table under fume hoods.

4.2.1.2 Infrared Heating

IR heating uses IR heat lamps, which emit invisible IR radiation and transmit it to the body that is being heated. IR heating lamps are commonly incandescent lamps. An IR heater transfers energy to a body with a lower temperature through electromagnetic radiation. Depending on the temperature of the emitting body, the wavelength of the peak of the IR radiation ranges from 700 nm to 1 mm (at frequencies between 430 THz and 300 GHz). No contact or medium between the two bodies is needed for the energy transfer. IR heaters can be operated in vacuum or atmosphere and satisfy a variety of heating requirements, including:

- Extremely high temperatures, limited largely by the maximum temperature of the emitter
- Fast response time, on the order of 1–2 seconds
- Temperature gradients, especially on material webs with high heat input
- Focused heated area relative to conductive and convective heating methods
- Non-contact, thereby not disturbing the product as conductive or convective heating methods do

Thus, IR heaters are applied for many purposes including heating systems, curing of coatings, plastic shrinking, plastic heating prior to forming, plastic welding, glass and metal heat treating, and cooking.

4.2.1.3 Heating with Heat Transfer Liquids (Hot Fluids)

Bulk heating of WPCBs for desoldering can be achieved in water-soluble ionic liquids ((BMIm)BF$_4$) and [20] dielectric liquids (transformer oils, mineral oils, methyl phenyl silicone oil, and cryogenic liquids O$_2$, N$_2$, H$_2$, He, and Ar). Ninety percent of the ECs were removed from waste PCBs at 250 °C with water-soluble ionic liquids. But most of them cannot be used again. These ECs are separately recycled according to their metal contents. Inert and stable molten salts (LiCl + KCl (58,2% + 41,8% mol) and NaOH-KOH eutectic composition (41% NaOH-59% KOH) ($T_{melting}$: 170 °C)) can be used as a heat transfer fluid to dissolve glasses and oxides and to destruct plastics present in PCBs without oxidizing most of the metals. At a range of 450–470 °C, metal products in either liquid (solder, Zn, Sn, Pb, etc.) or solid (Cu, Au, Fe, Pd, etc.) form can be recovered. PCB pyrolysis gives 70% solid (metal and glass), 23% oil, and about 6% gas product in average [16, 25]. Disposing the used hot fluid and cleaning of bare boards and components for further processing are disadvantages of using heat transfer liquids. Hot fluids have the disadvantages of generating large quantities of hazardous waste that need to be further disposed and bare boards and components that need to be further cleaned [1].

4.2.1.4 Heating with Industrial Waste Heat

Chen et al. (2013) used industrial waste heat (i.e., pulsating air jet) to melt solders and separate the ECs from base PCBs [1]. They tested preheat temperatures in increments of 20 °C between 80 and 160 °C and incubation times of 1, 2, 4, 6, or 8 min and heating source temperatures ranging from 220 to 300 °C in increments of 20 °C. The optimum preheat temperature, heating source temperature, and incubation time were 120 °C, 260 °C, and 2 min, respectively. The disassembly rate of small SMCs was 40–50%, and THC and other SMCs were more than 98%. Inlet pressure was 0.5 MPa.

4.3 Semiautomatic PCB Electronic Component-Dismantling Machines

PCB dismantling machines are used to remove the solder and ECs from mixed WPCB assemblies automatically in e-waste recycling. Thus, labor force is saved and only two people are enough for easy operation. Wang et al. (2016) used 245–265 °C heating temperature, 2–8 min. incubation time, and 6–10 rpm rotating speed for drum type of automated dismantling machine [21]. According to Boks and Tempelman (1998), the main obstacles preventing automatic disassembly of PCBs from commercially successful are [26]:

- Too many different types of products
- Very small products
- General disassembly-unfriendly product design
- Material logistics problems
- Control of functioning ECs

4.3.1 Drum-/Barrel-Type Dismantling Machines

The high-temperature and abrasion-resistant cylindrical drum/barrel is made of 6 mm thickness steel. This machine works with gas/dust cleaning system or under exhaust hood (Fig. 4.3). Dismantling machines should be safe and reliable, be operated easily, and have stable performance, high precision, and durability characteristics for WPCB recycling plants. Solder and dismantled ECs without damage are removed in one step by blowing hot air in the drum by a blower. WPCB dismantling machines have automatic temperature control system. Capacities ranges from 200 to

Fig. 4.3 Drum-type dismantling machine with air cleaning system and dismantled WPCBs and ECs developed in China

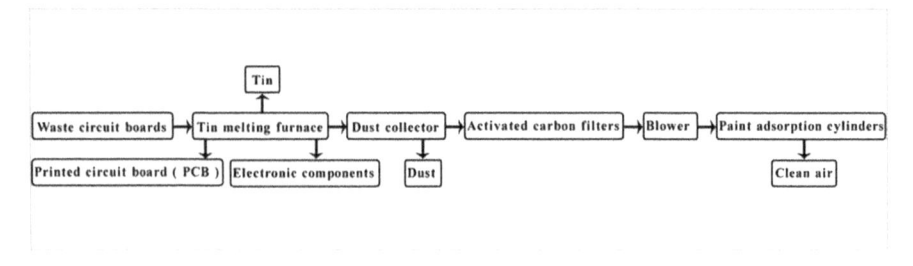

Fig. 4.4 Tunnel-type dismantling machine flowsheet

500 kg h^{-1}, motor power from 2.2 to 3.7 KW, and weight from 0.35 to 0.60 tons [27]. EC dismantling rate is claimed as high as 99% without dust in an eco-friendly way. The use of electrical resistances instead of hot air blowers is also possible for heating [28]. The off-gas purification with activated carbon is used to avoid hazardous gas discharge to the environment so as to protect workers' health and environment [21].

4.3.2 Tunnel-Type Dismantling Machines

This system consists of a horizontal solder melting furnace with IR heaters, automated dismantling, dust catching, conveying, and panel systems. Higher automation reduces dismantling labor and time. Dust catching prevents dust pollution in the environment. It requires less area in the plant. Dismantling machine technical block flowsheet is given in Fig. 4.4. From this operation solder/Sn, bare PCB assemblies, ECs, and dust are collected. Tunnel-type dismantling machine cross-sectional details and photos are shown in Fig. 4.5.

According to EPA, e-waste is the fastest-growing waste stream in the USA. Significant amounts are recoverable for reuse, resale, and refining for PMs [27]. The disassemblers send WPCBs to refineries/smelters for PM recovery. American-type e-waste disassemblers remove electronic parts from these WPCBs, so only certain parts are smelted. As such they are the key market for depopulator equipment. Refineries also have an incentive to remove the parts so that only certain parts are smelted. Both are WPCBs without using chemicals. The PM parts go to the smelter as opposed to the whole WPCBs. The weight reduction factor is 5:1 which reduces smelting-based pollution by 5:1. Figure 4.6 shows the tunnel-type WPCB depopulator developed in the USA. WPCBs on the conveyor are heated in the tunnel furnace, and ECs are fallen off with shaker from WPCBs at the end of conveyor.

200 kg h^{-1} capacity dismantling machine has external dimensions of 8135*1600*3260 mm. Total power supply is 15.2 KW and total weight is 1.37 tons. Dust-removing efficiency is more than 95% [31]. Table 4.2 shows technical data about tunnel-type industrial-scale machines. Dismantling machine heating

4.3 Semiautomatic PCB Electronic Component-Dismantling Machines

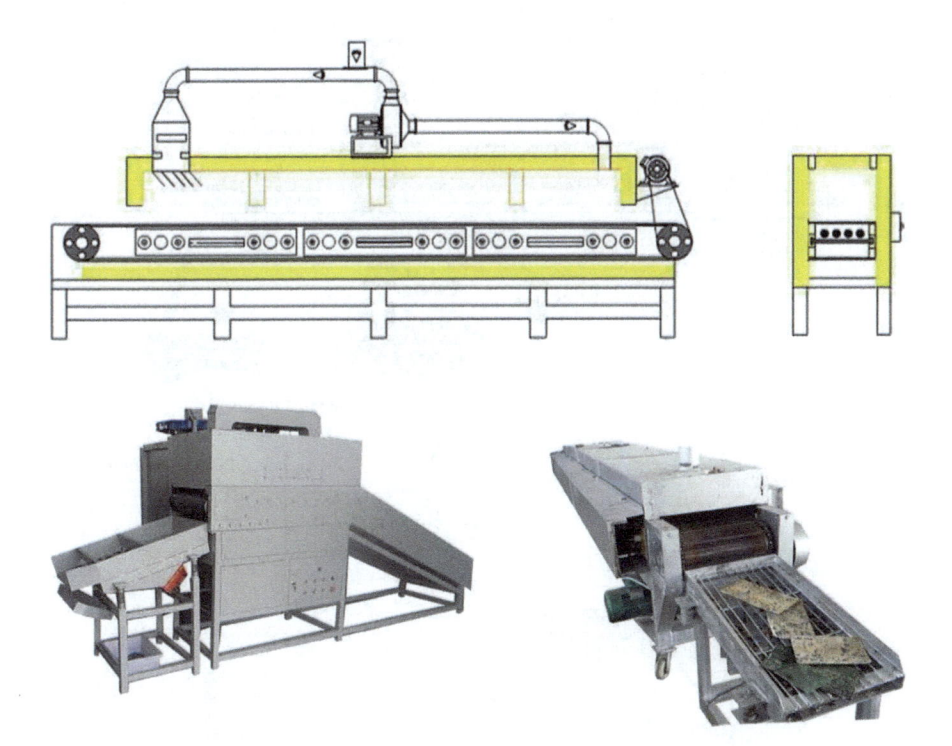

Fig. 4.5 Schematic view and photos of tunnel-type, heated dismantling machines [29]

resistance power is 18KW and feeder motor 0.75KW. Vibrating screen dimensions are 0.8∗1.2∗0.7 m.

4.3.3 Rod- and Brush-Type EC Disassembly Apparatus with IR Heater

Park et al. (2015) designed and tested automatic rotating rods and sweeping steel brush-type apparatus to dismantle ECs from WPCBs [32]. This apparatus used six IR heaters, three rows of rotating six-feeding rods, and six EC removing steel brushes (Fig. 4.7). Ninety-four percent disassembly ratio was obtained at a feeding rate of 0.33 cm s^{-1} and a heating temperature of 250 °C.

1. Control Panel (controllable factors: rotating speed of feeding rod, rotating speed of steel brush, heating temperature), 2. PCBA, 3. Feed hopper, 4. Feeding rod, 5. Steel brush, 6. IR heater, 7, 8. Trapezoidal gear change, 9. Product hopper, 10. Basket, 11, 12. Motors. (right) Detailed diagram of the disassembly modules. 13. PCBA, 14. Feed hopper, 15. Feeding rod, 16. Steel brush, 17. IR heater, 18. Product hopper.

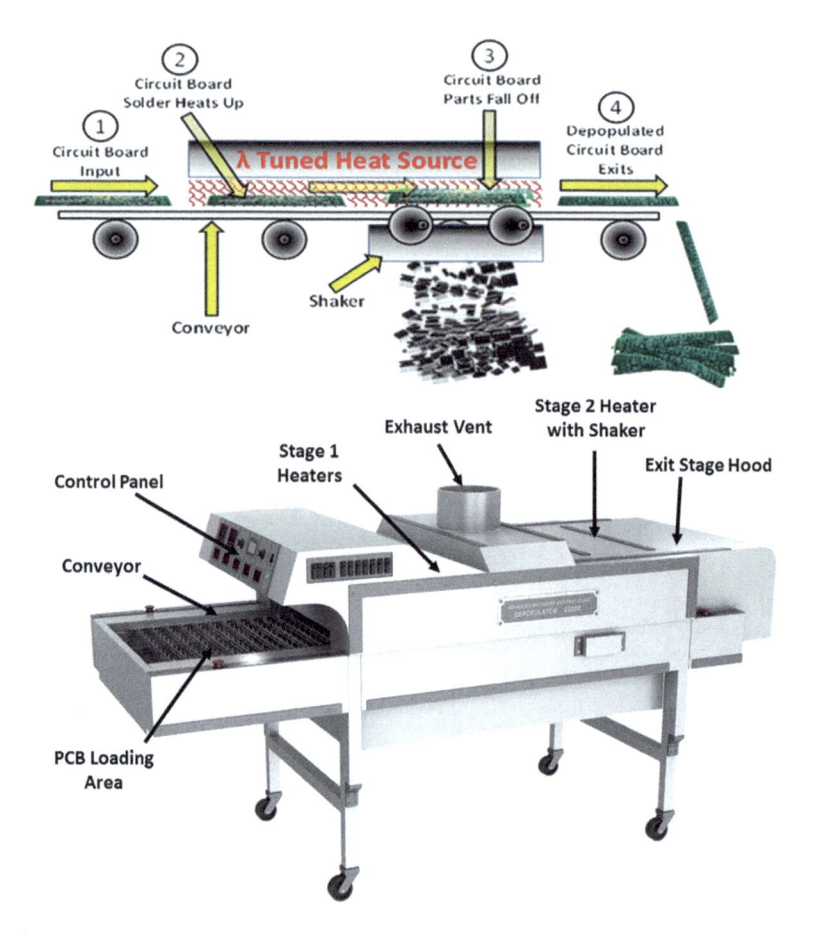

Fig. 4.6 Depopulator schematic diagram and rendering of the D2000 – the production depopulator [30]

Table 4.2 Technical parameters of tunnel-type dismantling machines

Name	Model	Power	Qty (pcs)	Dimensions
Melting furnace	SX-300	3000 W/ AC220 V	2	0.65 * 0.43 * 0.75 m
Automate dismantling machine	SX-600 SX-800	1.1 KW 1.1 KW	1 1	5 * 0.8 * 1.5 m 6 * 1 * 1.5 m

4.3 Semiautomatic PCB Electronic Component-Dismantling Machines 79

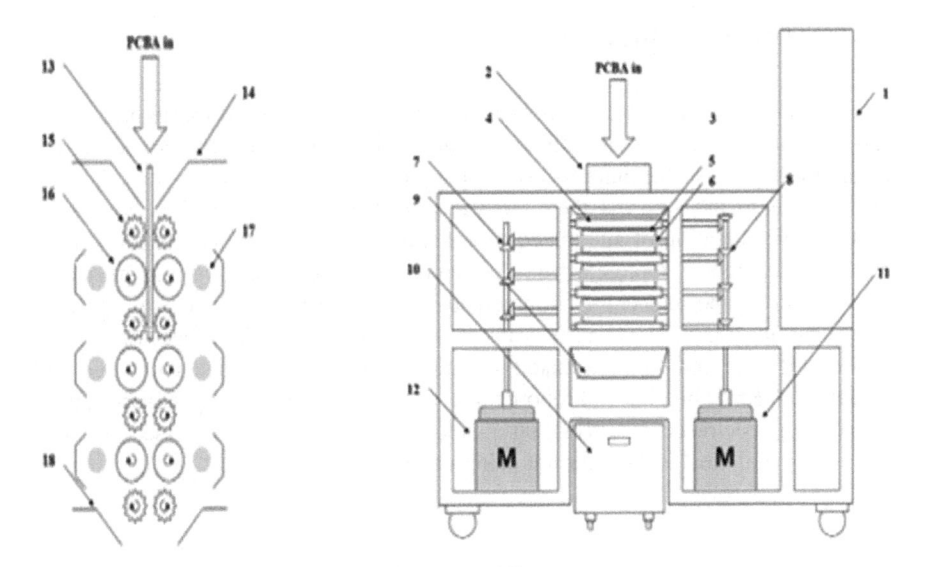

Fig. 4.7 Rod- and brush-type EC disassembly apparatus: (left) structure of the disassembly apparatus

4.3.4 Scanning and Laser Desoldering Automated Component-Dismantling Machine

From scrap, redundant, or malfunctioning PCB assemblies, an automated EC disassembly methodology was developed by the Austrian Society for Systems Engineering and Automation (SAT). While the existing production capacity deals with the recovery of relatively expensive ECs from faulty products and overcapacity manufacture from a number of German, Hungarian, and Austrian original equipment manufacturers (OEMs), the excess potential is used for disassembling WPCBs. SAT's technology essentially comprises automated component scanning to read all component identification data (ID) and dual-beam laser desoldering, with vacuum removal of selected components. The component disassembly operation comprises the following stages [33]:

- Scanning – read all component ID.
- Read stored component database – component cost data stored.
- Are the ID components soldered or surface mounted?
- If mounted, disassembly via robot in 3–5 s at a cost of 0.5 Euro.
- If soldered – three types – highest quality via laser with minimum thermal input (18–20 s/component) – lower quality and BGA (ball grid arrays) via IR heat input

4.4 Sensing Technologies

The most attractive research on disassembly process is the use of image processing and database to recognize reusable parts and toxic components. In automating the disassembly process, the main issues are imaging, recognition, and robotics. Sensing methods can greatly improve the effectiveness of WEEE recycling. They are crucial to the implementation of automated disassembly and can facilitate great improvements in separation. Opto-electronic sorters use conventional imaging device to discriminate shape and color. Augmentation by electromagnetic sensing permits identification of metals, as well as rubbers and plastics, allowing selective ejection of the identified items in automated separation processes. For better imaging and recognition, RFID tagging and robotics will be the future technology. Laser-induced breakdown spectroscopy (LIBS), laser-induced fluorescence, laser-based systems, and X-ray analytical techniques are being developed [33].

4.5 Eco-design/Design for Disassembly (DfD) Concept

DfD would be inherent from product conception and would involve the selection of ECs, to standardize material types and specifications and to both minimize and simplify fastener types. In order to facilitate disassembly [33]:

- Use biodegradable materials where possible.
- Provide accessibility to parts and fasteners to support disassembly.
- Use standardized joints to minimize number of tools for disassembling
- Modulate design for ease of part replacement
- Use connectors instead of hard wiring.
- Use thermoplastics instead of thermoset adhesives.
- Use snap-fit techniques to facilitate disassembly.
- Design product with weak spots to aid disassembly.
- Weight minimisation of individual ECs.

4.6 Active Disassembly (AD)

AD involves the disassembly of ECs using an all-encompassing stimulus, rather than a fastener-specified tool or machine. When designing for AD, we have to consider smart materials like shape-memory polymers (SMPs) and shape-memory alloys (SMAs). SMPs are polymeric smart materials that have the ability to return from a deformed state (temporary shape) to their original (permanent) shape induced by an external stimulus (trigger), such as temperature change, often in the form of screws, bolts, and rivets. AD fasteners change their form to a preset shape when exposed to a specific trigger temperature (65–120 °C) depending on the material.

References

1. Chen MJ, Wang JB, Chen HY, Oladele AO, Zhang MX, Zang HB, Hu JK (2013) Electronic waste disassembly with industrial waste heat. Environ Sci Technol 47(21):12409–12416. https://doi.org/10.1021/es402102t
2. Wang JB, Xu ZM (2015) Disposing and recycling waste printed circuit boards: disconnecting, resource recovery and pollution control. Environ Sci Technol 49:721–733. https://doi.org/10.1021/es504833y
3. Lee J, Kim YJ, Lee JC (2012) Disassembly and physical separation of electric/electronic components layered in printed circuit boards (PCB). J Hazard Mater 241–242:387–394. https://doi.org/10.1016/j.jhazmat.2012.09.053
4. Haruta T, Nagano T, Kishimoto T, Yamada Y, Yuno T (1991) Process for removing in and tin-lead alloy from copper substrates, US Patent, issued on July 30, 1991
5. Yokoyama SM (1997, May 5–7) Recycling of printed wiring boards with mounted electronic parts. In: IEEE international symposium on electronics and the environment, San Francisco, California. IEEE, New York, pp 109–114
6. Ding X, Xiang D, Lium X, Yang J (2008) Delamination failure in plastic IC packages disassembled from scrapped printed circuit boards engineering. In: Proceedings of the global conference on sustainable product development and life cycle sustainability and remanufacturing, Busan, Korea. Sep 29-Oct 1.
7. Gao P, Xiang D, Yang J, Cheng Y, Duan G, Ding X (2008) Optimization of PCB disassembly heating parameters based on genetic algorithm. Mod Manuf Eng 8:92–95, In Chinese.
8. Pan XY, Li ZL, Zhi H, Wang L (2007) Method and apparatus of separation for electronic components and solders from printed circuit boards. Chinese patent; 2007102015321
9. Li ZL, Zhi H, Pan XY, Liu HL, Wang L (2008) The equipment of dismantling for electronic components from printed circuit boards. Chinese patent; 2008103057561
10. Guo F (2007) Composite lead-free electronic solders. J Mater Sci Mater Electron 18:129–145. https://doi.org/10.1007/s10854-006-9019-1.
11. Zhang H, Song S, Liu Z, Zhang H et al (2008) Research on disassembling electronic components on discarded PCBs with hot fluid. Mach Des Manuf 4:101–103. (in Chinese)
12. Frank MC (2002) The recycling of computer circuit boards, http://p2pays.org/ref/02/01469.pdf.
13. Zhao Z, Wang Y, Song S, Liu G, Liu Z (2009) Development and application of unsoldering equipment for printed circuit board scraps. Modular Mach Tool Autom Manuf Technol 10:95–98. https://doi.org/10.3969/j.issn.1001-2265.2009.10.026
14. Huang K, Guo J, Xu Z (2009) Recycling of waste printed circuit boards: a review of current technologies and treatment status in China. J Hazard Mater 164:399–406. https://doi.org/10.1016/j.jhazmat.2008.08.051
15. Flandinet L, Tedjar F, Ghetta V, Fouletier J (2012) Metals recovering from waste printed circuit boards (WPCBs) using molten salts. J Hazard Mater 213–214:485–490. https://doi.org/10.1016/j.jhazmat.2012.02.037
16. Riedewald F, Gallagher MS (2015) Novel waste printed circuit board recycling process with molten salt. Methods X 2:100–106. https://doi.org/10.1016/j.mex.2015.02.010
17. Wang YL, Wang QY, Gao M, Chen FG, Liu FG (2013) A method for separation of components and solder from waste printed circuit boards, Chinese patent: 2013101808637
18. Zhou Y, Qiu K (2010) A new technology for recycling materials from waste printed circuit boards. J Hazard Mater 175(1–3):823–828. https://doi.org/10.1016/j.jhazmat.2009.10.083
19. Kaya M (2018) Waste printed circuit board (WPCB) recycling technology: disassembly and desoldering approach. In: Reference Module in Materials Science and Materials Engineering/Encyclopedia of Renewable and Sustainable Materials, Elsevier. https://doi.org/10.1016/B978-0-12-803581-8.11246-9
20. Zeng X, Jinhui L, Henghau X, Lili L (2013) A novel dismantling process of waste printed circuit boards using water-soluble ionic liquid. Chemosphere 93:1288–1294. https://doi.org/10.1016/j.chemosphere.2013.06.063

21. Wang J, Guo J, Xu Z (2016) An environmentally friendly technology of disassembling electronic components from waste printed circuit boards. Waste Manag 53:218–224. https://doi.org/10.1016/j.wasman.2016.03.036
22. Dodbiba G, Murata K, Okaya K, Fujita T (2016) Liberation of various types of composite materials by controlling underwater explosion. Miner Eng 89:63–70. https://doi.org/10.1016/j.mineng.2016.01.017
23. Duan H, Hou K, Li JH, Zhu XD (2011) Examining the technology acceptance for dismantling of waste printed circuit boards in light of recycling and environmental concerns. J Environ Manag 92:392–399. https://doi.org/10.1016/j.jenvman.2010.10.057
24. http://en.jxmingxin.com/circuit-board/33.html
25. European Parliament (2012) Directive 2012/19/EU of the European Parliament and of the Council of 4 July 2012 on waste electrical and electronic equipment (WEEE). Off J Eur Union L197:38–71
26. Boks C, Tempelman E (1998) Future disassembly and recycling technology –results of a Delphi study. Futures 30:425–442. https://doi.org/10.1016/S0016-3287(98)00046-9
27. https://www.alibaba.com/product-detail/Energy-Saving-PCB-Dismantling-MachineElectronic_ 6047480 1054.html
28. http://mtmakina.com/en/other/pcb-component-dismantling-machine.html
29. Kaya M (2018) Current WEEE recycling solutions, Chap. 3. In: Veglio F, Birloaga I (eds) Waste electrical and electronic equipment recycling, aqueous recovery methods, pp 33–93. https://doi.org/10.1016/B978-0-08-102057-9.00003-2
30. https://cfpub.epa.gov/ncer_abstracts/index.cfm/fuseaction/display.highlight/abstract/10489/report/F
31. http://www.ewasterecyclingmachine.com/pcb-dismantling-machine/
32. Park S, Kim S, Han Y, Park J (2015) Apparatus for electronic component disassembly from printed circuit board assembly in e-wastes. Int J Miner Process 144:11–15. https://doi.org/10.1016/j.minpro.2015.09.013
33. Kellner D (2009) Recycling and recovery. In: Hester RE, Harrison RM (eds) Electronic waste management, design, analysis and application. RSC Publishing, Cambridge, pp 91–110

Chapter 5
Traditional and Advanced WPCB Recycling

"Recycling is the Key to a Clean and Safe Environment"
Anonymous

Abstract Conventional and advanced/novel WPCB and e-waste recycling processes are introduced and compared in this chapter. Traditional uncontrolled incineration and mechanical separation (i.e., gravity separation) methods are covered. Novel pyrometallurgical methods for WPCB recycling include direct smelting, incineration, physico-mechanical separation, vacuum pyrolysis, and gasification. Limitations and emerging technologies in pyrometallurgy are also presented. Industrial pyrometallurgical processes for the recovery of metals from e-waste are stated. Hydrometallurgy, purification, solvent extraction, ion exchange, and electrowinning are explained in detail. Advantages and disadvantages of base metals, precious metals, brominated epoxy resin, and solder stripping solvents are compared. Possible chemical reactions between metals and reagents are presented. Finally, water treatment processes are mentioned shortly.

Keywords Landfill · Incineration · Mechanical separation · Pyrometallurgy · Pyrolysis · Hydrometallurgy · Bioleaching · Purification · Solvent extraction

5.1 Comparison of Traditional and Advanced WPCB Recycling Processes

WPCB treatment and recycling methods can be classified as thermal or nonthermal, chemical or mechanical, pyrometallurgical or hydrometallurgical, advanced or traditional/conventional, and primitive or direct or according to reaction atmosphere. In this book, WPCB treatment and recycling techniques will be classified as traditional and advanced recycling techniques. Table 5.1 shows the classification of traditional and advanced/novel WPBC recycling methods. Traditional recycling techniques include direct treatment, which includes uncontrolled incineration, and mechanical

Table 5.1 Classification of traditional and advanced methods for WEEE recycling

Traditional methods for WEEE recycling	Advanced methods for WEEE recycling

Direct treatment: uncontrolled incineration (i.e., burning in blast furnace)
Mechanical separation
   Wet gravity separation (i.e., shaking tables)
   Dry gravity separation (i.e., zigzag three-way air classifier)
Primitive acid leaching (solubilization + purification)
	Pyrometallurgical technology (smelting + converting + refining)
	Physico-mechanical separation (Gravity-magnetic-electrostatic)
	Hydrometallurgical processes
	Mild extraction technology
	   Bioleaching
	   Biosorption
	Electrochemical technology
	Supercritical fluid technology (For water: T, 374 °C; P, 218 atm.)
	Vacuum metallurgical technology (Vacuum evaporation/sublimation)
	Other novel technologies (Ultrasonic, mechanochemical technologies, DC arc plasma treatment, hydrothermal method)

Compiled from Zhang and Xu (2016) and Kaya (2018) [2, 3]

separation along with primitive crude acid leaching techniques. Pyrometallurgy, physico-mechanical separation (gravity/magnetic/electrostatic separations), hydrometallurgy, electrochemical, vacuum metallurgical, supercritical, and other novel technologies are advanced new technologies. In this book, all traditional technologies were covered in detail. In the conventional metal recovery process, whole WPCBs are typically crushed, ground, and smelted without disassembling the ECs [1]. Disadvantageously, when WPCBs are ground, only the plastic fraction can be effectively liberated from metals. However, this leads to the recovery of only Cu, Au, and Ag metals and loss of low concentration rare metals. These rare metals can only be recovered by hydrometallurgical processes. Therefore, it is imperative that ECs should be disassembled by desoldering in order to recover these valuable and scarce metals.

In the pyrometallurgy, the metal is separated from other impurities of the matte using differences in the melting points, densities, and other physical/chemical characteristics. In the hydrometallurgy, metals are separated using differences in solubilities and other electrochemical properties. There are two industrially driven ways of processing WPCBs, based on pyrometallurgical or hydrometallurgical routes or a combination between both. Both have their pros and cons, but especially for "low-grade" WPCBs, hydrometallurgy (i.e., leaching) offers distinct advantages, such as selectivity and reduced capital and operational costs. One way to explore the inherent advantages of the hydrometallurgical processing as downstream treatment is to prepare the WPCBs in an optimal way. This is realized through fragmentation followed by classification and physical sorting techniques aiming to selectively concentrate the valuable elements in specific streams. Further on, only the metal-bearing fractions are supposed to be delivered to leaching where metals are solubilized into a pregnant leach solution followed by their recovery using techniques such as solvent extraction (SX) and electrowinning (EW).

5.2 Traditional Processes for Mixed WPCB Recycling

Traditional methods are performed for partial recovery of MFs from WPCBs.

5.2.1 Uncontrolled Incineration

Incineration means the combustion of WPCB by converting its calorific value to energy and emitting the gas directly or after treatment with the purpose mainly to remove the nonmetallic fraction part (around 70 wt%). It has the advantages of significantly reducing the volume of WPCB by 50%, and also the calorific value of WPCB is relatively high compared to municipal solid waste, which is around $9.9 * 10^4$ kJ/kg. Therefore, it readily satisfies the minimum incineration calorific value for waste, which is roughly 5000 kcal/kg. Currently in the world, incineration is still widely used in America, Asia, and Europe due to the simplicity of the process. However, during the combustion of WPCB, toxic emissions including heavy metals, fly ash, PCDD/Fs, and PBDD/Fs are released into the atmosphere in the absence of post-purification. Cd, Cu, Ni, Pb, and Zn will be vaporized according to the order of their melting points and released into the atmosphere [4]. Therefore, incineration as a treatment method for WPCB is not an environmentally friendly option considering the toxic emissions. Also, the construction of an incineration plant is a major expenditure for local governments.

5.2.2 Mechanical Separation

Physical processes are commonly applied during the upgradation stage when various metals and nonmetals contained in e-waste are liberated and come apart by some means of shredding and crushing processes. The effort to recover the valuable metals in particular Au, Ag, Pd, and Cu has received enormous attention in recent years using extraction processes such as physical, chemical, and combined pyro-/hydroleaching separation routes.

Generic mechanical and physico-mechanical separators use the laws of physics and material properties for separating materials. Mechanical separation mainly includes sorting and gravity separation, while physico-mechanical separation contains gravity + magnetic + electrostatic separations together. Mechanical separation is traditional and physico-mechanical separation is advanced treatment method for WPCBs. Variables affecting separation include [5]:

- Fluid viscosity (ease of flow).
- Solid volume concentration (significantly affects viscosity).
- Flow turbulence (and turbulent energy dissipation).
- Boundary flow along the separator walls.
- Eddy flows, i.e., swirls that dissipate energy and efficiency.
- Particle shape affecting terminal velocity, e.g., platelike particles will fall differently than more spherical ones (shape factor).
- Hindered settling, i.e., particles affect each other's flow.
- Momentum transfer between phases and within a phase.
- Uneven force field separation, e.g., due to technological constraints.
- Zeta potential, e.g., the electrical surface charges on particles.
- pH of fluid.
- Physical distribution of particle properties, e.g., density, conductivity, and magnetism (and other properties than those used for separation).

None of the separation methods creates pure streams of material but only increases purity. Overall industrial recovery rate of mechanical separation changes from 80% up to 95% [5]. The physical/mechanical process comprises shredding and grinding of the whole mixed e-waste, followed by separation and concentration using gravity, magnetic, and/or electrostatic separators. Accordingly, mechanical methods do not result in high recovery rates, especially for PMs. Mechanical separation process flowsheet for a mixed WEEE metal recovery plant has a manual sorting line to remove batteries, toners/inks, paper, and external cables; shredding and manual picking line/band to remove large capacitors, motors, and transformers; overband magnetic separation to remove coarse ferrous metals; pulverizing followed by magnetic separation to remove again fine ferrous metals and steels; eddy current separators (ECS) to separate metallic (Cu and Al) and nonmetallic (plastics, rubbers, glass, wood, and stone); and finally Cu-rich and Al-rich material that can be separated by density separation [6]. Purely mechanical preprocessing leads to major losses of, especially, PMs in dust and ferrous fractions. With dry mechanical separation processes, the potential loss of PMs may be as high as 10–35% due to liberation problem between metals and plastics. To evade pollution with dust, a three-stage dust removal equipment is commonly used.

Preprocessing industry treats WEEE using a combination of manual and mechanical methods, with varying efficiency. Much innovation is possible for improving preprocessing performance, and incentives must be created for treating all WEEE with certified BAT technology. E-wastes are managed without proper handling by informal sectors in underdeveloped and developing countries. Since the informal sectors cannot invest in high-tech equipment and machines, it is important to provide a more economical yet environmentally friendly mechanical separation solution to this problem. Therefore, it is important to provide simple physical material recovery procedures, which can be applicable for informal sectors.

5.3 Advanced Methods for WPCB Recycling

5.3.1 Pyrometallurgy

Thermal/thermochemical processes at high temperature are currently the most important alternative solution for processing complex e-waste. They involve the simultaneous change of chemical composition and physical phase and are irreversible. The purpose of thermochemical treatment of e-waste is elimination of organic components (i.e., plastics) while leaving nonvolatile mineral and metallic phases in more or less original forms that could be recovered.

In pyrometallurgy, direct treatment (without size reduction and removal of attached ECs) targets Cu and PM recoveries. In this case, the initial pyrometallurgical step is generally followed by subsequent hydrometallurgical and electrometallurgical operations. During pyrometallurgical treatment, the polymers are used as reduction agent as well as energy source due to their intrinsic calorific value. Pyrometallurgical processes for recovering metals from various waste materials have been used during the last three decades. Smelting in furnaces, incineration, combustion, and pyrolysis are typical e-waste recycling processes. The high temperature in furnace or smelter is generated via the combustion of fuel or via electrical heating. Technical hardware includes submerged lance smelters, rotary furnaces, electric arc furnaces, etc. In contrast to WPCBs from computers, monitors and TV boards usually contain massive Fe and Al parts (cooling elements, transformers, frames, etc.) and possibly large condensers. It is recommended to remove these massive parts before sending the remaining boards to the integrated smelters for final processing. Benefits of such a removal are two folds: Firstly, Al and Fe parts can be valorized by sending to appropriate endprocessing facilities. Secondly, WPCBs freed from these massive parts are relatively upgraded in the Cu and PMs contents, which will generate better revenues obtained from smelter. Such state-of-the-art smelters and refineries can extract valuable metals and isolate hazardous substances efficiently. Such recycling facilities can close the loop of valuable metals and reduce environmental impact arising from large quantities of e-waste. Currently, e-waste recycling is dominated by pyrometallurgical routes [5, 7], whereas the steel industry embraces the ferrous fractions for the recovery of Fe, and the secondary Al industry takes the Al fractions. MFs separated during the preprocessing of e-waste are composed of Fe, Al, Cu, Pb, and PMs. After Fe and Al, Cu and Pb are the main constituents of a typical e-waste. Therefore, it is logical to send e-waste to smelters that accept Cu/Pb scrap. Currently, Cu and Pb smelters work as e-waste recyclers for the recovery of mixed Pb/Sn, Cu, and PMs. In these pyrometallurgical processes, e-waste/Cu/Pb scraps are fed into a furnace, whereby metals are collected in a molten bath and oxides form a slag phase. Cu smelting route is environmentally friendly and cheaper than Pb smelting route [8].

There are four different thermal processes used today in pyrometallurgy and waste recycling: direct smelting, incineration, pyrolysis, and gasification.

5.3.1.1 Direct Smelting

In direct smelting process, mixed WEEE is burned in the blast/shaft furnace to obtain 75–85% black Cu, then oxidized in converter and later reduced in the anode furnace. Impurities are mostly segregated into the vapor phase and are discharged in the off-gas. Anodic Cu can be purified with H_2SO_4 along with Ni, Zn, and Fe [2]. Thermal methods result in the emission of hazardous chemicals to the atmosphere and water as a result of degradation of epoxy and volatilization of metals (including Pb, Sn, As, and Ga). These processes consume high energy and require expensive exhaust gas purification systems and corrosion resistance equipment [1].

Direct smelting is melting the metals around 1200 °C in air-blown furnace. Isasmelt vertical furnaces and Kaldo rotary furnaces can be used for WPCB recycling. Crushed e-waste is charged in a molten bath to remove plastics and refractory oxides to form a slag containing valuable metals. Smelting leads to formation of hazardous by-product fumes and partial recovery of metals. Both ferrous and nonferrous smelters need to have state-of-the-art off-gas treatment in place to deal with VOCs, dioxins, and furans. Labels, plastics, and resins contain significant amount of flame retardants. For painted scrap, lacquer should be removed prior to smelting using appropriate technologies. For treatment of WPCBs, it is of utmost importance that smelter is equipped with gas treatment equipment, since otherwise dioxins will be formed and emitted. Process gases are cooled with energy recovery and cleaned. A number of technologies are available for destruction or capture of dioxins, furans, and other gases, for example, adiabatic coolers, scrubbers, bag house/electrofilters. Catalytic decomposition and off-gas treatment can be used together for optimum performance. Formation of dioxins during smelting can be prevented by sufficiently high temperatures and long residence time in the smelter. Excess heat from off-gas can be used in subsequent processes.

WPCBs and small devices are mixed with catalysts, by-products from nonferrous industries, or primary ore and directly (without further size reduction) treated in integrated smelter and refinery (ISR). Organic components are converted to energy during smelting. Mechanical upgrading other than disassembly, grading, and shredding for bulk volume reduction prior to smelting is not undertaken due to inherent yield loss, particularly of PMs. This loss may be typically in the order of 10% but may be much higher [6].

5.3.1.2 Incineration

Contrary to smelting, formation of melts is avoided during incineration. The purposes of incineration are simply to reduce the volume of WPCB and to achieve partial heat recovery through the incineration of WPCB when the heat released is adequate. Small-scale e-waste with high PM content can be incinerated in rotary furnaces at around 850 °C with around 6% oxygen. Around 20% weight loss occurs at the incinerated material. Incinerated product is brittle and pulverized to 0.2 mm.

This product is easily chemically and mechanically processed. Incineration is a classically open process with continuous flow of gases through the reactor: atmosphere → incinerator → atmosphere. Leaching incineration ashes with 95% H_2SO_4, almost complete Cu, and 97% Ag dissolutions are possible [9]. During incineration of WEEE and batteries, Hg is considered typically volatile. Ni, Mn, and Cd are also emitted if off-gas treatment systems are not adopted.

5.3.1.3 Pyrolysis

Pyrolysis is known as a promising technique for resource utilization of WPCB. The pyrolytic oil (PO) should be utilized according to the potential value of recycling. But it has not been applied in large scale due to its complicated chemical composition. Pyrolysis is a thermochemical decomposition of organic material (plastics/resin) at high temperature in the absence of oxygen/air. Pyrolysis involves changes of chemical composition and physical phase. Direct gas emissions are avoided in pyrolysis. The purpose of WPCB pyrolysis focuses more on resource recovery. Gases from pyrolysis may be considered as a potential source of energy or chemical products. Vacuum pyrolysis of organic materials produces combustible gases (rich in CO, CO_2, CH_4, H_2, etc.) (i.e., syngas) and liquid products (i.e., oil) and leaves a solid residue rich in carbon (char) (i.e., glass fiber, ECs, metals, other inorganic materials, etc.). After pyrolysis, WPCBs become brittle and easily undergo delamination which could be easily ground, while inorganic glass fiber remains fairly intense, which can be recycled into other composites or any other materials. Unlike incineration, the operation cost for pyrolysis is normally much higher than incineration due to maintaining the absence of oxygen from the process. Energy can be recovered from combustible gases, which make pyrolysis self-sustained process. Table 5.2 shows the comparison between incineration and pyrolysis of WPCBs. Pyrolysis is typically divided into two stages: In the first stage, the WPCBs start to decompose due to the intensive heat input and release of volatile organic compounds. The second stage is the formation of char due to the pyrolysis of polymer inside the structure. Up to 400 °C, liquid products are formed, and above 800 °C

Table 5.2 Comparison between incineration and pyrolysis of WPCB

Aspect	Incineration	Pyrolysis
Heavy metal emission	Results in Hg, Cd, or Pb	Reduced emission of heavy metals
Solder/Sn recovery	Is not possible	Possible
Br_2 recovery	Br_2/HBr emission	Br_2/HBr evolves in part in the gas and in the oil
Dioxins/furans emissions	Produces both gases and released in air and remains in the solid material	Reduced production of dioxins
Heat recovery	Generates more heat	Generates 40–70% of the heating value

breaks the high molecules, thus producing smaller organic molecules. The pyrolysis of WPCBs is dangerous while the temperature is lower than 800 °C and the absence of an inert atmosphere; since there will be toxic PBFF/Fs formation.

Pyrolysis, which belongs to pyrometallurgical route, has some obstacles to limit large-scale operations at present: (i) the yields of gas and liquid generated are low that indicate economic effect; (ii) further separation is essential for mixture of pyrolysis residues and metals (moreover, pyrolysis residues generally contain toxic materials, like PBDD/Fs and PCDD/Fs, and should be disposed of properly); (iii) the further process cleaning, which would lead to more complex of the process and facilities, is also needed for the metals obtained from the course of pyrolysis [10].

Vacuum Pyrolysis

Metals can be recovered from e-waste utilizing vacuum processes which have no wastewater pollution. Nonmetallic components from e-waste can easily be recovered. Metals separated out and recovered from WPCBs depend on their vapor pressures at the same temperature. Metals from WPCBs can be recovered in a two-step vacuum pyrolysis process. The initial process separates and recovers the solder alloy at temperature that ranges from 400 to 600 °C. In the ensuing process, WPCBs are pyrolyzed and the residue heated under vacuum to recover solder by centrifugal separation. Glass fiber and other inorganic metals and materials in the resulting residues still need additional treatment. Cd and Zn metals from WPCBs are recycled using vacuum process for their separation. Cd and Zn separation is easy due to vapor pressure differences. The separation of Pb from Pb-Bi alloy is more difficult. Indium (In) can be recovered from LCD panels by vacuum pyrolysis.

The effect of pyrolysis conditions on the products yield of epoxy resin in waste printed circuit boards was investigated using the vacuum pyrolysis oven heated by temperature controller by Qui et al. (2009) [11]. The effects of temperature, heating rate, pressure, and reaction time on the yield of vacuum pyrolysis production were analyzed. In addition, the compositions of liquid products were analyzed by FT-IR and GC/MS. The experimental results showed that temperature was the key factor in the vacuum pyrolysis process. At the same time, heating rate, pressure, and rest time could not be neglected. The optimization conditions for the liquid yield from the vacuum pyrolysis process were as follows: temperature 400–550 °C, heating rate 15–20 °C/min, pressure 15 kPa, and reaction time 30 min. The main composition in the product yield was phenolic organic compounds with the total of 84.08%, while considerable amount of brominated products was up to 15.34%, which lowered the value of liquid products.

5.3.1.4 Gasification

It is the process that enables complete elimination or organic fractions to the gaseous phase at low oxygen potential of the system. During gasification, steam, oxygen (air,

5.3 Advanced Methods for WPCB Recycling

oxygen enriched air), or CO_2 (much less) reagents are used. Same reactor can be used in pyrolysis, incineration, and gasification. The only difference consisted in gaseous phase introduced in the reactor – Ar/N_2 for pyrolysis, air for incineration, and distilled water for steam gasification. In laboratory-scale incineration tests, 1.2 m quartz tube reactor furnace can be used. Ar, air, or gases are introduced from one side of the reactor, and reaction gases are collected from the other side by Liebig condenser [9]. The change (or reduction) of WPCB sample mass is around 8% for incineration and pyrolysis and 18% for gasification. After thermochemical treatment, material becomes sensitive to mechanical damages (i.e., brittle), and color is changed to black, reddish, or gray. After thermal treatment, hydrometallurgy or smelting can be used. There are no halogens and organic fractions which are completely eliminated or transformed into char during thermal treatment.

Pyrolysis and gasification were not developed for metal recovery from e-waste; but, gasification was mentioned as accepted technology for municipal solid waste processing. Recently, there were presented laboratory experiments of WPCBs gasification in molten carbonates by steam [12–14]. There is also a paper reporting an experiment of processing of plastics from WPCB to hydrogen by pyrolysis and steam conversion [15]. Pyrolysis is widely used method for recycling synthetic polymers that are mixed with glass fibers. Laboratory-scale analytical pyrolysis kinetics under N_2 atmosphere was studied by many researchers. A kinetic analysis of the low-temperature (T: 200–600 °C) pyrolysis of WPCBs (3–5 mm) has been studied under N_2 and air atmosphere. The effect of thermal pretreatment on the leaching of Cu (C: 1 M HNO_3; T: 80 °C; t: 1 h; N: 300 rpm; and S/L: 25 g/L) from untreated, pyrolyzed, and air-burned (combustion) samples was also examined [16]. It was observed that weight loss increases with time at each temperature, initially rapidly, and then becomes constant after about 15 min. Moreover, the maximum weight loss increases with increasing up to the maximum temperatures of study which was observed as 25% at 600 °C at about 50 min of pyrolysis. The rate of removal of mass was slightly higher in air than in nitrogen.

The thermal treatment at the temperature of 300 °C does not have any marked influence on the release of plastics from WPCBs. With increase in temperatures, the amount of plastics removed increases. Temperature of about 500 °C and duration about 50 min are sufficient for the effective removal of VOCs of WPCB. Maximum pyrolysis (decomposition) was found in the temperature range of 400–500 °C. The pyrolysis behavior can be divided into three stages. The first stage is up to temperature of 296 °C where no loss in the weight of the PCB scrap was observed. The second stage was between 296 and 500 °C where rapid decomposition of the PCB scrap took place. The last stage was observed above 500 °C where the rate of decomposition became almost constant. The weight loss achieved under similar conditions by combustion was slightly higher than that of pyrolysis. Also the activation energy calculated indicated that combustion was more favorable than pyrolysis; however, the difference was not large. Cu recovery from the untreated sample was poor. However, Cu recovery obtained from combustion and pyrolysis was considerably higher indicating prior thermal treatment is necessary along with mechanical upgradation in order to improve the process efficiency, thus making pyrolysis as an economical alternative for recycling WPCBs [16].

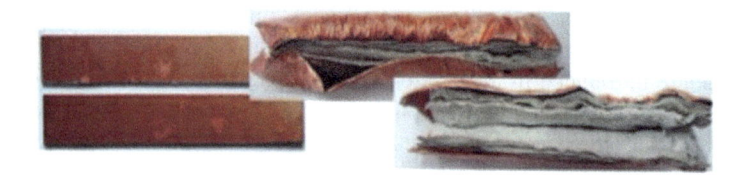

Fig. 5.1 Steam gasification of WPCBs

Allothermal steam gasification experiment was performed in a quartz reactor with samples of the raw FR-4 board at 860 °C for few hours by Gulgul et al. (2017) [9]. It may be noticed that the structure of the board was fully opened and the epoxy resin, which bonded fiber glass and Cu layers, was completely removed (Fig. 5.1). Steam gasification eliminates almost all organic materials and opens the interior of the WPCBs as compared to pyrolysis and incineration. This facilitates the penetration of lixiviants or gases during leaching.

5.4 Microwave Heating

Metals from WPCBs were separated utilizing a two-step microwave heating. Firstly WPCBs are heated to a temperature sufficient to combust the organic materials present, and the metals of low melting point (like Sn and Al) are removed simultaneously and secondly WPWBs are heated again to a higher temperature at which glass formers present in the WPCBs begin to vitrify, and then metals with higher melting points could be recovered from the solidified residues or might be removed at their respective melting points which are below the temperature of vitrifying. This approach is highly efficient; however, this recycling technology has not been transferred to industry because of its high economic cost [17]. In the processing the crushed scraps are burned in a furnace or in a molten bath to remove plastics, and the refractory oxides form a slag phase together with some metal oxides. Further, recovered materials are retreated or purified by using chemical processing. Energy cost is reduced by combustion of plastics and other flammable materials in the feeding.

Soare et al. (2016) used a laboratory furnace with microwave field in Ar inert atmosphere to melt WPCBs after comminution (0.5–2.0 cm) at 1000–1200 °C (2400 W) for 30 min [18]. A complete separation of the MF from the organic components was achieved. Multicomponent metallic alloy was processed through combined hydrometallurgical and electrochemical methods. Microwave melting presents a series of advantages, such as rapid heating cycles which save about 35% energy compared to conventional melting and improved process control, no direct contact with the heating materials, and the possibility of processing various nonferrous metals containing wastes.

5.5 Pyrometallurgical Processes for the Recovery of Metals from E-Waste

5.5.1 Ferrous Scrap

Ferrous metals (Fe, steel, etc.) have been recycled for well over a century, as steel does not lose its physical properties during the recycling process. It certainly is economically viable compared to mining Fe ore, handling it, and preparing it for the smelting process. The energy saving of recycling ferrous scrap is approximately 16% compared to mining and producing steel. Cu is an impurity for the steel recycling process.

5.5.2 Nonferrous Scrap

Metal wastes that do not contain large quantities of Fe are nonferrous scrap. These can be BMs such as Al, Cu, Pb, Ni, and Zn or alloys such as brass, and these metals or mixtures of metals can contain traces of PMs such as Ag, Au, or Pt or even rare metals.

5.5.3 Shredder Residues

These are fractions that are generated from shredding a variety of materials after the separation of the ferrous metals. These shredder residues can be the shredder heavies or the fines.

5.5.4 Metal-Containing Slags/Bottom Ash

Over the last few years, the incineration of e-waste has seen a fast growth. The incineration process converts the waste materials into slags, ash, and heat. Metals are recovered from these slags by pyrometallurgy.

5.5.5 Cu Smelters

Industrial processes for recovering metals from e-waste are based on combined pyrometallurgical, hydrometallurgical, and electrometallurgical routes (Table 5.3). In pyrometallurgical processes, e-waste is blended with other materials and

Table 5.3 Cu pyrometallurgical production layout from ores and WEEEs

	Primary Cu(S) ore	E-waste/WPCBs
Ave. grade	0.5–1.0% Cu	7.0–27.0%
Size reduction	with	without or with
Enrichment	Flotation (20.0–30.0% Cu)	without upgrading
Drying/roasting	Oxidized Cu	without roasting
Smelting	Cu mat (35.0–70.0% Cu)	Cu mat (35.0–70.0% Cu)
Conversion	Blister Cu (99% Cu)	Blister Cu (99% Cu)
Fire refinement	Anodic Cu (99.5% Cu)	Anodic Cu (99.5% Cu)
Electro refinement	Cathodic Cu (99.9% Cu)	Cathodic Cu (99.9% Cu)
Melting		
Casting	Final product	Final product

incorporated into primary/secondary smelting processes (e.g., into Cu or Pb smelters) at 1200 °C. Materials with a lower Cu content (like shredder Cu, Cu–Fe materials) are melted down with coke, quartz, and lime in the blast furnace. Cu smelting is the dominating route for e-waste recycling where PMs are collected in Cu matte or black Cu (approx. 75% Cu). The smelted metal in the blast furnace is processed in the converter with alloy materials such as brass, bronze, and red brass. In this process, using oxygen, Pb, Sn, and Zn are separated off as mixed oxides. The resulting slag is recycled in the blast furnace. The molten mass obtained from the converter consists of up to 96% Cu and is further refined in the anode furnace. Here, it encounters other feedstock such as scrap sheet metal, pipe, and wire as well as anode waste from the electrolysis. The finished molten mass from the anode furnace contains approximately 99% Cu and is molded into anode plates. In the final stage of Cu production, i.e., the electrorefining process, pure Cu metal is produced, and the PMs are separated into slimes where they are recovered using hydrometallurgical routes. Currently, various industrial processes are used globally for extracting metals from e-waste, including the Umicore Hoboken (Belgium) integrated smelting and refining facility (ISR), the Noranda process in Quebec (Xstrata), Rönnskär smelters in Sweden (Boliden), Kosaka's recycling plant in Japan (DOWA), the Kayser recycling system in Austria, and the Metallo-Chimique N.V plants operated in Belgium and Spain. SIMS Recycling Solution in Singapore and Roorkee, Attero Recycling (12,000 t/y) in India, and Aurubis Recycling Center in Lünen (Germany) are some other important e-waste recyclers.

In pyrometallurgical processing, generally WEEE is firstly dismantled, shredded, and ground for size reduction and liberation of components and then smelted in furnaces (i.e., blast furnace, reverberator furnace, imperial smelting furnace, Outokumpu flash smelting, Kaldo furnace, Mitsubishi continuous smelting,

5.5 Pyrometallurgical Processes for the Recovery of Metals from E-Waste

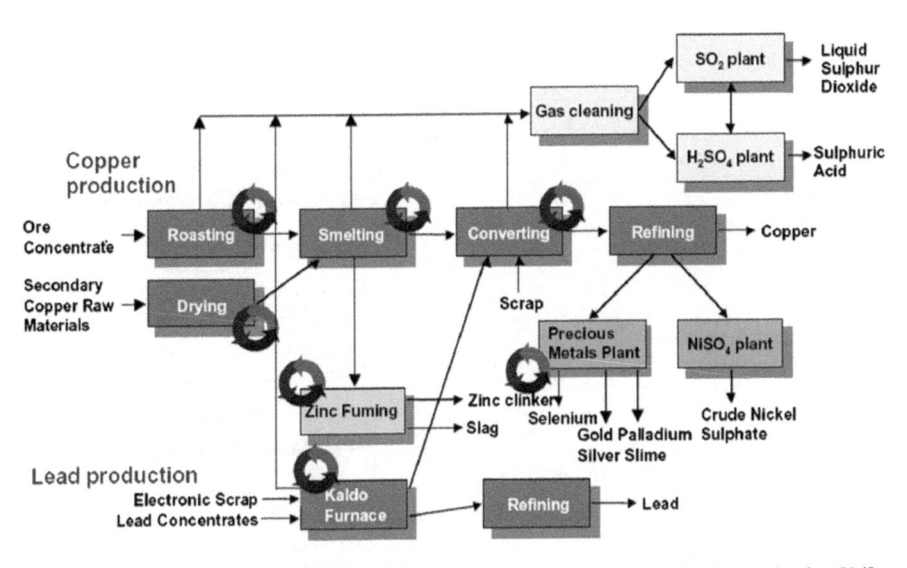

Fig. 5.2 Boliden's Rönnskar smelter (Skelleftehamn, Sweden) flowsheet for Cu production [14]

Isasmelt/Csiromelt, Noranda, CMT/Teniente, Vanyukov (Russia), Bayi (China), plasma arc furnace, etc.) to obtain coarse Cu sulfide/metal ingots. Noranda uses 10–14% WPCBs and 86–90% Cu concentrate as feed. The two basic and widely applied smelting processes include flash and bath smelting. The next step is converting process in Cu converter by blowing hot air from tuyeres in order to obtain molten Cu matte (Cu–Fe–S) (35–70% Cu). This step oxidizes iron sulfide and convert copper sulfide to metallic blister Cu (about 98.5% Cu). The last step is anodic Cu production in a furnace (99.5% Cu). Anodic Cu is refined by electrolysis to obtain 99.99% Cu ingots, and PMs are recovered from anodic mud. Figure 5.2 shows schematic diagrams of Rönnskar for Cu production from both ore concentrate and e-waste. Today, the smelter's annual capacity for recycling electrical material is 120,000 tons, including circuit boards from computers and mobile phones that are sourced primarily from Europe. The electronic material arrives at Rönnskar by train or truck from southern Sweden and has already been dismantled, with much of the plastic, Fe, and Al removed. The electronic material is sampled and shredded before being sent to a Kaldo furnace, which Boliden has specially adapted for smelting electronic material. The furnace consists of a leaning cylinder that rotates during the smelting process to ensure an even heat distribution. The smelted material, known as black Cu, then joins the facility's main smelter flow for further refining to extract Cu and PMs. WEEE often contains potentially hazardous substances and must be processed in a manner that ensures minimal environmental impact. The Rönnskar smelter is equipped with advanced systems to clean process gases and discharge water. Wet gas purification uses water to wash out dust particles, which are returned to the refining process. Rönnskar is also equipped with an additional purification stage for Hg. Plastic in the WEEE material melts during smelting, which acts as a source of energy and generates steam that is converted into electricity or district

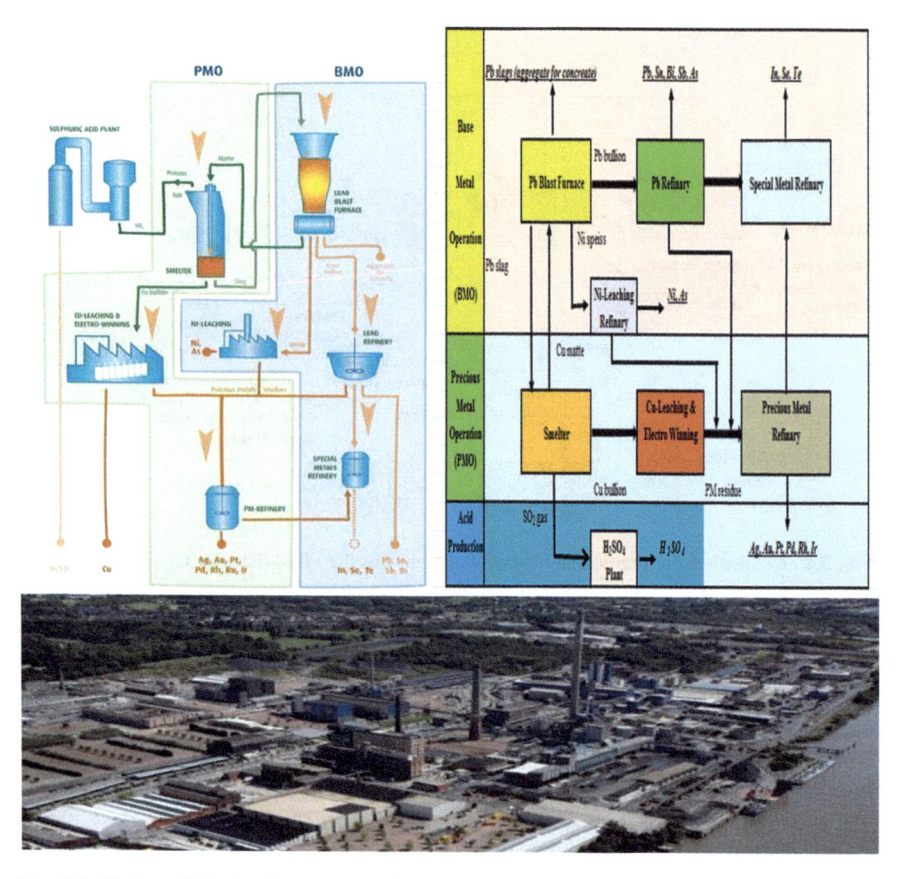

Fig. 5.3 Umicore Hoboken integrated smelter and refinery flowsheet and plant photo [21, 25]

heating. The heat is partially reused as district heating in the plant area, and the remaining heat is supplied to the local district heating system [19]. Zhou et.al (2010) used 12% NaOH as slag formation material which promotes the effective separation of metals from slag [20].

Another application for pyrometallurgical process was the plant of Umicore's ISR, which include Cu smelter, Pb blast furnace, and precious and special metal refineries along with H_2SO_4 plant (Fig. 5.3). Umicore has the world's largest waste recycling facility at Hoboken, Belgium, at a feeding capacity of 350,000 t/y and annual production capacity of over 50 tons of PGMs, 100 tons of Au, and 2400 tons of Ag [8, 21]. Umicore planned to expand its production capacity to 500,000 t/y [22]. Umicore's process mainly focused on recovery of PMs from WEEE including Ag, Au, Pt, Pd, Rh, Ru, and Ir. Its procedures included the following: firstly, WEEE was pretreated (i.e., dismantling, shredding, and physical processing) and then the precious metal operations (PMO) were smelted in an Isa Smelt furnace. Almost all other metals were concentrated in the slag after smelting; thirdly, the slag was further treated at the base metals operations (BMO). The BMs were the by-products from the PMO which were subjected to electrolytic refining to gain high-purity BMs, such as Cu [23–25].

5.6 Limitations of Pyrometallurgical Processes

Pyrometallurgical routes are generally more economical and eco-efficient and maximize the recovery of PMs; however, they have certain limitations that are summarized here [8]:

- Recovery of plastics is not possible because plastics replace coke as a reducing agent and fuel as a source of energy.
- Fe and Al recoveries are not easy as they end up in the slag phase as oxides.
- Hazardous emissions such as dioxins are generated during smelting of feed materials containing halogenated flame retardants. Therefore, special installations are required to minimize environmental pollution.
- A large investment is required for installing integrated e-waste recycling plants that maximize the recovery of valuable metals and also protect the environment by controlling hazardous gas emissions.
- Instant burning of fine dust of organic materials (e.g., NMFs of e-waste) can occur before reaching the metal bath. In such cases, agglomeration may be required to effectively harness the energy content and also to minimize the health risk posed by fine dust particles.
- Ceramic components in feed material can increase the volume of slag generated in the blast furnaces, which thereby increases the risk of losing PMs from BMs.
- Partial recovery and purity of PMs are achieved by pyrometallurgical routes. Therefore, subsequent hydrometallurgical and electrochemical techniques are necessary to extract pure metals from BMs.
- Handling the process of smelting and refining is challenging due to complex feed materials. The expertise in process handling and the thermodynamics of possible reactions will be difficult.

5.7 Emerging Technologies in Pyrometallurgy

Classical pyrometallurgical smelting processes are nowadays the major method of e-waste processing. Generally, these processes are focused on PMs and Cu recoveries. Organic materials (e.g., epoxy resins) are additional source of the heat; other components are collected in the slag or fly ash fractions. From this point of view, high concentration of chlorine or bromine is disadvantageous. However, there are undertaken efforts to develop alternative technologies, as incineration without smelting, pyrolysis, or gasification. It seems that the method of allothermal steam gasification is particularly interesting, because it enables complete elimination of organic fractions without high oxygen potential of the gaseous phase, collecting volatile contaminants (including halides) in the aqueous condensate, and production of H_2-rich syngas. It should be noted that the air is not used in this process.

Glass or binder encapsulation, where harmful contents are safely sealed in, to produce low-grade construction blocks, is a further thermal process being developed at the expense of resources lost. Vacuum, thermal plasmas, and lasers to enhance the thermal treatment are novel research topics being searched [6].

There are some attempts for selective transformation of PCBs into value-added metallic alloys instead of producing pure metals by using one-step or two-step pyrometallurgical process(es). In one-step thermal transformation process, the waste PCBs will be rapidly heated to high temperature, and metallic alloys will capture all the low-temperature melting elements. In the two-step process, the PCBs will first be exposed to lower temperature (<300 °C) to enable the low melting point metal alloys (Pb-Sn-based alloys) to form and to be separated out. The remaining material will then be rapidly heated to high temperature (>1200 °C) to produce the higher melting point metal alloy (Cu-based alloy).

5.8 Hydrometallurgy

5.8.1 Solvent Leaching

Generally, hydrometallurgical leaching technology is technically more exact, highly predictable, more flexible, and easily controllable in material extraction from resources and has already been applied in WPCB recycling with the assistance of mechanical crushing as pretreatment for almost three decades [2, 14, 26]. Hydrometallurgy uses various lixiviants/chemicals (such as strong acids, caustic watery solutions, halides, or microorganisms) to selectively dissolve and precipitate metals. Table 5.4 compares the advantages and disadvantages of leaching solvents/oxidants for base metals, precious metals, brominated epoxy resins, and solder stripping. Acids, bases, and cyanide are conventional leachants, and thiourea, thiosulfate, and halides are alternative leachants. Corrosion, toxicity, high consumption, slow leaching rate, high cost, low stability, and environmental irritations are main disadvantages for leaching solvents. Corrosion causes problems for the equipment when several kinds of leaching reagents are used. Also leaching process normally takes a long time to obtain metal-rich solution due to slow leaching rate, which makes the process a time-consuming one. Therefore, studies that reinforce the driving force and accelerate the leaching rate have been studied including using electrochemistry, pre-pyrolysis, supercritical extraction, and mechanical treatment.

Table 5.5 compares the most commonly used H_2SO_4 and HCl mineral acids, aqua regia and ionic liquids (IL), in leaching. HCl has a high dissociation constant and easily dissolves different base metal oxides. Neutralization with NaOH produces NaCl. Excessive corrosion and difficulty in Cu electrowinning and low quality of Cu

5.8 Hydrometallurgy

Table 5.4 Comparison of different kinds of leaching solvents

Base metal leaching solvents	Advantages	Disadvantages
Acids/mixed acids		
Inorganic/mineral acids (H_2SO_4, HNO_3, HCl, HClO, AR, etc.)	High dissolution rate	Corrosive, high concentration is needed pH: 1–2 is very low
Organic acids Methanesulfonic acid (MSA)	Environment-friendly, biodegradable, less toxic, less corrosive, no toxic gas, use mildly acidic conditions	Weak
(EDTA) + H_2O_2	Dissolve Cu with 0.1 M H_2O_2, 25 °C Solution potential >200 mV	
(Na-Citrate) + H_2O_2	Dissolve Cu with 0.1 M H_2O_2, dissolve BMs, less toxic, more stable and safe, Cu refining is simple, solution potential >300 mV, pH: 4.5	Low-solubility, high-temperature (100 °C) requirement
Acetic acid, oxalic acid, ascorbic acid	Environmentally benign, ease of biodegradation, and not corrosive	Low dissolution efficiencies
Bases/alkalines (KOH, NaOH, NH_4Cl, NH_4OH, NH_3)	Low corrosion, low toxicity, low pollution, $H_2(g)\uparrow$	
Acidic brines		
NaCl, $FeCl_3$ $(NH_4)_2SO_4$ $Fe_2(SO_4)_3$, $(NH_4)_2CO_3$	Used in Cu and Pd leach with $CuSO_4$ Used in Cu leach	Non-acid process
Precious metal leaching solvents	*Advantages*	*Disadvantages*
Cyanide NaCN/KCN	Low cost, high stability, low dosage, operating in alkaline solution, dissolve Au and Ag effectively	Difficult to process wastewater, high toxicity, slow leaching kinetics, bond with Au is strong, harmful to environment
Thiourea (+Fe^{3+}) $SC(NH_2)_2$	Low toxicity, noncorrosive, high dissolving power, selective for Au, less interference ions	High cost, difficulty in downstream metal recovery, bond with Au is strong, low stability, high consumption, poor stability
Thiosulfate $(NH_4)_2S_2O_3$ $Na_2(S_2O_3)\cdot 5H_2O$	High selectivity, nontoxic, noncorrosive, fast leaching rate, oxygen carrying catalyst is required	pH sensitive, low stability, high cost, high consumption, downstream metal recover
Potassium persulfate $K_2S_2O_8$	Nontoxic, strong oxidizing agent	Dissolve all BMs, only Au remains in solid residue
Halide/halogens NaCl	Non-acid, green, noncorrosive selective for Cu and Pd	Expensive, environmental irritation, consumption, corrosive

(continued)

Table 5.4 (continued)

Chlorine-chloride (Cl$^-$/Cl$_2$) NaCl, hypochlorite, HCl, HClO$_2$, NaClO Iodide (I$^-$/I$_2$), KI Bromide (Br$^-$/Br^2)	High leaching rate, reliable, safe, good selectivity, nontoxic, noncorrosive	chlorine gas, special reactor requirement, high consumption for iodine
Aqua regia (AR)	Fast kinetics, low reagent dosages	Strongly oxidative, corrosive, difficult to deal with downstream
Brominated epoxy resin leaching solvents	*Advantages*	*Disadvantages*
Dimethylacetamide (DMA)	High boiling point, relatively high viscosity, high thermal stability.	VOC
N-methyl-2-pyrrolidone (NMP)		VOC
Dimethyl sulfoxide (DMSO)		VOC, highly hygroscopic with specific heat, high viscosity, penetrate skin, vapor heavier than air
Dimethylformamide (DMF)	Aprotic, cheaper, colorless, strong hydrogen bond formation affinity between BER and DMF, breaks van der Waals bonds	VOC, temp. is high (135 °C)
Solder strip leaching solvents	*Advantages*	*Disadvantages*
HBF$_4$		Corrosive, strong acid
Oxidants	*Advantages*	*Disadvantages*
H$_2$O$_2$	Low toxicity, environment-friendly, extensively used, good for Cu with H$_2$SO$_4$, HCl, and HNO$_3$	Need acid assistance, high consumption due to its decomposition, high temperature
Fe$_2$(SO$_4$)$_3$ (Fe^{3+})	Dissolves Cu, Ni, low cost and regeneration, purified by goethite and jarosite precipitation before EW	Complicate Au recovery
Hypochlorite (NaOCl)	Easily decomposes	
Permanganate (KMnO$_4$)	Strong oxidizing agent, does not generate toxic by-products	
CuSO$_4$ (Cu^{2+})	Used in Cu and Pd leach, noncorrosive	
Ozone (O$_3$) injection	O$_2$–O$_3$ mixture generated by a UV lamp	Limited solubility in aqueous solutions Decrease solid-liquid contact area
Air (O$_2$) sparging		Low solubility of gas in aqueous solutions
Cl$_2$	Strongest oxidizing agent	Toxic

Compiled from [4, 27]

5.8 Hydrometallurgy

Table 5.5 Comparison of most commonly used two mineral acids, AR and ionic liquids, used in base metal leaching

Reagents	Pros	Cons
H_2SO_4	Strong proton donor Its cost is much less than other inorganic/mineral acids Handiness makes suitable lixiviant in the leaching process Highly hygroscopic Highly selective, well-established process for Cu	At elevated temperature Corrosive
HCl	Fast kinetics at room temperature High dissociation constant (k_a) High solubility and activity of base metals oxides Low toxicity Neutralized with NaOH to produce NaCl	Excessive corrosion Difficult electrowinning of Cu Poor quality of Cu
Aqua regia (AR) (3HCl + 1HNO$_3$ mixture)	Fast kinetics Effective	High reagent cost Highly corrosive Low selectivity
Ionic liquids (ILs)	Thermally stable Environmentally friendly	High cost Excessive dosage

are main drawbacks. H_2SO_4 is a strong proton donor. It is the cheapest mineral acid. Handiness makes it suitable for leaching process. H_2SO_4 is a highly selective and well-established process for Cu. Corrosiveness and elevated temperature requirements are some disadvantages. The use of AR in leaching has a fast kinetic and effectiveness. Reagent cost, corrosivity, and low selectivity are some drawbacks of AR. Ionic liquids are thermally stable and environmentally benign; but, expensive and excessive dosages are used. Mineral acids cause environmental pollution and may dissolve undesired impurities. Carbonaceous compounds using inorganic acids may lead to formation of CO_2 pressure due to fast dissolution which may cause risks.

Leaching selectivity and effectiveness of valuable metals are always important in hydrometallurgical processing. Chemical leaching is faster and more efficient; however, wastewater and waste gas are generated invariably during this process. Hydrometallurgy can be industrially applied to metal recovery for WPCB recycling due to its flexibility and environmentally benign and energy-saving features. Its corrosive and poisonous properties require corrosion-proof equipment as well as complicated process operation. Traditional hydrometallurgical extraction processes are acid leaching to recover BMs. $HCl/H_2SO_4/HNO_3/HClO_4$ can be used for extracting Cu, Pb, Zn, etc. Aqua regia can also be used for nonselective and aggressive digestion of base (Cu, Pb, and Zn) and precious (Ag and Au) metals from WEEE. The corrosivity toward Cu, Fe, and brass has been proposed in the following order $HNO_3 > H_2SO_4 > HCl$ [28]. The majority of hydrometallurgical routes for processing WPCBs use H_2SO_4 leaching in the presence of H_2O_2 as oxidizing agent, followed by solution refining. Other than H_2O_2, NaOCl and Cu^{2+} can also be used as oxidizing agent to promote the reaction. The use of HNO_3 and

HCl was studied extensively, but because of the environmental regulations and corrosive nature of these leaching agents, they are as appropriate as the less hazardous H_2SO_4 [29]. HNO_3 has the maximum environmental impact [30].

Before leaching, heat pretreatment under different atmospheres, roasting, and microwave-assisted digestion can be helpful. High-pressure oxidative leach (HPOL) with dilute acids results in removal of a significant amount of BMs from WPCB ash samples obtained by incineration at 800 °C. A sequential hydrometallurgical approach consisting of HPOL and thiourea leaching for recovery of both BMs and PMs from WPCB ash was studied by Batnasan et al. (2018) [31]. Havlic et al. (2010) also studied the HCl leaching of Cu and Sn from WPCBs after thermal pretreatment. After precombustion in air for 15–60 min in 500–900 °C, WPCB was leached in 1 M HCl. Precombustion to 900 °C significantly improved Cu leaching due to conversion of Cu to Cu_2O, which dissolves more preferable in HCl. However, the precombustion leads to the difficulty in extracting Sn, since it is oxidized to form SnO_2, which is very stable in acid with minimal leaching. Therefore, the efficiency of hydrometallurgy was seldom enhanced even without considering the emission during combustion process [32].

Generally, higher acid concentrations and higher temperatures enhance leaching rates. A smaller particle size and lower pulp density (S/L ratio) are desirable for metal leaching [33]. The particle size can influence the surface area exposed to the leachant. A low S/L ratio indicates that larger volumes of liquid are required. Stirring speed is essentially related to diffusion phenomena, since higher stirring speed usually promotes mass transfer of reactants and reaction products in solution as the diffusion layer at the S/L interface is reduced. Overgrinding should be minimized for economic reasons, since this operation can only be justified to facilitate material handling or for homogenization purposes. Organic acids, which undergo biodegradation and are not evolve toxic gases, are considered as green acids as they are less toxic and corrosive in comparison to mineral acids. For example, green strong methanesulfonic acid (MSA) (CH_4O_3S) (pK_a, 1.9, and low molecular weight, 96 g/mol) biodegrade by forming CO_2 and sulfate and leaches solder along H_2O_2 [34]. Thus, MSA + H_2O_2 aqueous system provides environment-friendly desoldering separation of Sn–Pb alloy for dismantling ECs from WPCBs. Aqueous bromide media (Br_2/KBr) were used for primary leaching and dismantling of PCBs in laboratory scale by Dorneanu (2017) [35].

Metal recoveries most of the time are higher than 90%. Leaching time changes from 60 to 180 min. Leaching can be accelerated with the use of mixing and supercritical fluids such as water or CO_2 with significantly increased power consumption [36]. Oxidation and reduction potential (ORP) and pH/ORP rate also affect metal dissolution. Thus, Pourbaix diagrams at determined temperature can be very helpful in dissolution studies [37]. The final free acidity of the leachates should be determined by titration with standard 1 M NaOH solution using methyl orange as indicator. Since the color change of the indicator occurs at relatively low pH, any precipitation should be prevented.

In general acid leaching reactions of metal extraction from WPCBs in the presence of oxidants can be expressed by the following equations:

5.8 Hydrometallurgy

$$Ag^0(s) + H^+(aq) \rightarrow Ag^+(aq) + 1/2H_2(g) \qquad (5.1)$$

$$Au^0(s) + 3H^+(aq) \rightarrow Au^{3+}(aq) + 3/2H_2(g) \qquad (5.2)$$

$$Cu^0(s) + 2H^+(aq) \rightarrow Cu^{2+}(aq) + H_2(g) \qquad (5.3)$$

$$Pd^0(s) + 2H^+(aq) \rightarrow Pd^{2+}(aq) + H_2(g) \qquad (5.4)$$

Except for Cu, Hg, Ag, Au, and Pt, other metals react with HCl and produce $H_2(g)$.

$$Zn + 2HCl \rightarrow ZnCl_2 + H_2(g) \uparrow \qquad (5.5)$$

$$2Al + 6HCl \rightarrow 2AlCl_3 + 3H_2 \qquad (5.6)$$

Cu, Ag, and Hg give reaction with oxygen containing oxidizing acids (i.e., H_2SO_4, HNO_3, and H_3PO_4). For Cu following balanced equations (5.7–5.14) may occur with H_2SO_4 in the absence or presence of oxidizing agents (such as $KMnO_4$ and H_2O_2):

$$Cu + H_2SO_4(aq) + 2H^+ \rightarrow Cu^{2+} + SO_2(g) \uparrow + 2H_2O(l) \qquad (5.7)$$

$$Cu + 2H_2SO_4(aq) = CuSO_4 + SO_2(g) \uparrow + 2H_2O(l) \qquad (5.8)$$

$$Cu + H_2SO_4(aq) \rightarrow CuSO_4 + H_2(g) \uparrow \qquad (5.9)$$

$$5Cu + 8H_2SO_4(aq) + 2KMnO_4 \rightarrow 5CuSO_4 + 2MnSO_4 + K_2SO_4 + 8H_2O \qquad (5.10)$$

$$Cu + H_2SO_4(aq) + H_2O_2 \rightarrow CuSO_4 + 2H_2O(l) \qquad (5.11)$$

$$Cu + H_2SO_4(aq) \rightarrow CuO + SO_2H_2O \qquad (5.12)$$

$$Cu + H_2SO_4(aq) \rightarrow CuO + SO_2(g) \uparrow + H_2O(l) \qquad (5.13)$$

$$3Cu + 3H_2SO_4(aq) + 2HNO_3(aq) \rightarrow 3CuSO_4 + 2NO(g) \uparrow + 4H_2O(l) \qquad (5.14)$$

Cupric ion (Cu^{2+}), copper sulfate ($CuSO_4$), and copper II oxide (CuO) compounds may occur.

For Cu following balanced equations (5.15–5.22) may occur with HNO_3, HCl, and H_3PO_4. $CuNO_3$, $Cu(NO_3)_2$, $Cu(NO_3)_2 * 3H_2O$, $CuCl_2$, and $CuPO_4$ compounds occur.

$$3Cu + 8HNO_3(aq) \rightarrow 3Cu(NO_3)_2 + 4H_2O(l) + 2NO(g) \uparrow \qquad (5.15)$$

$$Cu + 4HNO_3(aq) \rightarrow Cu(NO_3)_2 + 2H_2O(l) + 2NO_2(g) \uparrow \qquad (5.16)$$

$$Cu + 2HNO_3(aq) \rightarrow CuNO_3 + NO_2(g) \uparrow + H_2O \qquad (5.17)$$

$$3Cu + 8HNO_3(aq) \rightarrow 4H_2O(l) + 2NO + 3Cu(NO_3)_2 \qquad (5.18)$$

$$3Cu + 8HNO_3(aq) + 5H_2O \rightarrow 3Cu(NO_3)_2 * 3H_2O + 2NO(g) \uparrow \quad (5.19)$$
$$4Cu + 10HNO_3(aq) \rightarrow 4Cu(NO_3)_2 + N_2O(g) \uparrow + 5H_2O(l) \quad (5.20)$$
$$Cu + HCl(aq) + HNO_3(aq) \rightarrow H_2O(l) + NO_2(g) \uparrow + CuCl_2 \quad (5.21)$$
$$2Cu + 2H_3PO_4(aq) \rightarrow 2CuPO_4 + 3H_2 \uparrow \quad (5.22)$$

HCl and HNO_3 together react with Cu in the following way:

$$6HCl + 2HNO_3 + 3Cu \rightarrow 4H_2O + 2NO + 3CuCl_2 \quad (5.23)$$

Pb dissolution from solder with HNO_3 oxidizes Pb to PbO which subsequently lead nitrate as given below:

$$Pb(s) + 2HNO_3(aq) \rightarrow Pb(NO_3)_2(aq) + H_2(g) \uparrow \quad (5.24)$$
$$Pb(s) + 2HNO_3(aq) \rightarrow PbO(s) + H_2O(aq) + 2NO_2(g) \uparrow \quad (5.25)$$
$$PbO(s) + 2HNO_3(aq) \rightarrow Pb(NO_3)_2(aq) + H_2O(aq) \quad (5.26)$$

Sn in solder partially dissolves at low HNO_3 concentrations and forms metastannic acid above 4 M HNO_3 concentrations [38]. This precipitated metastannic acid dissolves at low HCl concentrations:

$$Sn(s) + 4HNO_3(aq) \rightarrow H_2SnO_3 \downarrow + 4NO_2(g) + H_2O \quad (5.27)$$
$$H_2SnO_3(s) + 6HCl(aq) \rightarrow H_2SnCl_6 + 3H_2O \quad (5.28)$$

Ni and Zn also react with HNO_3; Ni forms nickel (II) nitrate:

$$Ni(s) + 4HNO_3(aq) \rightarrow Ni(NO_3)_2 + 2H_2O + 2NO_2 \quad (5.29)$$

and Zn is oxidized to zinc nitrate:

$$3Zn + 8HNO_3(aq) \rightarrow 3Zn(NO_3)_2 + 4H_2O + 2NO \quad (5.30)$$

Aqua regia (AR) is a very strong acid formed by the combination of concentrated HNO_3 and concentrated HCl, both of which are strong acids. It is generally used to remove metals like Au and Pt from substances, especially in WEEEs.

$$HNO_3(aq) + 3HCl(aq) \rightarrow NOCl(g) + 2H_2O(l) + Cl_2(g) \quad (5.31)$$

The reason AR can dissolve Au (and metals like Pt and Pd) is that each of its two component acids (i.e., HCl and HNO_3 acid) carries out a different function. While HNO_3 acid is an excellent oxidizing agent, the chloride ions from the HCl form coordination complexes with Au ions, thereby removing them from the solution. The following are the chemical equations representing the reaction between Au and AR:

5.8 Hydrometallurgy

$$Au(s) + 3NO_3^-(aq) + 6H^+(aq) \leftrightarrow Au^{3+}(aq) + 3NO_2(g) + 3H_2O(l) \quad (5.32)$$

$$Au^{3+}(aq) + 4Cl^-(aq) \leftrightarrow AuCl_4^-(aq) \quad (5.33)$$

As you can see from the equation above, the reduction of the concentration of Au^{3+} ions shifts the equilibrium toward the oxidized form. The Au present in AR dissolves completely to form chloroauric acid ($HAuCl_4$).

For Au following balanced equations (5.34–5.40) may also occur with (HNO_3 + HCl) and ($NaNO_3$ + HCl):

$$Au + HNO_3(aq) + 3HCl(aq) \rightarrow AuCl_3 + NO(g) \uparrow + 2H_2O(l) \quad (5.34)$$

$$3Au + HNO_3(aq) + 3HCl(aq) \rightarrow 3AuCl + NO(g) \uparrow + 2H_2O(l) \quad (5.35)$$

$$Au + HNO_3(aq) + 4HCl(aq) \rightarrow HAuCl_4 + NO(g) \uparrow + 2H_2O(l) \quad (5.36)$$

$$Au + HNO_3(aq) + 4HCl(aq) \rightarrow H(AuCl_4) + NO(g) \uparrow + 2H_2O(l) \quad (5.37)$$

$$Au + NaNO_3 + 5HCl(aq) \rightarrow HAuCl_4 + NO(g) \uparrow + 2H_2O(l) + NaCl \quad (5.38)$$

$$3Au + NaNO_3 + 4HCl(aq) \rightarrow 3AuCl + NO(g) + 2H_2O(l) \uparrow + NaCl \quad (5.39)$$

$$3Au + NaNO_3 + 4HCl(aq) \rightarrow 3AuCl_3 + NO(g) \uparrow + 2H_2O(l) + NaCl \quad (5.40)$$

From the above reactions, AuCl, $AlCl_3$, $HAuCl_4$, $H(AuCl_4)$, and $Au_2(SO_4)_3$ may occur.

$H(AuCl_4)$ may react with $SnCl_2$ to form Au and HCl:

$$2H(AuCl_4) + 3SnCl_2 \rightarrow 2Au + 3SnCl_4 + 2HCl \quad (5.41)$$

$$2H(AuCl_4) + 3SnCl_2 + 6H_2O \rightarrow 2Au + 3SnO_2 + 14HCl \quad (5.42)$$

In general bases cannot give reaction with metals. But, amphoteric Al, Zn, Sn, Pb, and Cr metals react with bases and produce $H_2(g)$.

$$Zn + 2KOH \rightarrow K_2ZnO_2 + H_2(g) \uparrow \quad (5.43)$$

$$Al + 3NaOH \rightarrow Na_3AlO_3 + 3/2H_2(g) \uparrow \quad (5.44)$$

$$ZnO + 2NaOH \rightarrow Na_2O + Zn(OH)_2 \quad (5.45)$$

NaCN and KCN dissolve Au in the presence of air and water:

$$4Au + 8NaCN(aq) + O_2 + 2H_2O \rightarrow 4Na[Au(CN)_2] + 4NaOH \quad (5.46)$$

$$4Au + 8NaCN(aq) + O_2 + 2H_2O \rightarrow 4Na(Au(CN)_2) + 4NaOH \quad (5.47)$$

$$4Au + 8KCN(aq) + O_2 + 2H_2O \rightarrow 4KAu(CN)_2 + 4NaOH \quad (5.48)$$

Au is oxidized to +3 state ($Au^0 - 3e^- \rightarrow Au^{III}$). C and O are reduced ($C^{II} + e^- \rightarrow C^I$ and $O^0 + 2e^- \rightarrow O^{II-}$).

In MSA-H$_2$O$_2$ desoldering leaching, H$^+$ and CH$_3$SO$_3^-$ are ionized by MSA and the differences of the standard electrode potential between H$_2$O$_2$/Sn (1.912 V) and H$_2$O$_2$/Pb (1.902 V) are so close that they can be considered to be stripped down easily from the WPCBs with almost the same ratio at room temperature [31]. Moreover, H$_2$O$_2$ is relatively stable in the acidic medium, and its final products are environment-friendly, hardly affecting the recycling of the leachant. Thus, MSA-H$_2$O$_2$ aqueous solution system can be regarded as green desoldering separation of Sn-Pb alloy for simultaneous dismantling ECs from WPCBs. In MSA + H$_2$O$_2$ aqueous system Sn and Pb leached at room temperature in 45 min with negligible Cu recovery:

$$2H_2O_2 + Sn + Pb + 4H^+ = Sn^{2+} + Pb^{2+} + 4H_2O \qquad (5.49)$$

$$H_2O_2 + Cu + 2H^+ = Cu^{2+} + 2H_2O \qquad (5.50)$$

Full Sn and Pb dissolution was obtained at MSA and H$_2$O$_2$ concentrations of 3.5 and 0.5 mol/L, respectively, at 45 min leaching time [34].

5.8.2 Biometallurgical Leaching

Biotechnology/bioleaching is one of the most promising technologies in metallurgical processing. It is considered green technology having a lower operative cost and energy demands when comparing with chemical methods. Bioleaching is an excellent process for extracting metals from low-grade ores. It is technically practicable using bacteria-assisted reaction to extract BMs (Cu, Ni, Zn, Cr, etc.) and PMs (Au, Ag, etc.) from e-waste. Microbes have the ability to bind metal ions present in the external environment at the cell surface or to transport them into the cell for various intracellular functions. This interaction could promote selective or nonselective in recovery of metals. Bioleaching and biosorption are the two main areas of biometallurgy for recovery of metals. Biometallurgical leaching methods use microorganisms including bacteria or fungi to dissolve metals and are not currently applied in the e-waste chain. Microorganism-assisted reactions have been regarded as a potential major technology breakthrough and have already been applied to WPCB recycling in lab scale. However, low speed and high leaching duration in bioleaching make the process unviable for industrial scale, despite it is being an eco-friendly and cost-effective approach [33].

Recently, some studies for extracting metals from e-waste scrap have been done by many researchers. Heavy metals were excellently leached when acidophilic bacteria were used. Liang et al. (2013) used a mixed culture of *Acidithiobacillus thiooxidans* and *Acidithiobacillus ferrooxidans* to recover Cu from WPCB fine powder. The bacteria can oxide elemental S added to H$_2$SO$_4$ for bioleaching and then extract the metal from the broth. The highest Cu recovery rate (98.4%) was obtained at pH 1.56, elemental sulfur (S^0) 5.44 g/L, and 16.88 g/L FeSO$_4$.7H$_2$O

5.8 Hydrometallurgy

concentration. However, the increase of WPCB addition in culture from 18 to 32.4 g/L will cause a sharp drop in the Cu recovery rate from 98.3% to 87.2% [39].

Other commonly used bacteria include mesophilic, chemolithotrophic, cyanogenic, or moderately thermophilic bacteria. The mechanism of biometallurgy recycling techniques is similar to hydrometallurgy recycling techniques since they all incorporate the process of leaching. However, instead of adding leaching reagents, biometallurgy recycling techniques normally use the chemicals produced by the microorganism itself, including organic/inorganic acids, cyanide, or sulfate ions. After leaching, the metal ions will form complexes or precipitates, and thus they are separated from the culture broth for direct use or further refining. It has the advantages of only a small volume of wastewater discharge and is environmentally benign compared to hydrometallurgy, which generally requires a high dosage of toxic chemical reagents. Besides, certain kinds of bacteria ($2,4,5\text{-}2^1,4^1,5^1$-BB) are capable of reducing brominated flame retardants in the pathway, which is rarely achieved in other recycling techniques [4].

Ilyas et al. (2007) studied the bioleachability of metals from e-waste by the selected moderately thermophilic strains of acidophilic chemolithotrophic and acidophilic heterotrophic bacteria. These included *Sulfobacillus thermosulfidooxidans* and an unidentified acidophilic heterotroph (code A1TSB) isolated from local environments. At scrap concentration of 10 g/L, a mixed consortium of the metal adapted cultures was able to leach more than 81% of Ni, 89% of Cu, 79% of Al, and 83% of Zn at T, 45 °C; t, 8 days; pH, 2; and N, 180 rpm. Although Pb and Sn were also leached out, they were detected in the precipitates formed during bioleaching [40]. Ilyas et al. (2010) used moderately thermophilic bacteria to recover metals and conducted the process in a column test. The recovery rate for Zn, Al, Cu, and Ni is 80%, 64%, 86%, and 74%, respectively, which already meets the requirements of industrial-scale implementation for recycling of MF of WPCB. It is also noticed that NMF will contribute to the alkalinity which will affect the leaching of metal ions, requiring washing before conducting the culture stage [41].

Yang et al. (2009) used *A. ferrooxidans* to study the factors affecting the mobilization of Cu in the bioleaching process. The higher concentration of Fe^{3+} from 0.64 to 2.13 g/L in the stock solution will bring to the increase of Cu leaching in 12 h from 34.5% to 79.8%. A similar result was observed that Cu recovery decreased from 99.1% to 88.4% in 48 h when the pH value increased from 1.5 to 2.0. Therefore, it is noticed that the concentration of Fe^{3+} and the pH value have a very obvious effect on Cu leaching [42].

Ting et al. (2008) used two cyanide-producing bacteria, *Pseudomonas fluorescens* and *Chromobacterium violaceum*, to extract Au and Cu from WPCB, and the recovery rate was around 27% and 20% for Au and Cu. In a two-step extraction, the recovery rate for Au and Cu was increased to around 30% and 24%, respectively [43].

Bioleaching has been successfully applied for recovery of PMs and Cu from ores for many years. Although limited researches were carried out on the bioleaching of metals from e-wastes, it has been demonstrated that using *C. violaceum* [44] Au can be microbial solubilized from WPCB and using bacterial consortium enriched from natural acid mine drainage and Cu could be efficiently solubilized from WPCBs in about 5 days [45, 46]. The extraction of Cu was mainly accomplished indirectly through oxidation by Fe^{3+} ions generated from Fe^{2+} ion oxidation bacteria; a two-step process was necessary for bacterial growth and for obtaining an appropriate oxidation rate of Fe^{2+} ion.

The e-waste has no source of energy that is required for growing the bacteria, and therefore, an external supply of nutrients is a must for leaching metals from e-waste. However, biometallurgy recycling techniques require a lot of nutrients for microorganism enrichment and metal extraction. The addition of an Fe source is necessary for Cu extraction since the solubilization of Cu needs the presence of Fe^{3+} according to reaction (5.51). Also, the ferrous ion is the energy source for *A. ferrooxidans*, which is an aerobic and autotrophic bacterium according to reaction (5.52).

$$Cu^0 + 2Fe^{3+}(aq) \rightarrow 2Fe^{2+}(aq) + Cu^{2+}(aq) \qquad (5.51)$$

$$4Fe^{2+} + O_2 + 4H^+ \rightarrow 4Fe^{3+} + 2H_2O \qquad (5.52)$$

Also, the low extraction rate in high WPCB dosage due to the limitation in air distribution and oxygen mass transfer hinders the application of the biometallurgical treatment method [47]. Moreover, normally microorganisms are vulnerable to heavy metals; thus, the growth of them will be inhibited due to the toxicity of metals [48]. Although some bacteria or fungi can adjust to the condition after prolonged adaptation time and achieve a good recovery rate, the time required for this adaption is extremely long (more than a week) [49]. Furthermore, the recovery period for biometallurgical treatment is much longer than pyrometallurgy or hydrometallurgy recycling techniques, which affected the positive evaluation of process. Also, normally the WPCB feedstock needs to be a fine powder with a particle size around 100 μm or even lower to ensure adequate surface contact, which will consume a lot of energy in the early stage [50].

So far, very few studies reported the use of fungus for recovery of metals from e-waste. Because of constant supply of nutrients for fungal growth, handling of fungi in turnover and long processing time restricts the use of fungus. Besides these limitations, fungal bioleaching has several advantages over bacterial bioleaching: they can grow at high pH, which makes them efficient for alkaline materials bioleaching. They can leach metals rapidly and conceal organic acids that chelate metal ions, thereby being useful in metal-leaching process. Even with such advantages, still, there is an information shortage about using fungi to leach metals from e-waste. The e-waste has no source of energy that is required for growing the bacteria, and therefore, an external supply of nutrients is a must for leaching metals from e-waste. Although bioleaching process has many advantages, commercial performance of the process is still in the nascent stage. This is mainly slow kinetics of the process. Many bioleaching processes require long time ranging from 48 to 245 h to recover metals without recovering all the metals present in e-waste. Thus, there is an urgent necessity to develop a fast and economic bioleaching process that can be applied industrially.

Biosorption process is a passive physicochemical interaction between the charged surface groups of microorganisms and ions in solution. Biosorbents are prepared from the naturally abundant and/or waste biomass of algae, fungi, or bacteria. Physicochemical mechanisms, such as ion exchange, complexation, coordination, and chelating between metal ions and ligands, depend on the specific properties of the biomass (alive, or dead, or as a derived product). Compared with the conventional methods, biosorption-based process offers a number of advantages including low operating costs, minimization of the volume of chemical/biological sludge to be handled, and high efficiency in detoxifying. However further efforts are required because the adsorption capacities of PMs on different types of biomass are greatly variable and much more work should be done to select a perfect biomass from the billions of microorganisms and their derivatives. Most of the researches on biosorption mainly focused on Au more attentions should be taken into biosorption of Ag from solutions and on recovery of PMs from multielemental solutions [51].

5.9 Purification

The major aim of purification is to produce a pure electrolyte for electrolysis/solvent extraction. There are four main methods of metal recovery: precipitation, electrowinning, cementation, and ion exchange (Fig. 5.4).

5.9.1 Chemical Precipitation of Metals

When metals are precipitated out from aqueous solution, it is usual to adjust the pH of the solution or to add chemical precipitants and flocculants, but the solubilities of the metals in solution will determine the process used. Metals will often precipitate as hydroxides, sulfides, and carbonates. In order to recover Sn, Fe, Cu, and Zn from leach solutions by chemical precipitation, the pH of the solution was adjusted to 1.5, 3, 6, and 8, respectively. After precipitation, filtration can be performed to remove precipitated metals. NaOH and H_2SO_4 can be used to adjust pH.

5.9.2 Cementation

Cementation is a process where metal ions are reduced to their elemental state at a solid metal interface. It is a common process for removing one metal from solution that is not important enough to be electrowon. Cementation is a type of precipitation, a heterogeneous process in which ions are reduced to zero valence at a solid metallic

Fig. 5.4 Metal recovery processes for purification

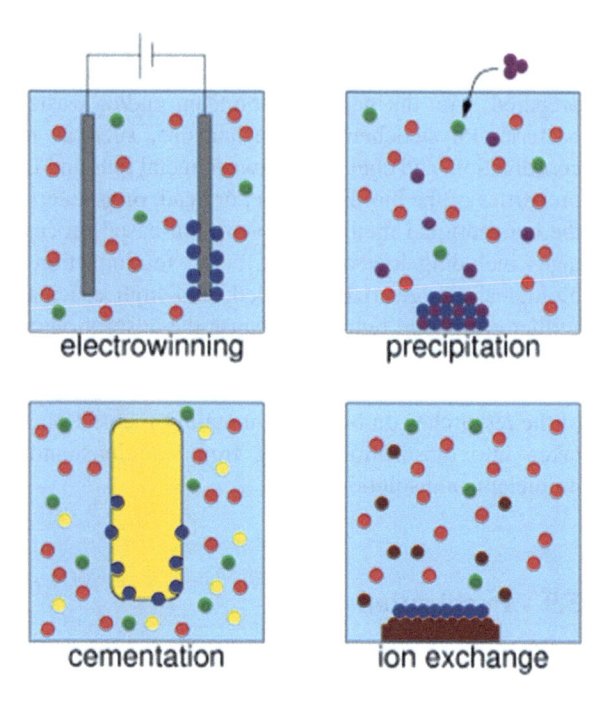

interface. The process is often used to refine leach solutions. Cu in solution can be cemented with scrap Fe or by Zn dusts (Eq. 5.53) and Ag in solution is precipitated with Zn powders.

$$Cu^{2+} + Zn \rightarrow Zn^{2+} + Cu \tag{5.53}$$

5.9.3 Electrowinning (EW)

Electrowinning is the recovery of metals, such as Au, Ag, and Cu, from aqueous solutions by passing a current through the solution with positive redox potentials. But, EW is less efficient for other metals such as Cr and is practically impossible for metals such as Al. Electrons from the current chemically reduce the metal ions, to form a solid metal compound on the cathode. Electrowinning is a widely used technology in modern metal recovery, mining, refining, and wastewater treatment applications. In EW, the electrolyte includes dissolved metals that have to be recovered. Another similar process is electrorefining which is strictly used in refining applications to improve the purity of the metals. Both processes use electroplating and are used to purify nonferrous metals such as Cu, Au, and Ag.

The electrowinning process is often sensitive to parameters such as pH change, as the pH of a solution affects its potential window and the solubilities of any metal

5.9 Purification

compounds dissolved in it. This is a well-studied area and pH-potential (Pourbaix) diagrams exist for most systems. Complexing agents, such as cyanide or ammonia, can be used to adjust metal speciation and therefore control the solubility and deposition properties of both the desired solute and any impurities present in solution. An overpotential in water can cause hydrogen gas evolution at the cathode and oxygen gas evolution at the anode, which can lead to embrittlement, a diminishing in the quality of the electrodeposit, and resulting in poor current efficiency. Metals such as Cr and Al must be chemically or electrochemically reduced in high-temperature processes, which often demand an aqueous pretreatment of the ore. Most of these procedures have extremely high energy demands and can produce large volumes of waste.

Electrowinning is performed in a divided cell with anode and cathode in the presence of solutions. Ion concentrations, temperature, current density, acidity, and cell voltage are important variables. Figure 5.5 shows industrial-scale Cu EW plant. Precipitation on cathode surface may be powdery, dendritic, smooth, rough, etc. Smooth surface with metallic luster is preferred in precipitation. Both quality and adherence of metals on cathode are important. Cathodic current efficiency and cell voltage are main cell performance parameters. Higher current efficiencies are better. In the catholyte, as the dissolved ion concentration increases current efficiencies increase. In the catholyte solution, acid and/or salt concentration is also important. Salt increases the conductivity of the solution during EW. Lower $H_2(g)$ evolution is desired. Higher temperatures result in water and acid vaporization.

The reactivity of metals depends on how easily they can lose their outer shell electrons to form a stable electron configuration. Let's assume Metal X is higher in reactivity series (more reactive) than Metal Y. This means that Metal X can react with Metal Y and reduces the salt or oxide of Metal Y. In other words, Metal X will displace Metal Y from its compound because Metal X is more reactive than Metal Y. For example, Fe, Zn, and Mg are more reactive than Cu, so

$$Fe(s) + CuCl_2(aq) \rightarrow FeCl_2(aq) + Cu(s) \downarrow \quad (5.54)$$

$$Zn(s) + Cu(NO_3)_2(aq) \rightarrow Zn(NO_3)_2(aq) + Cu(s) \downarrow \quad (5.55)$$

$$Mg(s) + Cu^{2+}(aq) \rightarrow Mg^{2+}(aq) + Cu(s) \downarrow \quad (5.56)$$

Oxidation and reduction can be defined in terms of oxygen and electron transfers. Oxidation is gaining of oxygen and reduction is loss of oxygen. Because both **red**uction and **ox**idation are going on side by side, this is known as a **redox** reaction.

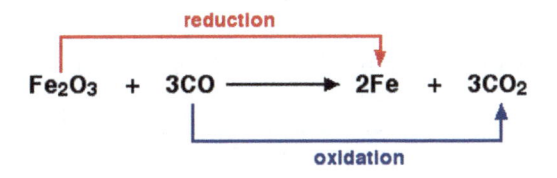

$$Fe_2O_3 + 3CO \xrightarrow{\text{reduction}} 2Fe + 3CO_2 \xleftarrow{\text{oxidation}}$$

Fig. 5.5 Industrial-scale Cu electrowinning plant

An oxidizing agent is substance which oxidizes something else. In the above example, Fe^{3+} oxide is oxidizing agent. A reducing agent reduces something else. In above equation, CO is the reducing agent. Oxidizing agents give oxygen to another substance and reducing agents remove oxygen from another substance.

Oxidation is loss of electrons and reduction is gain of electrons. Magnesium reduces Cu and loses two electrons and become reducing agent. In the following example, Cu is oxidizing agent. An oxidizing agent oxidizes something else and this means that the oxidizing agent is reduced. Reduction is gain of electrons; thus, an oxidizing agent gains electrons. Equations (5.57) and (5.58) show oxidation reaction of Sn metal and reduction of Pb^{2+} ions. Table 5.6 shows the reduction half-reaction potentials in acidic solutions.

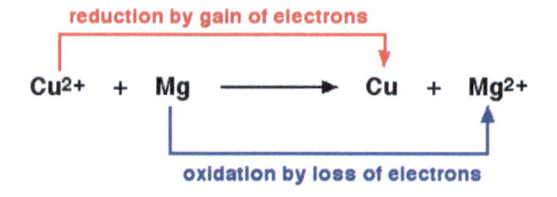

$$Cu^{2+} + Mg \longrightarrow Cu + Mg^{2+}$$

reduction by gain of electrons

oxidation by loss of electrons

5.9 Purification

Table 5.6 Reduction half-reaction in acidic solutions (25 °C, 101 kPa, 1 M) [51]

Half reaction potentials	V
$2H^+(aq) + 2e^- \to H_2(g)$	0
$Pb^{2+}(aq) + 2e^- \to Pb$	−0.125
$Sn^{2+}(aq) + 2e^- \to Sn(s)$	−0.137
$Fe^{2+}(aq) + 2e^- \to Fe(s)$	−0.440
$Zn^{2+} + 2e^- \to Zn(s)$	−0.763
$Al^{3+}(aq) + 3e^- \to Al(s)$	−1.676
$Mg^{2+}(aq) + 2e^- \to Mg(s)$	−2.356
$Na^+(aq) + e^- \to Na(s)$	−2.713
$Ca^{2+}(aq) + 2e^- \to Ca(s)$	−2.84
$K^+(aq) + + e^- \to K(s)$	−2.924
$Li^+(aq) + e^- \to Li(s)$	−3.040
$Sn^{4+}(aq) + 2e^- \to Sn^{2+}(aq)$	+0.154
$Cu^{2+}(aq) + 2e^- \to Cu(s)$	+0.340
$Ag^+(aq) + e^- \to Ag(s)$	+0.800
$Fe^{3+}(aq) + e^- \to Fe^{2+}(aq)$	+0.771
$O_3(g) + 2H^+(aq) + 2e^- \to O_2(g) + H_2O(l)$	+2.075
$Br_2(l) + 2e^- \to 2Br^-(aq)$	+1.065
$SO_4^{2-}(aq) + 4H^+(aq) + 2e^- \to 2H_2O(l) + SO_2(g)$	+0.17
$NO_3^-(aq) + 4H^+(aq) + 3e^- \to NO(g) + 2H_2O(l)$	+0.956
$O_2(g) + 4H^+(aq) + 4e^- \to 2H_2O(l)$	+1.229
$O_2(g) + 2H^+(ag) + 2e^- \to H_2O_2(aq)$	+0.695
$PbO_2(s) + 4H^+(aq) + 2e^- \to Pb^{2+}(aq) + 4H_2O(l)$	+1.455
$H_2O_2(aq) + 2H^+(aq) + 2e^- \to 2H_2O(l)$	+1.763
$S(s) + 2H^+(aq) + 2e^- \to H_2S(g)$	+0.14

$$\text{Oxidation}: Sn_{(s)} \to Sn^{2+}{}_{(aq)} + 2e^-{}_{(aq)} \tag{5.57}$$

$$\text{Reduction}: Pb^{2+}{}_{(aq)} + 2e^-{}_{(aq)} \to Pb_{(s)} \tag{5.58}$$

The major issues in the electrowinning are selecting the right electrodes and also the voltage or the potential required to separate the metals from AR medium. The electrodes are chosen based on the galvanic series, and the potential can be calculated using the Nernst equation in electrochemistry. The electrodes are to be decided based on the reactivity or the galvanic series. The active metal acts as anode and the passive or inert metal acts as cathode. The series show the tendency of different materials, which are tending to be more active or more inert. The anodic material gives the electron and the metal precipitated from the solution gets deposited on cathode. Thus, the reactivity of anode material should be more than the cathodic material. Table 5.7 shows reactivity order of some metals, reaction with O_2, dilute acids and water, ease of oxidation, oxidation reactions, and standard half-cell potential values. So, according to galvanic series, for Cu recovery, Al foil can be anode and pure Cu rod can be cathode electrode. For Au, Ag, and Pd recoveries, stainless steel mesh can be anode and stainless steel rod can be cathode electrode.

Table 5.7 Reactivity order of some metals, reaction with O_2, dilute acids and water, ease of oxidation, oxidation reactions, and standard half-cell potential values

								Oxidation reaction	Half reaction	Volts
MOST REACTIVE		Li	Lithium			Can not be extracted with C			$Li^+ + e^- \to Li$	-3.04
More cathodic (inert)	Extraction by elecro winning expensive	Li	Potassium		React with water		Strongly reducing	$K \to K^+ + e^-$		
		Ba	Barium							
		Ca	Calcium		React with dilute acids					
		Na	Sodium							
		Mg	Magnesium							
Increasing reactivity		Al	Aluminum	React with oxygen			Ease of oxidation increase	$Al \to Al^{3+} + 3e^-$	$Al^{3+} + 3e^- \to Al$	-1.68
		C	Carbon							
	Extraction by metal oxide reduction with C/CO₂ cheap	Zn	Zinc			To form oxides			$Zn^{2+} + 2e^- \to Zn$	-0.76
		Fe	Iron (Steel)						$Fe^{2+} + 2e^- \to Fe$	-0.44
		Ni	Nickel					$Ni \to Ni^{2+} + 2e^-$	$Ni^{2+} + 2e^- \to Fe$	-0.26
		Sn	Tin							
More anodic (active)		Pb	lead			Can be extracted with C	Weakly reducing		$Pb^{2+} + 2e^- \to Pb$	-0.13
		H	Hydrogen						$2H^+ + 2e^- \to H_2$	0.00
		Cu	Copper					$Cu \to Cu^{2+} + 2e^-$	$Cu^{+2} + 2e^- \to Cu$	0.34
		Hg	Mercury		Zn, Fe, Ni, Sn & Pb are BMs					
		Ag	Silver						$Ag^+ + e^- \to Ag$	0.80
		Au	Gold		Au, Ag & Pt are PMs			$Au \to Au^{3+} + 3e^-$	$Au^{3+} + 3e^- \to Au$	1.52
		Pt	Platinum		C & H are not metals					
LEAST REACTIVE										

Cu can be used as cathode; Cu extracted will be coated on the pure Cu rod, enhancing it, which is more economical. Other metals use stainless steel (SS) as both anode and cathode as they are PMs, and using it as cathode would not be economical. As for the other metals, the metal powder deposited on SS rod can be scrapped out. The anode would be lined on the surface of EW chamber in order to increase the contact surface with the liquid and SS mesh used. When voltage is passed through this arrangement, the anode loses electron which are gained by metallic ions in the liquid and gets deposited on the cathode. Thus, the extraction of metal depends on the voltage required to lose an electron and also converting the metals in the liquid into ions.

The electric potential required to reduce the metal is known as standard half-cell potential and the series is called EMF series. Further this standard half-cell potential is used to calculate the actual voltage required to deposit metal on cathode. This is done by Nernst equation, which is given below [51]:

$$E = E^\circ - \frac{2.303 * R * T}{z * F} * \log(a) \qquad (5.59)$$

where:

E: half-cell potential of the reaction in volts
E°: standard half-cell potential at STP and molarity of 1
R: universal gas constant, 8.314 J/mole-K
z: number of moles of electrons transferred in the reaction
F: Faraday's constant, 96,485 Coulombs/mole of electrons
a: chemical activity of the species

Here the following assumptions were made before calculating the potential [51]:

5.9 Purification

Table 5.8 Calculated electrolysis data

	Cu	Au	Ag	Pd
a (ion activity)	0.000754	0.000000518	0.00000382	0.000000212
z (ionization state)	2	3	1	4
$E°$ (standard half potential)	0.3419	1.498	0.7996	1
$E = V_I$ (standard potential)	0.2498	1.3741	0.4793	0.9014

1. $V_I = E$ (standard potential).
2. Assume activities equal to concentration in mole/L

Table 5.8 shows a, z, $E°$, and half-cell potential values of Cu, Au, Ag, and Pd metals [51].

In EW process, firstly Cu and then Ag, Pd, and Au are recovered, respectively. As the DC current is supplied, the metal from the solution gets collected at the cathode. Bubbling depicts that the reaction is going on. Eventually bubbling stops, which means that the reaction is being completed. The powder deposited on the cathode end can be scrapped out. This powder can be melted to form ingots of the metal. After complete recovery of the metals, the waste stream contains insoluble metals and the acid slurry. These waste streams are to be treated for environment, health, and safety norms. Generally, acids are neutralized and the insoluble contents are separated using a most common method membrane separation technique.

5.9.4 Solvent Extraction (SX)

Solvent extraction is a process that allows the separation of two or more components due to their unequal solubilities in two immiscible liquid phases. It is an important method in hydrometallurgy in separation of metal ions from their solution. SX consists of transferring one (or more) solute(s) contained in a feed solution to another immiscible liquid (solvent). The solvent that is enriched in solute(s) is called extract, and the feed solution that is depleted of solute(s) is called raffinate (Fig. 5.6).

Solvent extraction, also called liquid-liquid extraction (LLE) and partitioning, is a method to separate compounds based on their relative solubility in two different immiscible liquids, usually water and organic solvents (Fig. 5.7). Immiscible liquids are ones that cannot get mixed up together and separate into layers when shaken together. LLE is an extraction of a substance from one liquid into another liquid phase. The most common use of the distribution principle is in the extraction of substances by solvents, which are often employed in a laboratory- or in large-scale manufacturing. Organic compounds are generally much more soluble in organic solvents, like benzene, chloroform, and ether, than in water, and these solvents are immiscible with water. Organic compounds are then quite easily separated from the mixture with inorganic compounds in aqueous medium by adding benzene, chloroform, etc. Upon shaking, these separate into two layers. Since organic compounds have their distribution ratio largely in favor of the benzene phase, more of them

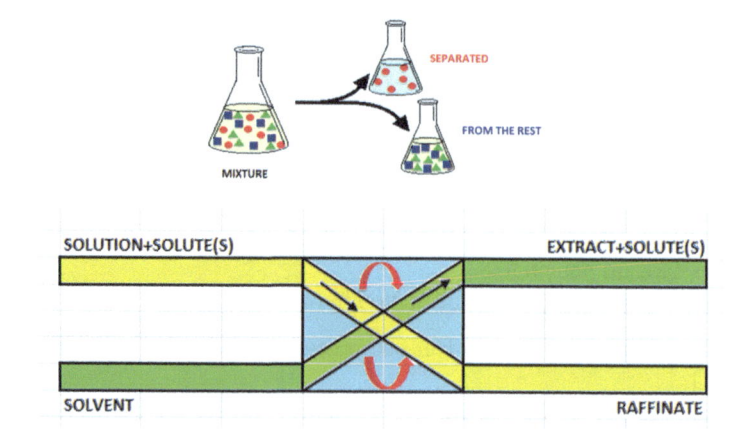

Fig. 5.6 Solvent extraction process feeds and products

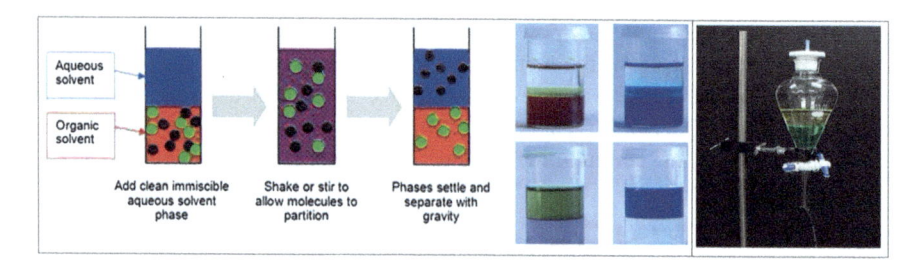

Fig. 5.7 Principles of solvent extraction and phases in SX separation

would pass into a nonaqueous layer. Finally, this nonaqueous layer is removed and distilled to obtain the purified compound. In solvent extraction, it's advantageous to do extraction in successive stages using smaller lots of solvents rather than doing extraction once using the entire lot. Solvent extraction tests can be done in separation vessels and mixer-settlers.

SX, which is an important part of hydrometallurgy, includes extraction and stripping steps after leaching and before reduction. SX flowsheet is shown in Fig. 5.8. Leaching dissolves mixed metal(s). Extraction step is performed in a mixing chamber, where selected metal is loaded in organics and raffinate is sent back to the leaching step. In stripping step, selected metal is extracted to the aqueous phase and spent electrolyte is recycled to the extraction step. After stripping metal (s) are reduced in the reduction step and spent electrolyte is recycled to the stripping stage. SX tests are performed in mixer-settler chambers in pilot and plant scales (Fig. 5.8). Table 5.9 shows types of solvent extractions along with extraction solvents and application areas. Organic solvents in SX are selected depending on the following properties: high extraction capacity, selectivity, easily stripped, immiscible in aqueous phase, sufficient density difference with the aqueous phase, low viscosity, nontoxicity, nonexplosiveness, and cheapness. Aliphatic or aromatic hydrocarbons or combination of both are used as diluent for diluting the organic

5.9 Purification

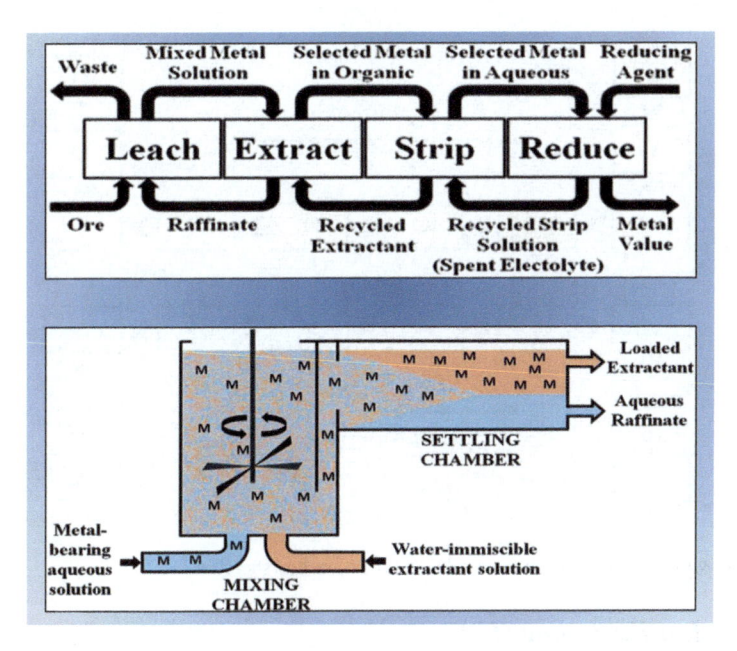

Fig. 5.8 Solvent extraction flowsheet and mixer-settler setup

Table 5.9 Types of solvent extraction, solvents used, and application areas

	Solvation extraction	Cationic exchange	Anionic exchange	Chelating exchange
Solvents used	Tributyl phosphate (TBP)	Di-(2-ethylhexyl)phosphoric acid (D2EHPA)	Primary amines (RNH$_2$)	Lix 63
	Trioctylphosphine Oxide (TOPO)	Naphthenic acid	Secondary amines (R$_2$NH)	Lix 65
	Methyl isobutyl ketones (MIBK)	Versatic acid	Tertiary amines (R $\geq$ C$_{12}$–C$_{14}$)	Kelex 100
Applications	Pb, Zn, U, Fe, Cd, Hf, Zr, Pu	Cu, Zn, Ni, Co, Ag	U, Th, V, Co	Cu, Ni, Co

extractant so that its physical properties like viscosity and density become more favorable for better mixing for two phases and their separation. Some common diluents (such as kerosene, benzene, chloroform, xylene, naphtha, or toluene) affect the extraction, scrubbing, stripping, and phase separation process quite significantly [52].

Much of the metals move from aqueous phase to organic phase in a single-stage contact (in laboratory scale) or multiple-stage counter-current contacts (in industrial scale). Multistage counter-current SX flowsheet along with organic and aqueous phase flow is given in Fig. 5.9 [52].

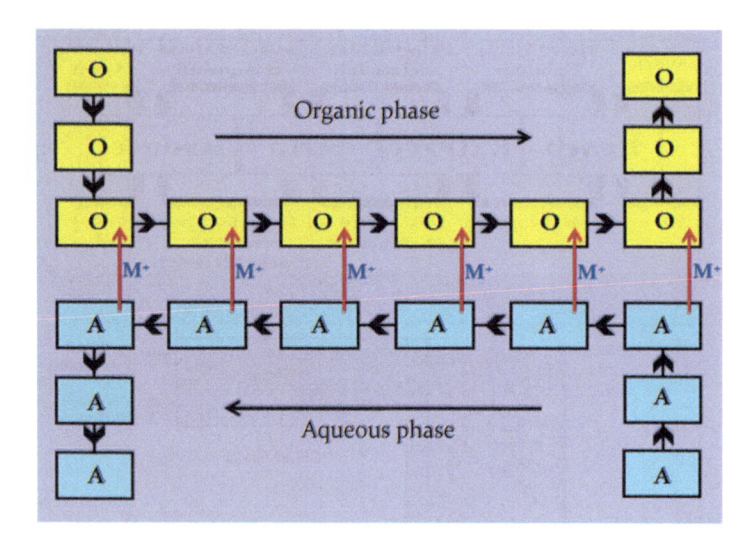

Fig. 5.9 Multistage counter-current solvent extraction flowsheet

5.9.5 Ion Exchange (IX)

In ion exchange, ions are exchanged between two electrolytes or between an electrolyte solution and a solute or complex. The sacrificial metal is usually one of low value. Ion exchange processes are reversible chemical reactions for removing dissolved ions from solution and replacing them with other similarly charged ions. Ion exchange is a reversible electrostatic adsorption phenomenon between two oppositely charged surfaces (solute molecules and resin) [53]. The positive and negative ions are exchanged by the medium ions. There are two types of ion exchangers: cation and anion exchangers. Cation exchangers exchange positive ions or cations. Cation exchangers can be inorganic or organic. Inorganic and organic cationic exchangers may be natural or synthetic. Ion exchange resins or polymers are used. Figure 5.10 shows the classification of ion exchangers. Ion exchange brings total environment solutions – water treatment, liquid waste treatment, recycle, and solid waste management. Applications of IX include water hardness removal (Ca and Mg exchange with Na and H); Fe and Mn removal from groundwater; recovery of Ag, Au, and U from wastewater products; demineralization; and removal of NO_3, NH_4, and PO_4.

Au recovery by IX process was investigated using three resins (Amberlite XAD-7HP, Bonlite BA304, and Purolite A-500) by Kim et al. (2011) [54]. Purolite A-500, a macroporous strong base resin, was supplied by Purolite Company, whereas Bonlite BA304 (a gel strong base resin) and Amberlite XAD-7HP (a nonionic resin) were obtained from Born Chemical Company and Rohm and Haas, respectively. The resins were thoroughly washed with water and dried in oven

5.10 Water Treatment

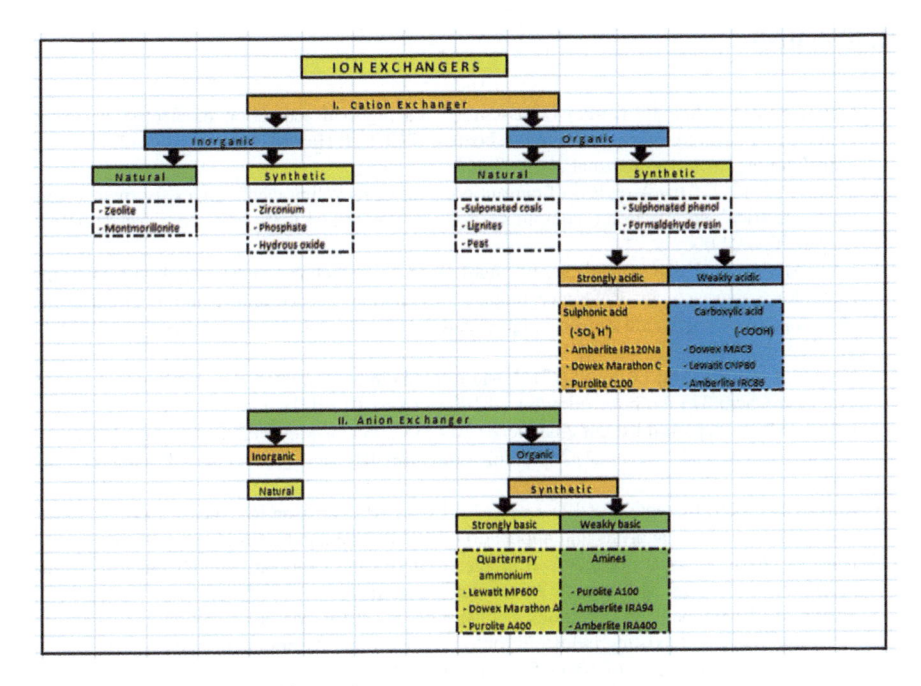

Fig. 5.10 Classifications of ion exchangers

at 50 °C. A glass column (ID: 1 cm * height: 25 cm) with a stopcock and a porous disk affixed at the bottom were used in all experiments. The glass column was loaded with 1.0 g (1 bed volume of resin, BV = 4.6 mL) of the resins. After each experiment, the resin filled in the column was washed with a large volume of distilled water and stored in water for the next set of experiment. For each experiment, a 5 BV of leach liquor was prepared through two-stage leaching by electrogenerated chlorine. The solution was passed through the column at a rate of 1.0 mL/min. The loaded resin was eluted by HCl in acetone.

5.10 Water Treatment

Not only waste process water but also cooling water, all rainwater, and sprinkling water should be treated in BAT in wastewater treatment plant. Acids should be neutralized; metals, sulfates, and fluorine should be removed by physical chemistry processes. Generally, two-thirds of treated water are reused internally, while one-third is discharged.

References

1. Lee J, Kim YJ, Lee JC (2012) Disassembly and physical separation of electric/electronic components layered in printed circuit boards (PCB). J Hazard Mater 241–242:387–394. https://doi.org/10.1016/j.jhazmat.2012.09.053
2. Zhang L, Xu Z (2016) A review of current progress of recycling technologies for metals from waste electrical and electronic equipment. J Clean Prod 127:19–36. https://doi.org/10.1016/j.jclepro.2016.04.004
3. Kaya M (2018) Waste printed circuit board (WPCB) recycling: conventional and emerging technology approach. In: Reference module in materials science and materials engineering/encyclopedia of renewable and sustainable materials. Elsevier. https://doi.org/10.1016/B978-0-12-803581-8.11246-9
4. Ning C, Lin CSK, Hui DCW (2017) Waste printed circuit board (PCB) recycling techniques. Top Curr Chem 375:43. https://doi.org/10.1007/s41061-017-0118-7
5. http://wedocs.unep.org/bitstream/handle/20.500.11822/8423/-Metal%20Recycling%20Opportunities%2c%20Limits%2c%20Infrastructure-2013Metal_recycling.pdf?sequence=3&isAllowed=y
6. Kellner D (2009) Recycling and recovery. In: Hester RE, Harrison RM (eds) Electronic waste management, design, analysis and application. RSC Publishing, Cambridge, pp 91–110
7. Kaya M (2016) Recovery of metals from electronic waste by physical and chemical recycling processes. Int J Chem Mol Eng 10(2). scholar.waset.org/1999.2/10003863.
8. Khaliq A, Rhamdhani MA, Brooks G, Masood S (2014) Metal extraction processes for electronic waste and existing industrial routes: a review and Australian perspective. Resources 3(1):152–179. https://doi.org/10.3390/resources3010152
9. Gulgul A, Szczepaniak W, Zablocka-Malicka M (2017) Incineration, pyrolysis and gasification of electronic waste, E3S Web of conferences, 22, 00060. https://doi.org/10.1051/e3sconf/20172000060
10. Wang JB, Xu ZM (2015) Disposing and recycling waste printed circuit boards: disconnecting, resource recovery and pollution control. Environ Sci Technol 49:721–733. https://doi.org/10.1021/es504833y
11. Qui K, Wu Q, Zhan Z (2009) Vacuum pyrolysis characteristics of waste printed circuit boards epoxy resin and analysis of liquid products. J Cent South Univ/Sci Technol 5 (in Chinese). http://en.cnki.com.cn/Article_en/CJFDTOTAL-ZNGD200905009.htm
12. Zhang S, Yoshikawa K, Nakagome H, Kamo T (2013) Kinetics of the steam gasification of a phenolic circuit board in the presence of carbonates. Appl Energy 101:815–821. https://doi.org/10.1016/j.apenergy.2012.08.030
13. Salbidegoitia JA, Fuentes-Ordonez EG, Gonzalez-Marcos MP, Gonzalez-Velasco JR, Bhaskar T, Kamo T (2015) Steam gasification of printed circuit board from e-waste: effect of coexisting nickel to hydrogen production. Fuel Process Technol 133:69–74. https://doi.org/10.1016/j.fuproc.2015.01.006
14. Zhang S, Yu Y (2016) Dechlorination behavior on the recovery of useful resources from WEEE by the steam gasification in the molten carbonates. Procedia Environ Sci 31:903–910. https://doi.org/10.1016/j.proenv.2016.02.108
15. Acomb JC, Anas Nahil M, Williams PT (2013) Thermal processing of plastics from waste electrical and electronic equipment for hydrogen production. J Anal Appl Pyrolysis 103:320–327. https://doi.org/10.1016/j.jaap.2012.09.014
16. Mankhand TR, Singh KK, Gupta SK, Das S (2012) Pyrolysis of printed circuit boards. Int J Metall Eng 1(6):102–107. https://doi.org/10.5923/j.ijmee.20120106.01.
17. Jones DA, Lelyveld TP, Mavrofidis SD, Kingham SW, Miles NJ (2002) Microwave heating applications in environmental engineering: a review. Resour Conserv Recycl 34(2):75–90. https://doi.org/10.1016/S0921-3449(01)00088-X

18. Soare V, Burada M, Dumitrescu DV, Costantian I, Soare V, Popescu ANJ, Carcea I (2016) Innovation approach for the valorization of useful metals from waste electrical and electronic equipment (WEEE). IOP Conference Series, Materials Science and Engineering. https://doi.org/10.1088/1757-899X/145/2/022039; http://researchgate.net/publication/304310103
19. www.boliden.com/sustainability/case-studies/largest-electronic-material-recycler-in-the-world
20. Zhou G, He Y, Luo Z, Zhao Y (2010) Feasibility of pyrometallurgy to recover metals from waste printed circuit boards. Fresenius Environ Bull 19(7):1254–1259. ISSN: 03043894
21. Hagelüken C (2006) Recycling of electronic scrap at Umicore's integrated metals smelter and refinery. World Metall ERZMETALL 59(3):152–161
22. Scott A (2014) Innovations in mobile phone recycling: biomining to dissolving circuit boards. Available from: http://theguardian.com/sustainable-business/2014/sep/30/innovations-mobile-phone-recycling-biomining-dissolving-circuit-boards
23. Tuncuk A, Stazi V, Akcil A, Yazici EY, Deveci H (2012) Aqueous metal recovery techniques from E-scrap: hydrometallurgy in recycling. Miner Eng 25:28. https://doi.org/10.1016/j.mineng.2011.09.019
24. Le Ret C, Briel O (2011) Umicore manufacturing precious metal based catalysts and APIs... and more! Chim Oggi/Chem Today 29:2–3
25. Kaya M (2016a) Recovery of metals and nonmetals from electronic waste by physical and chemical recycling processes. Waste Manag 57:64–90. https://doi.org/10.1016/j.wasman.2016.08.004
26. Baba H (1987) An efficient recovery of gold and other noble metals from electronic and other scraps. Conserv Recycl 10(4):247–252. https://doi.org/10.1016/0361-3658(87)90055-5
27. Kinoshita T, Akita S, Kobayashi N, Nii S, Kwaizumi F, Takahashi K (2003) Metal recovery from non-mounted printed wiring boards via hydrometallurgical processing. Hydrometallurgy 69:73–79. https://doi.org/10.1016/S0304-386X(03)00031-8
28. Samina M, Karim A, Venkatachalam A (2011) Corrosion study of iron and copper metals and brass alloy in different medium. E J Chem 8:344–348. https://doi.org/10.1155/2011/193987
29. Yang JG, Tang CB, He J, Yang SH, Tang MT (2011) A method of extracting valuable metals from electronic wastes, State Intellectual Property Office of the People's Republic of China, ZL 200910303503.5.
30. Iannicelli-Zubiani EM, Giani MI, Recanati F, Dotelli G, Puricelli S, Cristiani C (2017) Environmental impacts of a hydrometallurgical processes for electronic waste treatment: a life cycle assessment case study. J Clean Prod 140:1204–1216. https://doi.org/10.1016/j.jclepro.2016.10.040
31. Batnasan A, Haga K, Shibayama A (2018) Recovery of precious and base metals from waste printed circuit boards using a sequential leaching procedure. JOM 70(2):124–128. https://doi.org/10.1007/s11837-017-2694-y
32. Havlik T, Orac D, Petranikova M, Miskufova A, Kukurugya F, Takacova Z (2010) Leaching of copper and tin from used printed circuit boards after thermal treatment. J Hazard Mater 183:866–873. https://doi.org/10.1016/j.jhazmat.2010.07.107
33. Yang C, Li J, Tan Q, Liu L, Dong Q (2017) Green process of metal recycling: coprocessing waste printed circuit boards and spent tin stripping solution. ACS Sustain ChemEng 5:3524–3535. https://doi.org/10.1021/acssuschemeng.7b00245
34. Zhang X, Guan J, Gua Y, Cao Y, Gua J, Yuan H, Su R, Liang B, Gao G, Zhou Y, Xu J, Guo Z (2017) Effective dismantling of waste PCB assembly with methanesulfonic acid containing hydrogen peroxide, AIChE. Environ Prog Sustain Energy 36(3). https://doi.org/10.1002/ep.12527
35. Dorneanu SA (2017) Electrochemical recycling of waste printed circuit boards in bromide media. Part 1: Preliminary leaching and dismantling tests. Stud Univ Babes Bolyai Chem LXII (3):177–186. https://doi.org/10.24193/subbchem.2017.3.14
36. Calgaro CO, Schlemmer DF, da Silva MDCR, Maziero EV, Tanabe EH, Bertuol DA (2015) Fast copper extraction from printed circuit boards using supercritical carbon dioxide. Waste Manag 45:289–297. https://doi.org/10.1016/j.wasman.2015.05.017

37. Silvas FPC, Correa MMJ, Caldes MPK, Moraes VT, Espinosa DCR, Tenorio JAS (2015) Printed circuit board recycling; physical processing and copper extraction by selective leaching. Waste Manag 46:503–510. https://doi.org/10.1016/j.wasman.2015.08.030
38. Mecucci A, Scott K (2002) Leaching and electrochemical recovery of copper, lead and tin from scrap printed circuit boards. J Chem Technol Biotechnol 77:449–457. https://doi.org/10.1002/jctb.575
39. Liang G, Tang J, Liu W, Zhou Q (2013) Optimizing mixed culture of two acidophiles to improve copper recovery from printed circuit boards (PCBs). J Hazard Mater 250–251:238–245. https://doi.org/10.1016/j.jhazmat.2013.01.077
40. Ilyas S, Anwar MA, Niazi SB, Ghauri MA (2007) Bioleaching of metals from electronic scrap by moderately thermophilic acidophilic bacteria. Hydrometallurgy 88:180–188. https://doi.org/10.1016/j.hydromet.2007.04.007
41. Ilyas S, Ruan C, Bhatti HN, Ghauri MH, Anwar MH (2010) Column bioleaching of metals from electronic scrap. Hydrometallurgy 101:135–140. https://doi.org/10.1016/j.hydromet.2009.12.007
42. Yang T, Xu Z, Wen J, Yang L (2009) Factors influencing bioleaching copper from waste printed circuit boards by Acidithiobacillus ferrooxidans. Hydrometallurgy 97:29–32. https://doi.org/10.1016/j.hydromet.2008.12.011
43. Ting Y-P, Tan CC, Pham VA (2008) Cyanide-generating bacteria for gold recovery from electronic scrap material. J Biotechnol 136:S653–S654. https://doi.org/10.1016/j.jbiotec.2008.07.1515
44. Faramarzi MA, Stagars M, Pensini E, Krebs W, Brandl H (2004) Metal solubilization from metal-containing solid materials by cyanogenic Chromobacterium violaceum. J Biotechnol 113(1–3):321–326. https://doi.org/10.1016/j.jbiotec.2004.03.031
45. Xiang Y, Wu P, Zhu N, Zhang T, Liu W, Wu J, Li P (2010) Bioleaching of copper from waste printed circuit boards by bacterial consortium enriched from acid mine drainage. J Hazard Mater 184(1–3):812–818. https://doi.org/10.1016/j.jhazmat.2010.08.113
46. Marhual NP, Pradha N, Kar RN, Sukla LB, Mishra BK (2008) Differential bioleaching of copper by mesophilic and moderately thermophilic acidophilic consortium enriched from same copper mine water sample. Bioresour Technol 99:8331–8336. https://doi.org/10.1016/j.biortech.2008.03.003
47. Sampson MI, Phillips CV (2001) Influence of base metals on the oxidizing ability of acidophilic bacteria during the oxidation of ferrous sulfate and mineral sulfide concentrates, using mesophiles and moderate thermophiles. Miner Eng 14:317–340. https://doi.org/10.1016/S0892-6875(01)00004-8
48. Brandl H, Bosshard R, Wegmann M (1999) Computer-munching microbes: metal leaching from electronic scrap by bacteria and fungi. Process Metall 9:569–576. https://doi.org/10.1016/S1572-4409(99)80146-1
49. Zhu N, Xiang Y, Zhang T, Wu P, Dang Z, Li P, Wu J (2011) Bioleaching of metal concentrates of waste printed circuit boards by mixed culture of acidophilic bacteria. J Hazard Mater 192:614–619. https://doi.org/10.1016/j.jhazmat.2011.05.062
50. Luda MP (2017) Chapter 15: Recycling of printed circuit boards. In: Kumar S (ed) Integrated waste management, vol 2. InTech, Rijeka, pp 285–298. https://doi.org/10.5772/17220
51. Memon AH, Patel RL, Pitroda DJ (2017) Design for recovery of precious and base metals from e-waste using electrowinning process. Int J Adv Res Eng Sci Technol 4(5):579–586. e-ISSN: 2393-9877, p-ISSN: 2394-2444
52. https://www.slideshare.net/DilipSaha1/solvent-extraction-51718683?from_action=save
53. https://slideplayer.com/slide/3503176/
54. Kim EY, Kim MS, Lee JC, Pandey BD (2011) Selective recovery of gold from waste mobile phone PCBs by hydrometallurgical process. J Hazard Mater 198:206–215. https://doi.org/10.1016/j.jhazmat.2011.10.034

Chapter 6
Size Reduction and Classification of WPCBs

"Think Smart, Think Green – Recycle!"

Anonymous

Abstract This chapter explains the aim and importance of size reduction in e-waste recycling. Shredding, crushing, pulverizing, and grinding are general comminution methods used in WPCB recycling. Industrial-size shredders, which are hammer mill pulverizers, are covered in detail. Emerging fractionation and electrodynamic fragmentation with high-voltage pulse technologies are introduced for WPCB recycling. Industrial plant application flowsheets are demonstrated. Industrial-scale material classification methods and equipment are presented. Plant-scale conveyor belts for material transport between equipment and working principles of filtration and dust collection systems for high filtration efficiency are also covered in this chapter.

Keywords Size reduction · Shredder · Pulverizer · Fractionator · Delamination · Electrodynamic fragmentation · Classification · Conveying · Filtration

6.1 Size Reduction of PCBs

The reutilization of WPCB is a focused topic in the field of environment protection and resource recycling. Presently, WPCB is mainly processed using physical methods, chemical methods, biological methods, or combinations of these approaches. The purpose of the crushing process is to liberate metals from nonmetals of WPCBs, which are a mixture of woven glass fiber-reinforced resin and multiple kinds of metals and have high hardness and tenacity. In the process of reclaiming WPCB, crushing is the crucial technique. Its importance for chemical and biological methods is that the crushing process must make the surface of the resulting particles appropriate for subsequent contact with the chemical or biological medium. Considering physical methods, the process of crushing may directly affect the efficiency of successive separation steps, which then can affect the recovery rate and the purity of the metal. If the size reduction process is done under dry conditions, it can produce local high temperature, which makes the organic matter in the WPCB pyrolysis, and produce harmful gases, causing air pollution. If size reduction is done under wet conditions, it produces wastewater that contains a large number of suspended particulate matters, which is an environmental pollutant. Before the water can be discharged or reused, specialized wastewater treatment facilities must be set up, which increases construction and operation costs. One problem with current crushing methods is that they cannot achieve the complete dissociation of metal and base plate. Other problems include large energy consumption, high economic costs, and environmental pollution [1].

WPCB comminution, which is performed for material liberation, to −5 mm generates 96–99% metallic liberation [2]. Duan et al. (2009) found that most WPCBs were liberated in the 2–5 mm size [3]. Ninety-five percent liberation was obtained at −1 mm size for WPCBs. Total liberation for metals, particularly for Cu prints on WPCB, was not achieved even at −212 + 75 μm. At the finest fraction (−75 μm), unliberated particles still can be found [4]. At coarse-size fractions, metals were mostly locked with plastics/resins, while needle-shaped metals in general were liberated. Agglomerated particles hinder the efficiency of following physical separation. Therefore, removing ECs prior to size reduction and physical separation of WPCBs should be suitable. Zhang and Forsberg (1999) suggested to remove Al foils with cyclones prior to physical processing [5]. Material loss during size reduction with hammer mills was 25–29 wt. % in Morales (2011) and Yamane (2012) studies [6, 7]. But, Silvas et al. (2015) reduced this loss less than 6 wt. % with combination of knife and hammer mills [8]. Avoid dissipation of trace elements, Au losses of up to 75% if PC-motherboards are not removed prior to shredding and to 100% in a car shredder [9].

Shredding (fragmenting), crushing, pulverizing, grinding, and ball milling are relatively conventional methods for reducing particle size. Recently, the use of cryogenics is being tested, and it is believed that it will be used for size reduction in the future due to high cost. Here only shredding and pulverizing will be covered in detail. Shredding is a process in which feed material is fragmented, ground, ripped, or torn into small pieces. Then, shredded materials (3–20 mm) are granulated into fine particles in pulverizers. Leak tightness of size reduction machines and pipelines must be ensured. Dry crushing generates fugitive odor and dust during crushing. Wet impact crushing to −1 mm is possible. This prevents odor and dust problems, and then wet gravity separation using Falcon concentrator can be used for metal recoveries [3]. Wet impact crushing has the advantages of higher crushing efficiency, less over-crushing, and no secondary pollution. Water can be recycled in the process wherein only a small amount of fresh water need be supplied.

6.2 Shredders

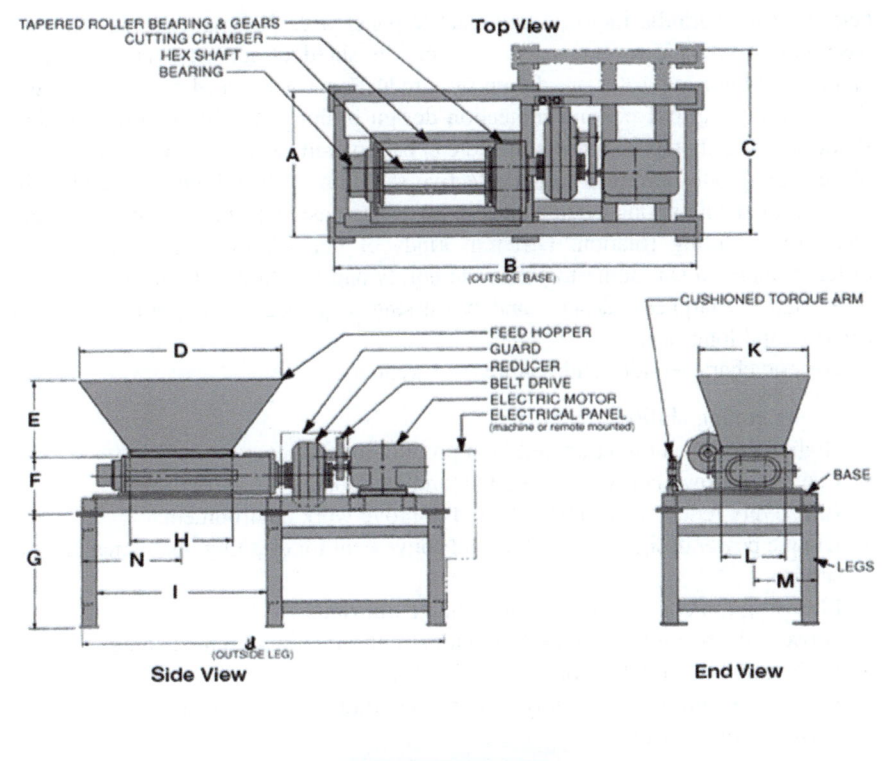

Fig. 6.1 Shredder machine details and cutter knife discs/blades for shredders

6.2 Shredders

Shredders are size-reducing, crushing, and volume-reducing machines for e-waste/e-scrap. There are three types of industrial shredders: single-shaft, double-shaft, and four-shaft. Figure 6.1 shows the top, side, and end views of a typical one motor-driven shredder along with blade details. Some high-capacity shredders have two motors. Shredder machines employ a low-speed and high-torque mechanism at low noise, energy, and dust. They have a wide range of applications because they can shred many different materials. Shredders include cutting chamber, bearings and gears, hexagonal shaft, feed hopper, guard, reducer, cushioned torque arm, drive belt, electric/hydraulic motor(s) and electric panel with PLC control, cabinet, and base unit. Automatic reverse sensors protect the shredder against overloading and jamming. Motor and rotor are driven by a reducer for low speed and high torque. Water- and dustproof bearing protection design prolongs the life of the shredder. Blade wearing should be minimum and can be repaired easily at a low cost. Helical blades are spirally arranged to get effective shredding. Blade inner hole and main rotor adopted with strong hexagonal column design ensure the blade stability and get equal force during rotation. Different kinds of blades/knives are available for different material shredding to ensure durability and flexibility. Rotary knife thickness, shape, sharpness, design, and processing sequence are important for high capacity and long-lasting.

General characteristics and features of shredders:

- Strong cutting ability.
- High degree of automation and high production efficiency.
- Low noise, low energy consumption, and less dust.
- Effectively reduce labor intensity and improve work environment.
- Unique power design, detachable, and convenient for cleaning, maintenance, and service.
- PLC program control for safe and reliable utilization.
- Advanced structure and drive design for strong shredding/high crushing capacity.
- High and steady quality for long-lasting life.
- Easy to use under heavy-duty working conditions and maintain to enable the recycling of e-waste.
- Program control, overload protection, and automatic reset.
- Rotary knife thickness, shape, sharpness, design, and processing sequence.
- Making up cutting tools from special hardened alloy steel (SKD-11) with a hardness of HRC55-58 for durability and wear resistance.
- Large hopper chamber dimensions which change from 300 to 4500 mm for high capacities.

Industrial shredder features are summarized in Table 6.1 [10]. Blade diameters change from 200 mm up to 800 mm, and blade thicknesses range from 20 mm to 80 mm. Thicker blades are used for higher diameter blades. As the blade diameter increases, the number of blades increases from 16 to 50. Shredder capacities change from 300 to 40,000 kg h^{-1}, and power consumption changes from 7.5 to 264 KW.

Table 6.1 Industrial shredder characteristics

Blade diameter (mm)	Blade thickness (mm)	Number of blades	Capacity (kg/h)	Total power (KW)
200	20	16–32	300–1200	7.5–22.0
300	40	16–36	800–5500	30.0–110.0
400	40	20–36	1500–7000	44.0–150.0
600	50	30–50	3000–22,000	74.0–220.0
800	80	30–50	4500–40,000	150.0–264.0

6.2 Shredders

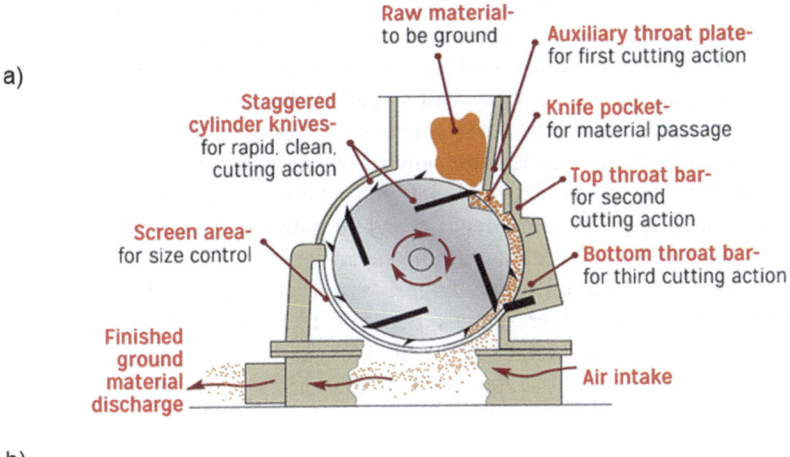

a)

- Raw material- to be ground
- Auxiliary throat plate- for first cutting action
- Staggered cylinder knives- for rapid, clean, cutting action
- Knife pocket- for material passage
- Top throat bar- for second cutting action
- Screen area- for size control
- Bottom throat bar- for third cutting action
- Finished ground material discharge
- Air intake

b)

Fig. 6.2 Single-shaft shredder details (**a**) and industrial-scale machines (**b**)

6.2.1 Single-Shaft Shredders

They consist of a knife disc and a static knife to complete shredding material. The cutter is composed of a base shaft and several shape quadrilateral knife blocks. The knife block has three, four, or more blades which can be replaced/sharpened after wearing. The knife block along the axis of the base is in multiple rows into type V, fixed with screws in the radical axis, rotated with the based shaft together, and finally comprised of a knife disc. The two static knives are fixed on the machine frame. The material is sent to the cutter by the horizontal hydraulic cylinder. The speed is stable and adjustable, and the propulsive force is large and uniform. It is applicable to the recovery of a variety of bulk solid materials, refractory materials, irregular plastic containers and plastic barrels, tubes, films, fibers, paper, etc. Spindle speed is between 45 and 100 rpm, which has a stable work and low noise. Figure 6.2 shows the details of the single-shaft shredders and industrial-scale machines [11].

Single-shaft shredder is mainly used for wood, large plastic waste recycling, large caliber pipe, and die material. Single-shaft shredders should be strong and durable, have low noise, and be an energy saver. Single-shaft shredders crush material down to 20 mm. In single-shaft shredders, the number of rotary blades is between 24 and 57; shaft diameter between 220 and 350 mm; rotor length between 500 and 2000 mm; rotating shaft speed between 75 and 90 rpm; fixed knife between 1 and 2; pressure between 6 and 12 MPa; crushing capacity between 250–400 kg h^{-1} and 600–1200 kg h^{-1}; feeding caliber/hopper size between 550–650 mm and 1280–1300 mm; motor power outputs between 18.5 and 55 KW; mesh size between 20 and 40 mm; weight between 3.5 and 9.5 tons; and dimensions (L*W*H) between 3000*1600*2200 mm and 4500*2500*2800 mm.

6.2.2 Double-Shaft Shredders

The uniquely designed machines feature low noise, high torque, and wide applications; they serve the functions of shearing, tearing, fracturing, breaking, and auto-discharging. Double-shaft PCB shredders are mainly used for crushing packing belt, tire, film, woven bag, fishing nets, and other plastic waste utilization. The biaxial independent drive, the unique knife shaft structure, and a four-angle rotary knife are used in the production process of high torque in low speed. In double-shaft shredders, the number of rotary blades is between 60 and 150; rotating shaft speed between 65 and 87 rpm; fixed knife between 4 and 8; blade thickness between 30 and 75 mm; crushing capacity between 500 and 2000 kg h^{-1}; feeding caliber between 700–800 mm and 1650–2000 mm; motor power between 18.5*2 and 45*2 KW; and weight between 4.0 and 9.5 tons. Product size changes from 5 to 20 mm. Figure 6.3 shows a double-shaft shredder in an industrial application.

Fig. 6.3 Double-shaft shredders

Fig. 6.4 Industrial-scale four-shaft shredders and obtained product

6.2.3 Four-Shaft Shredders

These machines are heavy-duty machines with modular design and good compatibility. The cutting tool is made of special alloy steel after special craft, with many characteristics (such as good wear resistant and highly repairable) and other characteristics. Driving part adopts four drives, with overloading planet gear reducer transmission and differential operation between main shaft. Electrical part of the application of PLC and touch screen control, and with a video monitoring system, the machine also has the overload protection function and so on. The machine has certain advantages such low rotating speed, high torque, and low noise, which are widely applicable to the recycling industry in the process. Materials suitable for crushing with four-shaft shredder are plastics, metals, glasses, paper, and e-wastes. Figure 6.4 shows four-shaft shredders, industrial-size equipment, and product sizes. In four-shaft shredders, the number of the main/assistant knife rotary diameter is between 150 and 400 mm; feeding caliber between 1000–800 mm and 2000–1800 mm; rotating speed between 25 and 43 rpm; main knife between 24 and 64; and main power between $11*2$ and $45*2$ KW. Some companies are producing double-story/flat double-shaft shredders. Double-story shredder and pulverizer on a single chassis are available on the market today. A precrushing shredder is accumulated on the same monoblock body with a pulverizer in some industrial equipment [12].

6.3 Blade/Hammer Mill Pulverizers/Granulators

Granulator machines are designed with high speed, medium inertia, open rotor body for fine grinding, and two, three, or five hardened steel knives. Granulators can grind material down to 0.185 mm (80 meshes) or up to 5 cm in size. Generally resulting particles can vary in size from 3 to 20 mm. Interchangeable qualifying screens with various diameter holes determine the final reduction size. With decibel ratings of less than 65 dB, these units are ideal for placement at individual work stations. Granulators are sized by the dimensions of the cutting chamber and range in size from $20*25$ to $40*88$ cm. Motor sizes range from 5 HP to 40 HP. Complete systems can

Fig. 6.5 Two different industrial-scale pulverizer machines along with screens and dust collection cyclones

include air discharge units or conveyors and can easily be integrated with existing shredder or grinder systems. Granulators have a smaller footprint than a full-size grinder but can still handle high volumes of product in the granulation process (Fig. 6.5) [13]. Hammer mills accomplish size reduction by impacting at rates of typically 7000 rpms and higher [2]. These granulators are used for the sizing of plastics, nonferrous metals, and heterogeneous materials and enable to reach controlled output size in the recycling process with the use of classifier screens starting from 2 mm diameter. Granulator sizes are $(1060-1800) * (1700-1800) * (2000)$ mm; power changes from 8 and 90 KW and weight from 700 to 4200 kg.

Because of the strength and tenacity of PCB material, much dust and harmful gas could be produced in the process of dry pulverizing at normal temperatures. On the other hand, after a very long operation, very high temperatures can develop in the equipment causing melting and jamming of the particles and, hence, reduced efficiency of crushing/pulverizing and environmental pollution too. Enhancement of crushing efficiency and reduction of secondary pollution during WPCB processing are important issues. One typical method is to cool the WPCB below the embrittlement temperature using liquid nitrogen and then crush it. The high cost of low-temperature crushing limits its further industrial application. To achieve high crushing efficiency and to resolve the secondary pollution problems of dust and gas

6.4 Fractionator Technology

from the crushing process, a wet impacting crusher can be used to perform comminution of WPCB in a water medium [3].

Fractionation and electrodynamic fragmentation (EDF) with high-voltage pulse (HVP) technologies, which will be covered in this chapter, are emerging technologies for e-waste/WPCB comminution for size reduction.

6.4 Fractionator Technology

The fractionation process technology is a breakthrough process to recycle metals from wastes in a purely mechanical way. The underlying physical principle is to use the different properties of the materials (density, plasticity, ductility, etc.) to separate the layers by creating huge accelerations and decelerations [14].

6.4.1 Physical Principles

In a fractionator, a composite material (like WPCBs) with heterogeneous composition and size is submitted to a high-frequency series of accelerations and decelerations (impacts). The phases of the material will react differently due to their heterogeneous densities and plasticities (Fig. 6.6):

- During accelerations (respectively, decelerations), the lighter phases will tend to go faster (respectively, slower), which will create shearing forces at the boundaries.
- Impacts will deform the phases in different ways according to their plasticities. The metals, which are ductile, will tend to aggregate into compact shapes; the minerals (i.e., fibers), which are brittle, will be fragmented while the plastics will tend to stay in their original shape.

6.4.2 Material Preparation

WPCBs are shredded and granulated to less than 22 mm. During the process, ferrous materials are easily separated by magnetic separation.

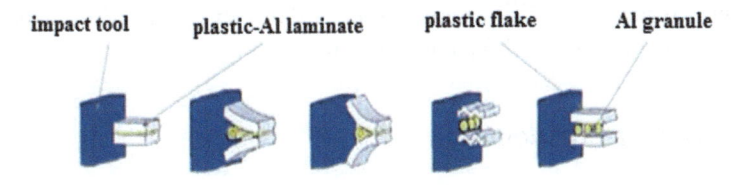

Fig. 6.6 Delamination process in a fractionator machine [14]

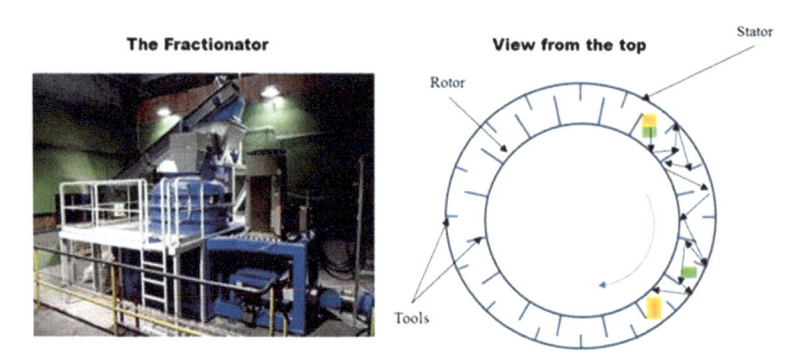

Fig. 6.7 Industrial-size fractionator and its cross section

6.4.3 Delamination

The series of high-frequency impacts are achieved in a fractionator machine which features an annular space between a rotor and a stator equipped with radial tools. Materials are fed from the top and fall by gravity. The high-speed rotation of the rotor drives the materials which then impacts with a high frequency the static tools of the stator as shown in Fig. 6.7 (view from the top) [14].

6.4.4 Separation into High-Purity Output Fractions

At the bottom of the fractionator, the various materials (metals, plastics, minerals) are liberated but still need to be separated. The size distribution is much differentiated at the exit of the fractionator:

- Plastics are found in relatively coarse particle size distributions (i.e., big pieces).
- Metals (Al and Cu) are found as very pure granules with sizes between 50 μ and 5 mm (the size is a function of the thickness and surface of the metal layer in the input).
- Precious metals are in the fine sizes (Fig. 6.8).

This makes the two-stage separation process (by size and density) very efficient:

- Metals and nonmetals are separated by size.
- Light (Al) and heavy (Cu) metals are separated by density (spherical shapes make it easy) in a fluidized bed separator (Fig. 6.9).

6.4.5 Output Fractions

The process generates up to 22 different fractions:

6.4 Fractionator Technology

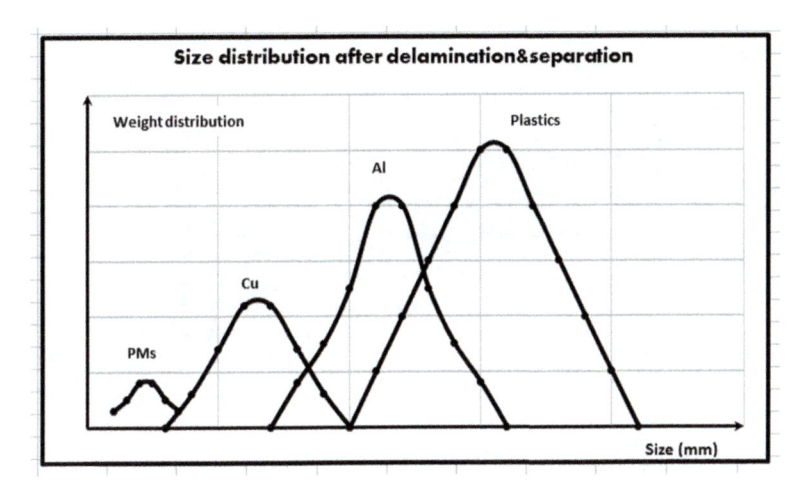

Fig. 6.8 Size distributions of materials after delamination and separation into fractions [14]

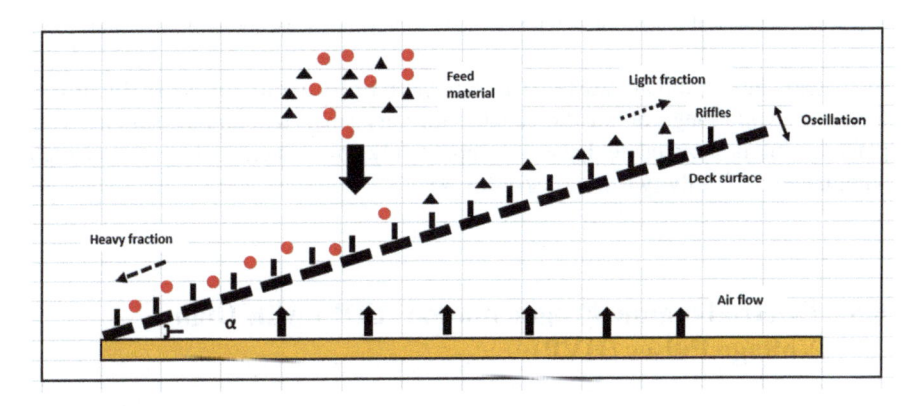

Fig. 6.9 Fluidized bed separator operation principles

- Several (plastics, resins, fibers, etc.) nonmetallic fractions with sizes.
- A ferrous fraction.
- Several light metal (typically Al) fractions.
- Several heavy metal (PMs, Cu and Cu alloys, Ni, Pb, etc.) fractions.
- Three dust fractions (the three stages – granulator, fractionator, and separator – are equipped with aspiration systems and filters).

Figure 6.10 shows industrial-scale plant flowsheet for the mechanical separation with fractionator and fluidized bed separator for delamination. This process has been used in more than 20 power plants built worldwide since 1995 with the previous generations. Metal recovery rates are claimed to be above 97%, and the output fractions have a high purity. This process is environmentally friendly, due to dry mechanical treatment, low temperature, no gases or fluids, and very limited

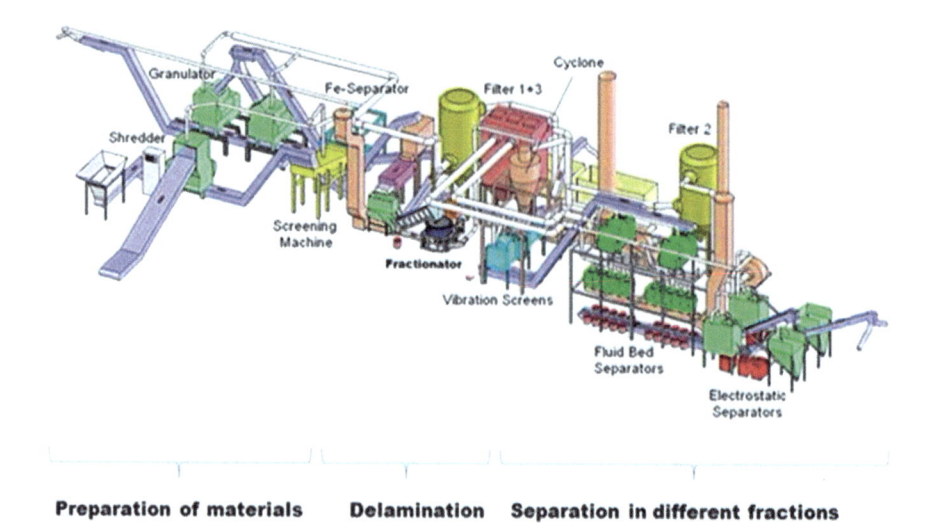

Fig. 6.10 Mechanical separation with fractionator and fluid bed separator for delamination

(potentially zero) ultimate wastes. This flexible process has a capacity of 3 tons h^{-1} for a very wide range of wastes, suitable for very small lots and campaigns. This plant layout is claimed to have a competitively low capex and apex and high automation with only two workers/shift [14].

6.5 Electrodynamic Fragmentation (EDF) with High-Voltage Pulses (HVP)

Novel comminution technologies are developing for breaking ores and secondary materials. Some of them are patented and some of them (i.e. industrialized high-voltage pulse (HVP) and electrodynamic fragmentation (EDF) technologies) are commercialized by Selfrag AG (establihed in 2007) in recycling (Fig. 6.11). HVP uses high-voltage pulse power technology which allows liberation or weakening of material along natural boundaries. It also allows for controlled crushing without contamination due to a combination of pulse power technology, physical (electrical) material discontinuities, and high-voltage and mechanical engineering skills.

6.5.1 Pulse Power Generation

Research into HVP technology began in the 1930s using capacitors to discharge electricity for producing X-rays. Since the 1960s, HVP technology rapidly developed. HVP creates repetitive electrical discharges to materials immersed in a process

6.5 Electrodynamic Fragmentation (EDF) with High-Voltage Pulses (HVP)   135

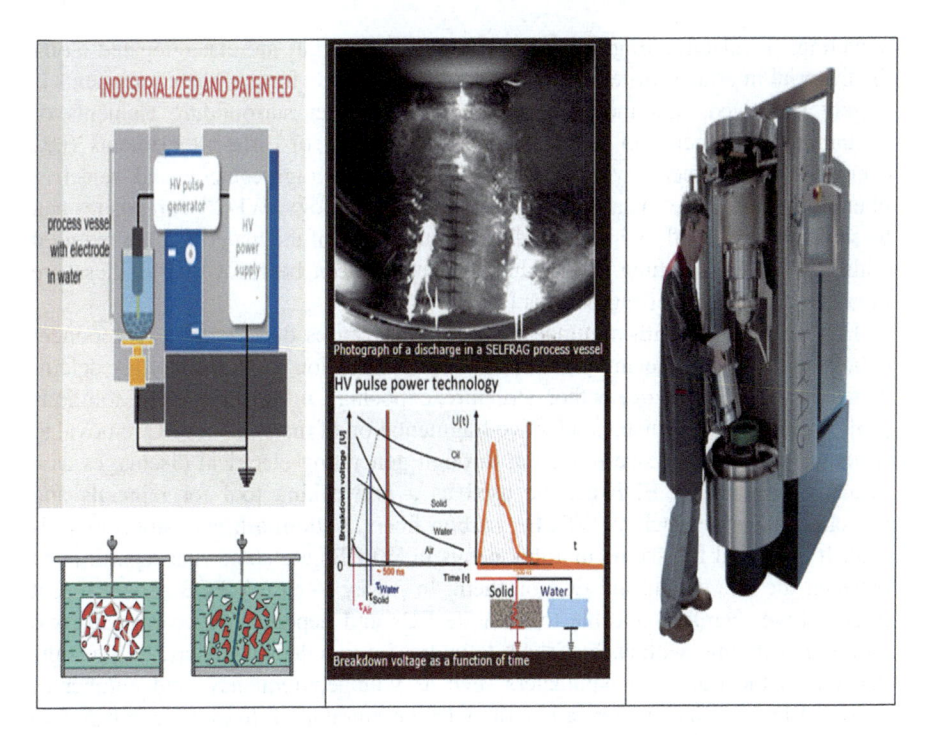

Fig. 6.11 HVP laboratory-scale equipment and pulse power generation [15]

liquid, like water, which has a high dielectric strength when voltage rise time is kept below 500 ns. As a result, discharges are forced through the immersed material. The introduced electrical energy is then transformed into an acoustical shockwave resulting into a huge tensile stress regime within the material.

6.5.2 Material Treated

Particles inside a composite material (WPCBs) or defects in a crystalline material lead to discontinuity in the electrical and acoustical properties. The discontinuity in the dielectric permittivity enhances the electrical field at the grain boundaries and forces the discharge channels to the boundaries. The sudden expansion of the created plasma produces a shock wave with local pressures up to 10,000 bar. The interaction of a shock wave and an acoustical discontinuity concentrate tensile stress at these interfaces. The selective fragmentation technology of Selfrag uses these effects to liberate material along the material boundaries, to weaken material along particle boundaries, or to diminish the size of material without introducing contamination.

Recovery is directly related to the selectivity of the process. Being able to separate all components without damage at a coarse fraction is the future of recycling. In selective fragmentation, composite materials are disaggregated along the material interfaces instead of being comminuted. The form of the components is largely preserved, and they are separated from other surrounding elements or contaminants. Benefits of HVP are coarse liberation of different elements (i.e., energy savings), recovery of original-size elements (aggregates), and selective liberation of valuable metals. With the Selfrag technology, WPCB composites can be separated efficiently without the fine comminution of the material often required with mechanical crushing. The valuable materials can be recovered as these are liberated during the Selfrag process [15].

EDF technology with minimal size reduction causes depopulation (component detachment) of ECs, delamination, and board destruction. EDF technology delaminates the WPCBs and opens their structure, exposing Cu foils to leaching solutions [16]. The use of EDF enables selective fragmentation of materials for EC removal to structure opening and size reduction through generating electrical discharges as a means of fracturing. EDF can be used as a preweaking tool for minerals and recycling materials such as WPCBs, carbon fibers, bottom ash, and siliceous rods [16]. It is logical to assume that the metals in WPCBs are somewhat repartitioned between the boards and the ECs on them. In a way to ease WPCB processing, it might prove therefore feasible to separate ECs and depopulated boards into two individual streams, with each of them to be treated in a dedicated way. EDF results depend on the operating parameters such as voltage, frequency, and number of pulses. EDF results in generation of a lesser amount of fines as compared to conventional comminution [16]. Liberated materials can be thus processed downstream in a more efficient way especially when value-added EoL electronic equipment is recycled. In EDF technique, there are three stages: depopulation, delamination, and entire fragmentation. Depopulation (for removal of ECs) followed by delamination is important for WPCB recycling. Finally, the entire WPCBs are physically fragmented for size reduction. A 150 kV voltage has been used for depopulation and 180 kV for delamination and entire fragmentation. Table 6.2 shows the comparison of energy consumptions for conventional hammer milling and EDF fragmentation. Liberation-oriented leaching results showed that EDF treatment should be used as a pretreatment stage for depopulation of WPCBs. Delamination and fragmentation stages consume more energy. Thus, EDF depopulated WPCBs can be further size-reduced to render Cu available to hydrometallurgical downstream processing.

Table 6.2 Energy consumption comparison for shredding and EDF technology [16]

Shredding energy consumption	EDF preprocessing energy consumption
Hammer-knife milling: 23 kWh/t	Depopulation: 132 kWh/t (150 kV, 5 Hz, 40 mm electrode gap)
	Delamination: 877 kWh/t (170–180 kWh, 5 Hz, 40 mm electrode gap)
	Fragmentation: 1485 kWh/t (170–180 kWh, 5 Hz, 40 mm electrode gap)

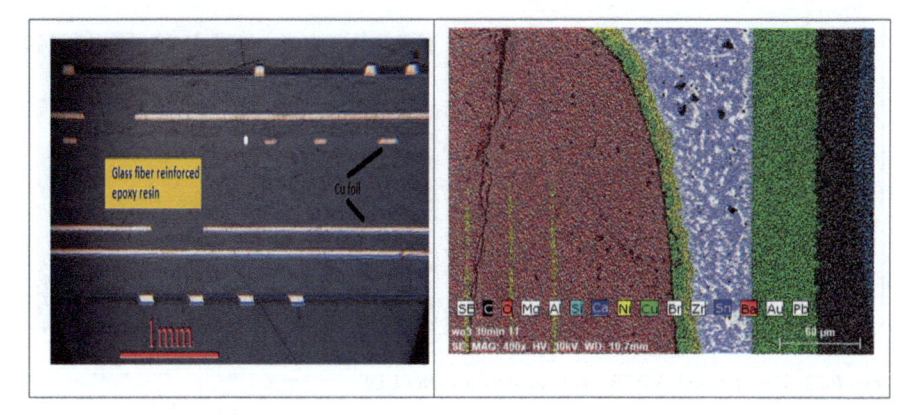

Fig. 6.12 Cross-sectional view of the internal structure of WPCBs. SEM view of the components fixed on the PCB surface

To sum up, the increased number of EDF pulses, with rest operating parameters kept constant, has resulted in immediate EC removal, i.e., depopulation. Then, when the depopulated boards are impacted through further pulses, the inner layers of the boards start to delaminate. Provided that after this step the number of pulses is further increased, the boards start to fragment having their size reduced. In that context, stage 2 could offer an ideal compromise between energy consumption and resulting delamination effects bringing an optimal liberation of Cu foils and threads without excessively reducing the size of the fragments [17]. EDF is a structure opening along interfaces which enables metals leaching from WPCBs at a relatively coarse size with minimum metal losses.

Figure 6.12 shows the cross section of an internal structure of a WPCB and SEM view of components fixed on the WPCB surface. Cu foil layers in glass fiber-reinforced epoxy resin can easily be seen. SEM views of the components fixed on PCBs show different elements from Cu, Sn, Ni to Br, Au, Ba, etc. Depopulated PWBs and removed components after EDF can be seen in Fig. 6.13. Figure 6.14a, b shows a cross-sectional view of the internal structure of a WPCB after depopulation treatment. Figure 6.14c, d shows cross-sectional pictures of PCBs after entire fragmentation [16].

Laboratory-scale batch EDF systems, which can take 200–300 g of materials, have been on the market since 2006. But, the cost of the laboratory equipment is very high. A fully automated pilot plant processing 2 t h^{-1} is also available in Selfrag AG.

6.6 Classification Screens

Screen is a machine with surface(s) used to classify materials by size. Screening is defined as "The mechanical process which accomplishes a division of particles on the basis of size and their acceptance or rejection by a screen surface". Material

Fig. 6.13 Depopulated WPCBs and components after EDF

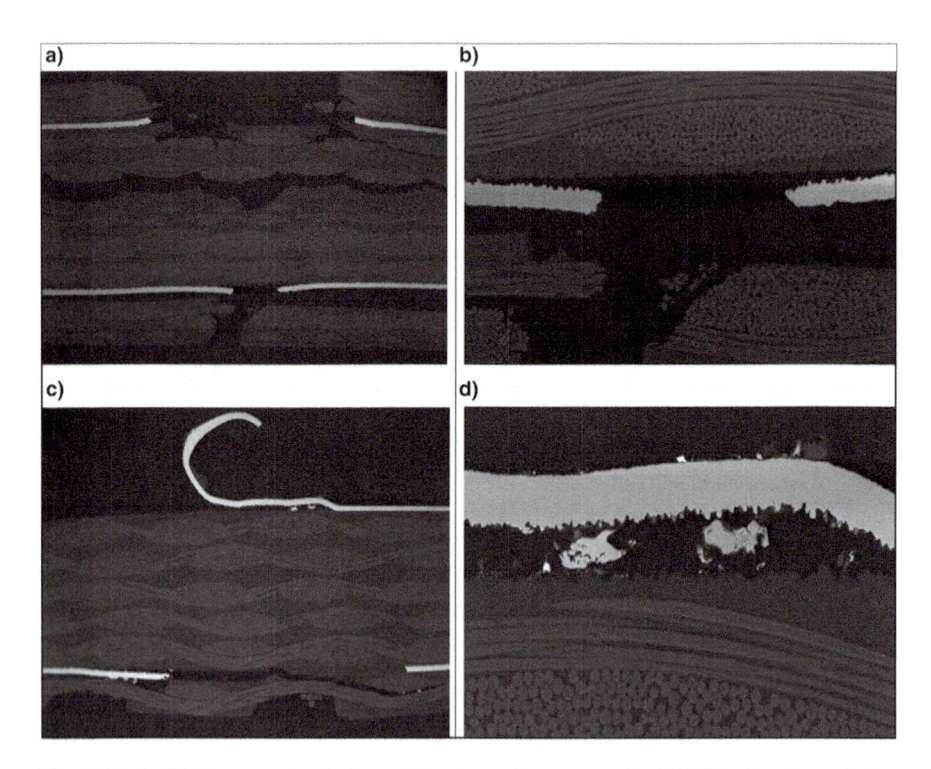

Fig. 6.14 (**a, b**) Cross-sectional view of the internal structure of a WPCB after depopulation treatment. (**c, d**) Cross-sectional pictures of WPCBs after entire fragmentation

larger than the hole size is oversize/coarse fraction and smaller than the hole size is undersize/fine fraction. Separation size is the particle size at which feed (t h^{-1}) separates into two products (undersize and oversize). Screening efficiency in industrial-scale varies between 90% and 95%.

6.6 Classification Screens

Grizzly Screen   **Shaking Screen**   **Vibrating Screen**   **Revolving Screen**

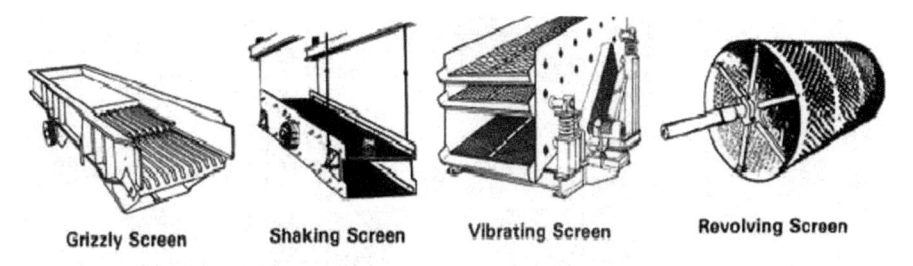

Fig. 6.15 Industrial-scale screen types

Fig. 6.16 Vibrating screen for sorting and classification [11]

Round and rectangular deck-shaken sieves and drum sifters are used to classify and sort materials according to their size. Grizzly screen, shaking screen, vibrating screen, and revolving screens (trommel) can be used industrially (Fig. 6.15). Trommel screens are large drum-shaped devices with a grate-like surface with large openings. Trommels are used to separate very coarse materials from bulk materials such as coarse plastics from finer Al-recycled material. Figure 6.16 shows industrial-scale round vibrating screen for sorting and classification. Screens can be classified according to opening size (coarse, medium, and fine), according to configuration (bar and mesh), according to methods used to clean the entrapped materials (manually, mechanically, raked, or water jet), and according to screen surface (fixed or moving).

Air/water classifiers, cones, rake classifiers, or cyclones can also be used for size classification. Cyclones are essentially settling chambers where the effect of gravity has been replaced by centrifugal acceleration. Screening is industrially used at Galloo, Salyp process, Stena, R-plus (WE-SA-SLF), and VW-Sicon and trommel separation at Argonne, Galloo, Salyp process, and Stena for end-of-life vehicle (ELV) recycling.

Fig. 6.17 Conveyor belts used in e-waste recycling

6.7 Conveyor Belts

Conveyor belts transport material between equipment. Conveyor belts in various sizes and configurations with or without magnetic rollers for removing ferrous fragments from the material stream during a recycling process are used. Conveyor belt length changes between 2000 and 3500 mm; width between 500 and 800 mm; height between 1500 and 2000 mm; weight from 120 to 400 kg; roller diameter size from 100 to 170 mm; roller height from 100 to 530 mm; and motor power from 0.5 KW to 1.1 KW. Figure 6.17 shows industrial-scale inclined conveyor belts used in e-waste/WPCB recycling [18].

6.8 Filtration Systems

6.8.1 Pulse-Jet Bag Filter Dust Collection

Pulse-jet baghouse is an improved new type and high-efficiency dust collector (99%) on the basis of fabric filters. Advantages of using pulse-jet baghouse are high filtration efficiency, stable performance, small footprint, large amounts of filtration, etc. Different from traditional dust collectors, pulse-jet fabric collector improves the structure and pulse valve and solves the problem of open air placement and compressed air source. This dust collector inlet and outlet are compact and air resistance is little. Its frame adopts sealing material, which reduces air leakage and enhances good tightness. This dust collector collects tiny dust with 0.3 μm particle size with 99% dusting efficiency.

Pulse-jet fabric collector is mainly composed of dust hopper and upper, middle, and lower boxes. When working, dust enters into the dust hopper through the air inlet firstly, coarse particles of dust directly drops into the bottom of the dust hopper,

6.8 Filtration Systems

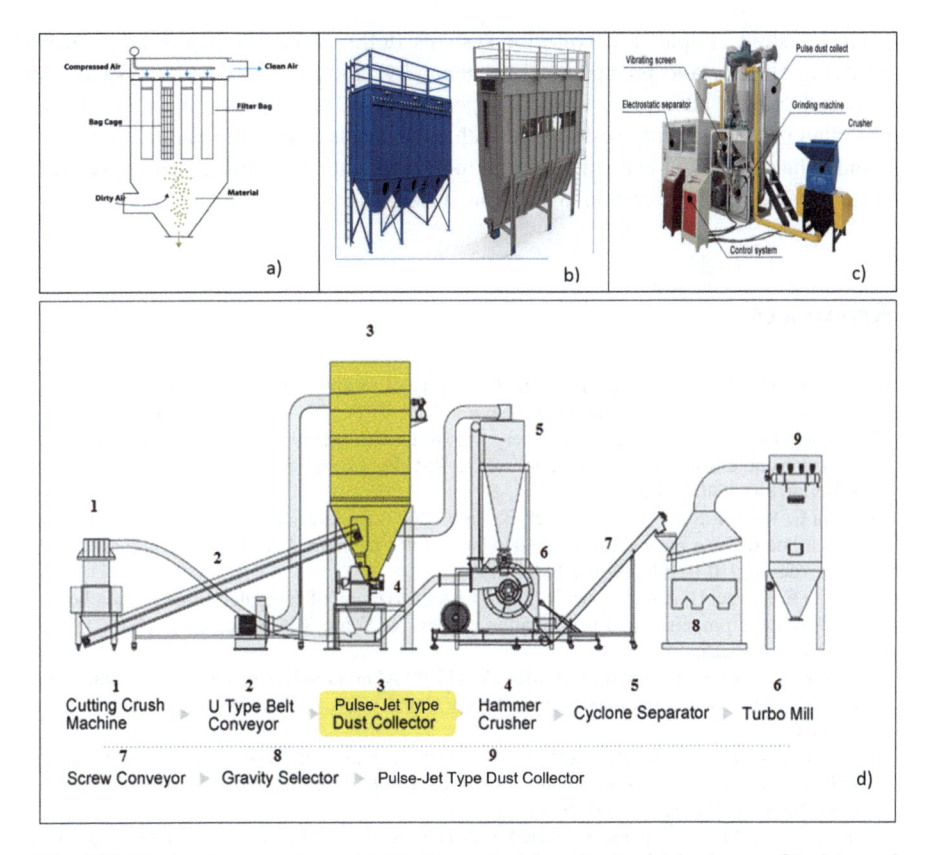

Fig. 6.18 Dust collector systems. (**a**) Working principles of pulse-jet baghouse, (**b**) picture of pulse-jet baghouse picture, and (**c**) e-waste recycling flowsheet with pulse dust collector and (**d**) e-waste crushing and milling flowsheet with cyclone and pulse-jet type dust collection system

and, at the same time, fine particles of dust enters into the middle and lower boxes with the help of airflow. Then, dust attaches to a filter bag for filtration; besides, the filtered air enters into the atmosphere by an exhaust blower. Figure 6.18a shows working principles of pulse-jet baghouse. Figure 6.18b shows the picture of pulse-jet baghouse picture. The resistance of machine rises to set value when dust on the surface of filter bag is increasing, and, at the same time, time relay sends out a signal that means the programmer has to work. Turn on the pulse valve one by one so that compressed air will clean the filter bag through jet. Finally, filter bag begins to swell; meanwhile, dust separates from filter bag quickly and falls into dust hopper under the effect of reverse airflow. Dust will be discharged from a dumper after blowing. Figure 6.18c shows full e-waste recycling flowsheet with shredder, pulverizer, electrostatic separator, and pulse dust collection system.

Filtration systems, operating with sleeve filters, offer a full range of solutions for the dedusting of WPCB recycling process improving the air quality and increasing the efficiency of plants. Reliable and robust SM filters can provide air cleaning with control system. Triple dust collector systems (tetra cyclone + bag + air cleaner/ purifying electrostatic precipitator) are generally practiced. Dedusting system and activated carbon adsorption tower realize that there is no pollution in the whole recycling process. Plants are designed with exhaust gas treatment system. In order to avoid exhaust gas and bad smell for the high temperature, plants are equipped with water spray and active carbon absorption system.

References

1. Zhou G, He Y, Luo Z, Zhao Y (2010) Feasibility of pyrometallurgy to recover metals from waste printed circuit boards. Fresenius Environ Bull 19(7):1254–1259.ISSN: 03043894
2. Kellner D (2009) Recycling and recovery. In: Hester RE, Harrison RM (eds) Electronic waste management, design, analysis and application. RSC Publishing, pp 91–110. ISSN: 1350-7583, ISBN: 978-0-85404-112-1
3. Duan C, Wen X, Shi C, Zhao Y, Wen B, He Y (2009) Recovery of metals from waste printed circuit boards by a mechanical method using a water medium. J Hazard Mater 166:478–482. https://doi.org/10.1016/j.jhazmat.2008.11.060
4. Yazici E, Deveci H, Alp I, Yazıcı R (2010) Characterization of computer printed circuit boards for hazardous properties and beneficiation studies, XXV Int. Min. Proc. Cong. (IMPC), B10, Brisbane, Australia
5. Zhang S, Forssberg E, Arvidson B, Moss W (1998) Aluminum recovery from electronic scrap by high-force eddy-current separators. Resour Conserv Recycl 23:225–241. PII S0921-3449 (98)00022-6
6. de Moraes VT (2011) Recuperação de metais a partir do processamento mecânico e hidrometalúrgico de placas de circuito impresso de celulares obsoletos. Engenharia Metalúrgica e de Materiais, Universidade de São Paulo
7. Yamane LH, Moraes VT, Espinosa DCR (2011) Recycling of WEEE: characterization of spent printed circuit boards from mobile phones and computers. Waste Manag 31:2553–2558. https://doi.org/10.1016/j.wasman.2011.07.006
8. Silvas FPC, Correa MMJ, Caldes MPK, Moraes VT, Espinosa DCR, Tenorio JAS (2015) Printed circuit board recycling; physical processing and copper extraction by selective leaching. Waste Manag 46:503–510. https://doi.org/10.1016/j.wasman.2015.08.030
9. Hagelüken C (2014) High-tech recycling of critical metals: opportunities and challenges, American Association for the Advancement of Science (AAAS) Annual Meeting, Chicago
10. https://www.alibaba.com/product-detail/2015-Circuit-board-shredder-PCBboard_60203968862.html?spm=a2700.7724857.0.0.PX1fPN
11. http://www.stokkermill.com/en/recycling-plants/shredder-residues-refining-plant/
12. http://mtmakina.com/en/shredder/double-storey-shredders.html
13. http://www.jordan-reductionsolutions.com/product-grinder.html
14. http://ecovaluemetal.com/technology.htm
15. http://www.selfrag.com/technology.php
16. Martino R, Iseli C, Gaydardzhiev S, Streicher-Porte M, Weh A (2017) Electro dynamic fragmentation of printed circuit wiring boards as a preparation tool for recycling. Miner Eng 107:20–26. https://doi.org/10.1016/j.mineng.2017.01.009
17. Chancerel P, Meskers CEM, Hagelüken C, Rotter VS (2009) Assessment of precious metal flows during preprocessing of waste electrical and electronic equipment. J Ind Ecol 13:791–810. https://doi.org/10.1111/j.1530-9290.2009.00171.x
18. http://www.stokkermill.com/en/company/

Chapter 7
Sorting and Separation of WPCBs

"No pollution is the best solution"

Anonymous

Abstract Manual and automatic sensor-based sorting and separation systems are used in WPCB recycling industries. Physical properties of WEEE for separation method selection are introduced here. Wet and dry gravity separation equipment (such as rising current separators, cyclone, triboelectric cyclone, Kelsey centrifugal jigs, heavy media separators, jigs, shaking tables, Falcon separators, and Mozley separators) are widely used in the crude metal and nonmetal separation for e-waste recycling. These processes produce huge amount of wastewater, dust, and residues which may create serious secondary environmental problems. Dry gravity separators (such as zigzag separators, fluidized bed separators, air tables, etc.) do not have water problem but require expensive dust collection systems. The advantages of density-based systems are simplicity, low energy consumption at a low product purity, and loss of some valuable metals. Electrostatic and magnetic separation types and industrial-scale equipment are covered in detail. Froth flotation and mixed separation method usage for e-waste recycling are also explained in this chapter as well.

Keywords Sorting · Wet/dry gravity separation · Electrostatic separation · Magnetic separation · Flotation

7.1 Sorting Systems

Manual and automatic sensor-based sorting systems can be used in e-waste recycling. People are probably the best sorters due to the learning capability of the brain fed by all human senses and experience. This is true where there are visual differences between materials and while their concentration lasts. However, there are limits, i.e., an eye cannot read an RFID tag or barcode but a machine can. Product design can prohibit manual sorting, by designing in components that cannot be separated by hand.

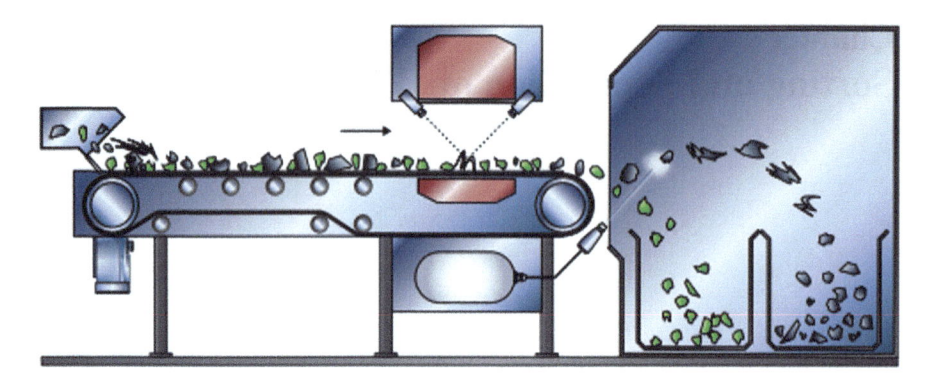

Fig. 7.1 Sensor-based automatic sorter for metal-glass/plastic separation

Automatic sorting with different detection systems (XRT/XRF) can substitute for manual sorting. For example, a camera, used as a detector in combination with a computer (Fig. 7.1), scans a product on a belt conveyor before processing each frame by the computer. After processing and identification, for instance, based on color, the computer can activate air valves that shoot identified particles into collection bins. This type of sorter can detect over 99% of all metal particles larger than 4 mm in an input stream, and, depending on the settings, 95–98% of the metal will be removed. An electrical signal produced by receiver coils can also be processed in different ways to determine metals from nonmetallic materials. The difference between stainless steel and nonferrous metals can also be determined [1].

7.2 Physical Properties of WEEE

Figure 7.2 shows both physical properties of metals and plastics in WEEE and potential physical separation method selection for metal recovery from WEEE. Nonmetallic fraction of WPCB has significantly low densities as compared to metallic fraction. Densities of plastics are between 0.9 and 2.0 g cm^{-3}. In WEEEs, epoxy resin has a density between 1.1 and 1.4 g cm^{-3}. Polyester resin has a density between 1.2 and 1.5 g cm^{-3}. Fiberglass epoxy composite has a density between 1.9 and 2.0 gr cm^{-3}. Glass fiber has a density of 2.5 g cm^{-3}. Al is the lightest metal in WEEE. Densities of Mg, Al, and Ti are between 1.7 and 4.5 g cm^{-3}. Six metals (Cr, Sn, Fe, Ni, Zn, Co, and Cu) have densities between 7 and 9 g cm^{-3}. Pb, Ag, and Mo have density between 10 and 12 g cm^{-3}, and four metals (Pd, W, PGMs, and Au) have between 19.3 and 21.4 g cm^{-3}. For density separation of polymers between each other in WEEE, at least 0.15 g cm^3 density difference should exist. However, laminated materials are to be avoided because of separating the materials [2]. For metals, the use of surface coatings/treatments (such as steel → Cu, Sn, Zn, Pb, Al; Al → cast Fe, steel, Cr, Zn, Pb, Cu, Mg; Zn → Cast Fe, steel, Pb, Sn, Cd; Au coatings, etc.) reduces the recyclability of materials.

7.2 Physical Properties of WEEE

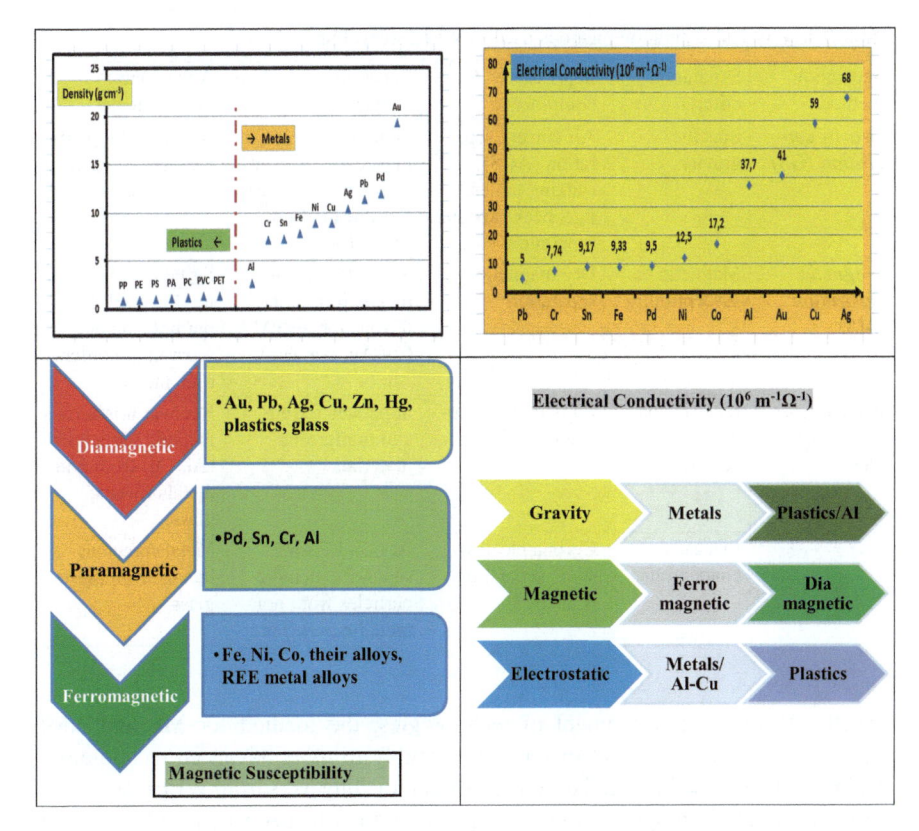

Fig. 7.2 Physical properties of metals and plastics in WEEE and potential physical separation method selection for metal recovery from WEEE

Physico-mechanical recycling steps included selectively dismantling, crushing, and physical separation methods. Physical separation methods, criteria, possible equipment, and separation characteristics are summarized in Table 7.1 along with advantages and disadvantages [3]. Plastics and glass are nonconductor. Five metals (Pb, Cr, Sn, Fe, and Pd) have electrical conductance less than 10; three metals (Ni, Co, and Al) have between 10 and 40, and three metals (Au, Cu, and Ag) have between 40 and $70*10^6$ m^{-1} Ω^{-1}. Au, Ag, Pb, and Cu metals, plastics, and glass are diamagnetic; Pd, Sn, Cr, and Al are paramagnetic; and Fe, Ni, and Co are ferromagnetic metals. Based on these data, plastics and Al metal can be separated from other metals by gravity separation; ferromagnetic (Fe, Ni, and Co) can be separated from diamagnetic (plastics, glass, Au, Pb, Ag, and Cu) by magnetic separation, and metals and Al-Cu can be separated from plastics by electrostatic separation methods [4].

WEEE is a complex system, containing about 70% NMF (various organic substances) and 30% MF. Therefore, separation of MF and NMF without damage to the structure of NMF is the precondition for NMF recycling and itself is also regarded as one type of recycling since MF separated from feedstock can be sold to the market

Table 7.1 Physical separation methods used in WEEE recycling

Separation method	Separation criteria	Separation equipment	Separation character	Advantages/disadvantages
Gravity separation (GS)	Specific gravity	Air current separation (ACS), shaking tables, jigs, heavy media separation	Separates light plastics/glass from heavy metals, separates resins from fiber glass	Air velocity, particle size, and density affect separation
Magnetic separation (MS)	Magnetic susceptibility	Dry magnetic separation	Separates ferrous metals from non-ferrous metals (Al, Cu, Au, Ag, etc.)	Suitable for separating Fe, steel, Ni, Cr but not suitable separating nonferrous metals
Electrostatic separation (ES)	Electric conductivity/density	Eddy current separation (ECS)	Separates ferrous and nonferrous materials	Recover nonferrous metallic particles (Al, Cu) from nonmetals (plastic, glass)
	Electrical conductivity	Corona electrostatic separation (CES)	At 0.2–1.2 mm, separates metallic particles from nonmetallic particles	Metals/precious metals from plastic/glass

directly. With the development of technologies, the method for MF and NMF separation has changed from manual disassembly to more advanced technologies. Additionally, the methods and technologies of recycling MFs from WEEE are totally different from technologies of mineral separation. In order to develop and implement both environmentally friendly and economically viable recycling processes of metals, a large number of novel technologies are proposed in recent years and mainly include pyrometallurgical technology, mild extracting technology, electrochemical technology, vacuum metallurgical technology, etc. [5, 6]. Over the past decades, the investigations on integrated recycling processes [7–9] for waste desktop computers, mobile phone, CRT TVs, and so on have achieved great progresses. Nevertheless, some technical obstacles also existed that limit the industrial application of WPCB recycling. Hence, up to now, the e-waste recycling should be developed toward more depth and refinement to promote industrial production of e-waste resource recovery [10]. In this book, the traditional and some advanced recycling processes and techniques of MFs and NMFs from WPCBs with high efficiency and good purity are mainly focused on. Therefore, it makes it possible to treat MF and NMF separately, and it brings about the selling of high purity MF to the market and also directing recycling and modification of NMF from WPCB as two potential ways to treat the NMF part of WPCB. Also, the separation of MF and NMF avoids potential interactions of NMF in further treatment stages because the metals in MF can catalyze unwanted side reactions of NMF.

7.3 Gravity/Density Separation

Gravity separation is based on the specific density differences. Concentration criterion (CC) can be used for gravity separation possibility. CC = $(\rho_h-\rho_{fluid})$ / $(\rho_l-\rho_{fluid})$, where ρ_h is the density of heavy material, ρ_l is the density of light material, and ρ_{fluid} is the density of fluid medium (water/air). If CC > ± 2.5, gravity separation is relatively easy. If 2.5 < CC < 1.25, efficiency of gravity separation decreases. If CC < 1, gravity separation can be possible under careful density control conditions.

For plastics and Al/glasses	CC = (2.65–1)/(2–1) = 1.65	(Gravity separation is possible but not easy)
For plastics and metals	CC = (7–1)/(2–1) = 6.0	(Gravity separation is easy)
For Al/glasses and metals	CC = (7–1)/(2.65–1) = 3.64	(Gravity separation is easy)

After required liberation, plastics and Al/glasses can easily be separated from metals by gravity separation methods. Plastics and Al/glasses gravity separation can be possible under careful specific gravity control conditions. Gravity concentration methods separate materials of different specific gravity by their relative movement in water or air. Nonetheless, this separation is not only dependent on the density of the components but also on their size and shape. Besides gravity one or more of the other forces, like force exerted by the viscous liquids, can serve as the separation medium. Density-based separation uses the differences of particles size and density to separate MF and NMF, and various equipment were developed. Water or airflow tables, heavy media separation, and sifting are common gravity separators used in e-waste recycling. Rectangular riffled air tables can be used for e-waste sorting. By using different heavy liquids (such as tetrabromoethane (TBE) (2967 gr/cm^3) and acetone, which lowers density and viscosity), the metals can be separated from the plastics or ceramics. Al went to the light fraction due to the lower density. Cu was more effectively separated at size 149 µm [11]. Different metal particles can be further separated. For this purpose, the WPCB material is processed on concentrating tables. The shaking tables exploit the difference in the specific gravity and particle size to achieve desired separation.

Density separation techniques which are well-known in the mineral processing industry have found their way into e-waste recycling based on the fact that e-scrap consists essentially of plastics, with a density less than 2.0 g/cm^3; light metal, primarily Al and glass, with a density of 2.7 g cm^{-3}; and heavy metals, predominantly Cu and ferromagnetics, with a density more than 7 g cm^{-3}. In sink-float separation, both PC and PCB scraps ~50% (weight) of floats which is primarily plastics can be separated out at the specific density of 2.0 g cm^{-3} [12].

In the past, wet gravity methods (i.e., hydraulic shaking tables) were widely used to obtain only crude Cu particles. This process produces huge amount of wastewater and residues which may create serious secondary environmental problems. Other metals and nonmetals cannot be recycled [13]. Recently, air classification is a popular method based on the fact that the particles suspended in the gas stream will be separated due to the gravity and drag forces experiencing in opposite directions. The heavier particles will gain larger terminal settling velocity and move toward the bottom of the air stream, while light particles go to the opposite side, which is the top of the column. Zigzag three-way air classifiers were used for dry separation (pneumatic) to overcome above problems along with magnetic and electrostatic separators.

7.3.1 Wet Gravity Separation

WPCB recycling lines with gravity separating type use wet gravity separators (i.e., shaking tables, jigs, rising-current separators, hydrocyclones, sink-float methods, Falcon centrifugal separator, etc.) to recover metals from WPCB boards. Metals and nonmetal (i.e., plastic and glass fiber) materials have different densities (Fig. 7.2). If there is a large difference in densities, gravity separation is more economical than electrostatic separation. Figure 7.3 shows a typical wet gravity separation flowsheet for WPCB recycling. Cu metal can be separated from plastics. Ferrous material is removed after shredding by magnetic separators. Wet separation reduces air and noise pollutions, recycles water, and reuses limited resources. Duan et al., (2009) used a Falcon SB40 centrifugal separator to recycle metals from WPCBs crushed to less than 1 mm. This separator separates materials by density using up to 300 g acceleration [14].

Industrial-scale plants have a capacity of 0.8–1.0 t d^{-1}, power requirement 160 KW, water requirement 20–30 t d^{-1}, and floor space requirement 150 m^2. Dimensions are 26.3∗6.1∗6.4 m. Fe recovery rate is claimed to be 99% and Cu and epoxy resin recovery rates are more than 95% [15]. Sink-float separation is industrially used at Galloo, Salyp Process, Stena, and VW-Sicon, and heavy media separation is industrially used at Stena plants.

7.3.1.1 Rising-Current Separators

In the rising-current method, a continuous rising water column is projected through a tank/ column that receives the feed material. Material that sinks faster than the rising water column falls to the bottom of the separator. The material carried up by the column of water can be separated from the water by a screen or a sieve (Fig. 7.4). A complication is that the water must be constantly cleaned from dissolved compounds [1].

7.3 Gravity/Density Separation

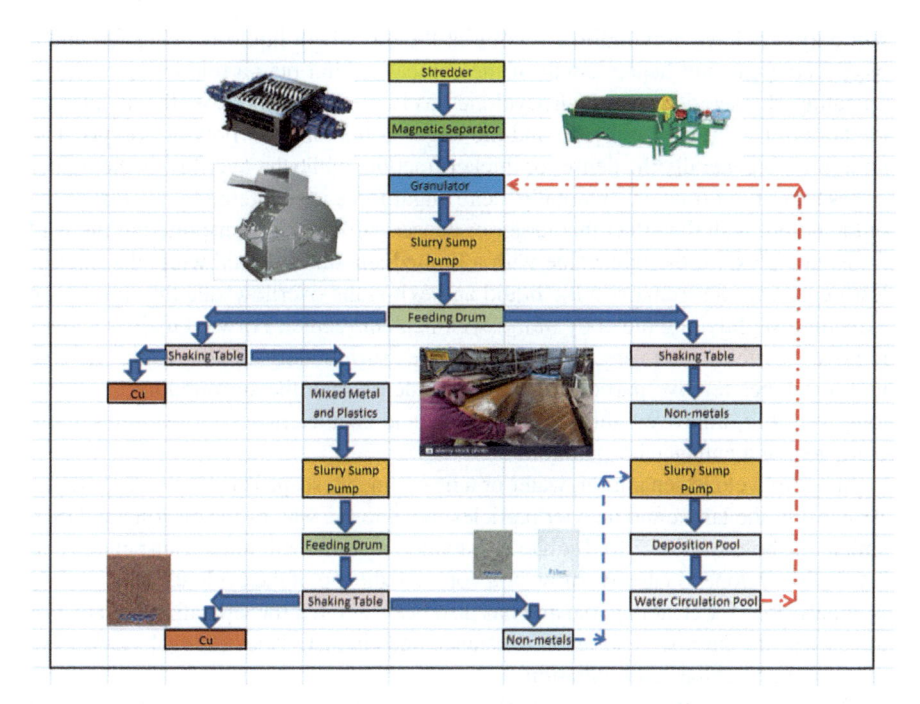

Fig. 7.3 Flowsheet of PCB recycling with wet gravity separation and industrial-scale shaking tables

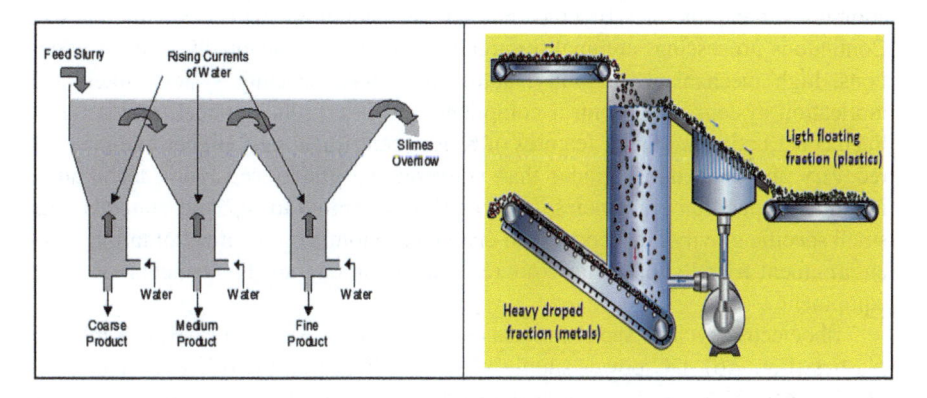

Fig. 7.4 Rising-current separator

7.3.1.2 Hydrocyclone/Triboelectric Cyclone and Kelsey Centrifugal Jig Separation

A hydrocyclone is a device to classify/separate or sort particles in a liquid suspension based on the densities of the particles. A hydrocyclone may be used to separate solids from liquids or to separate solids/liquids of different density. A hydrocyclone will normally have a cylindrical section at the top where liquid is being fed and a base. The angle and hence length of the conical section play a role in determining operating characteristics. A hydrocyclone has two exits on the axis: the smaller on the bottom (underflow or reject) and the larger at the top (overflow or accept). The underflow is generally the denser or thicker fraction, while the overflow is the lighter or has more fluid fraction. Internally, centrifugal force is countered by the resistance of the liquid, with the effect that larger or denser particles are transported to the wall for eventual exit at the reject side with a limited amount of liquid while the finer or less dense particles remain in the liquid and exit at the overflow side through a tube extending slightly into the body of the cyclone at the center. Forward hydrocyclones remove particles that are denser than the surrounding fluid, while reverse hydrocyclones remove particles that are less dense than the surrounding fluid. For long wear life, improved abrasion resistance and characteristics for cyclones allow consistent and enhanced separation.

In hydrocyclone separation, water or a heavy media suspension is made to rotate. As a result, the larger and heavier particles move to the wall of the cyclone and then sink to the bottom (Fig. 7.5). Lighter, or smaller, material will remain suspended and leaves the cyclone via the overflow or a vortex at the top of the cyclone. This method is widely used for cleaning coal and has also been adapted for separating metals. For long wear life, improved abrasion resistance and characteristics for cyclones allows consistent and enhanced separation.

When you need very fine material recovery and concentrate grades beyond the capacity of conventional gravity separation, the Kelsey Centrifugal Jig extends the efficient size recovery range of fine mineral separation processes down to 10 μms by combining the principles of conventional jig technology with centrifugal force. Continuous processing, nominal throughputs to 50 t h^{-1}, adaptation to feed variations, high mechanical reliability, automatic screen cleaning system, integrated lubrication system, PLC control compliant, and heavy duty, rubber, and ceramic lined construction are major features of Kelsey centrifugal jigs. Higher fine material recovery and concentrate grades than conventional gravity separation techniques, improved separation efficiencies down to 10 μms, ability to separate minerals with small specific gravity differences, and enabled economic re-treatment of tailings and environment-friendly operation (no reagents required) are the benefits of these equipment.

Triboelectric cyclone separator was used to separate ABS from ABS-PS mixture plastics (Fig. 7.6). DC power supply was used. Based on the different charges of plastic types, ABS goes to negative with 100% recovery and PS goes to positive side with 74% recovery.

7.3.1.3 Sink-Float Methods (Heavy Media)

Sink-float methods separate materials based on whether they float or sink in liquid. For example, wood and insulation (paper, PUR foam, board, etc.) will float on water and can be removed from the surface. Figure 7.7 shows sink-float drum separation

7.3 Gravity/Density Separation

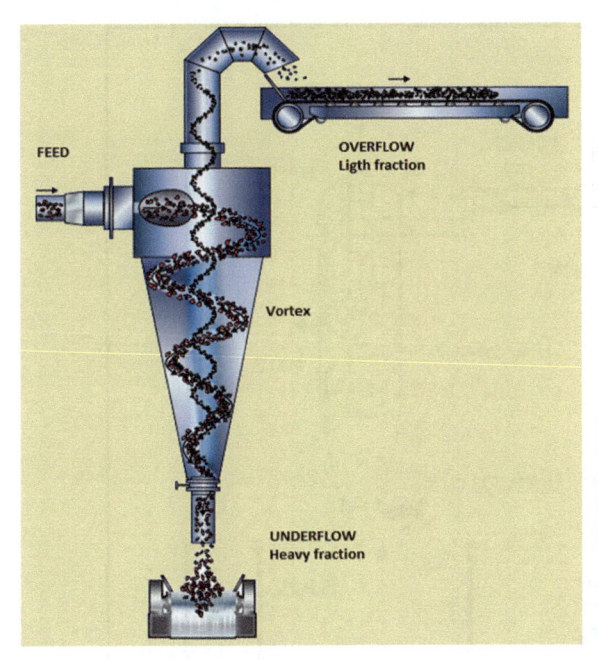

Fig. 7.5 A cyclone separating mixed feed into light and heavy fractions along with industrial-scale hydrocyclone batteries working together

system cross section and industrial-scale equipment details. The method is used in plastic separation, making use of the fact that PVC has a specific weight of >1 g cm^{-3} (heavier than water) while other plastic types, such as PE, PUR, PP, and PS, usually are <1 g cm^{-3}. Sink-float methods can use water, another liquids or a suspension of water and a solid material (heavy media) (e.g., ferrosilicon (FeSi) or magnetite (Fe_3O_4)) to vary the density of the liquid medium, adapting to which materials will float and sink. This variation is used particularly for the separation of shredded Al and Mg from a nonferrous mix. Sufficient FeSi or Fe_3O_4 is added to water for making a constantly agitated suspension with a specific density of anywhere between 1.8 and 3.3 g cm^{-3}. This enables, for instance, Al with a density of 2.7 g cm^{-3} to sink and Mg with a density of 1.74 g cm^{-3} to float. Sink-float method +flotation was used for PET and PE separation with 90.3% PET recovery and at 99.7% PET grade (Fig. 7.8) [16].

7.3.1.4 Jigs

Jigging is one of the oldest methods of sink-float separation. In a piece of equipment called a jig, input material is formed into a thick layer/bed. This is fluidized by pulsating a water current through it. On pulsation the bed of input material is lifted as a mass; then, as the velocity of the water stream decreases, the particles fall with

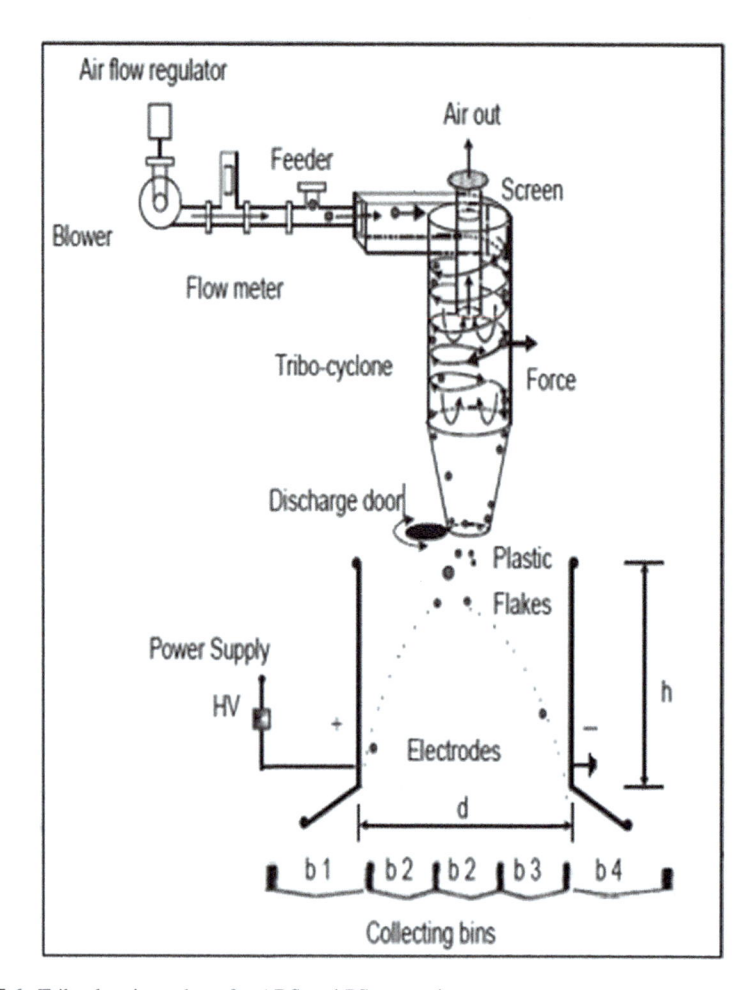

Fig. 7.6 Triboelectric cyclone for ABS and PS separation

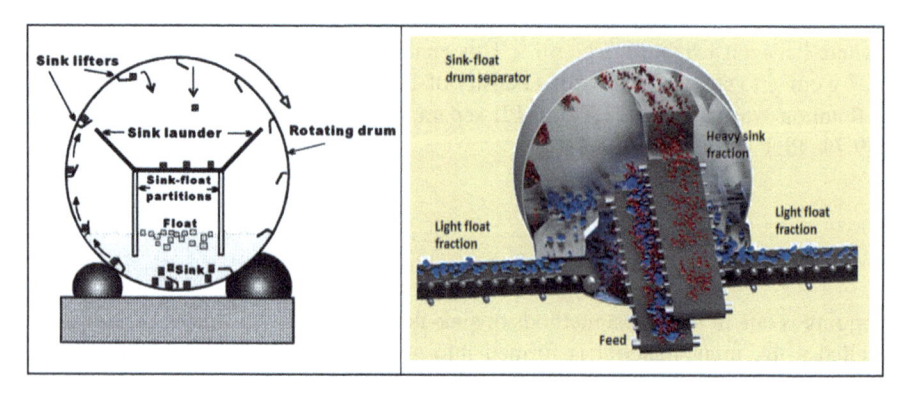

Fig. 7.7 Sink-float drum separation system

7.3 Gravity/Density Separation

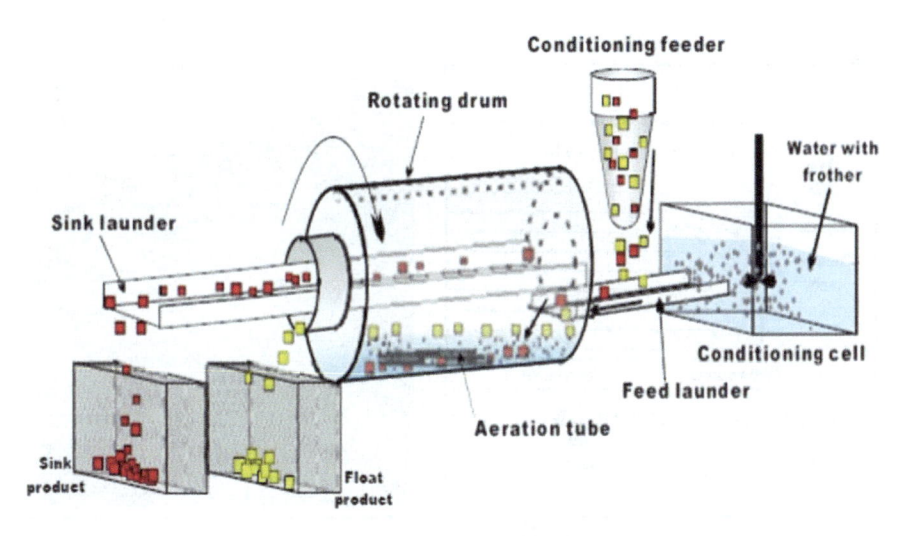

Fig. 7.8 PET and PE separation using heavy media and flotation

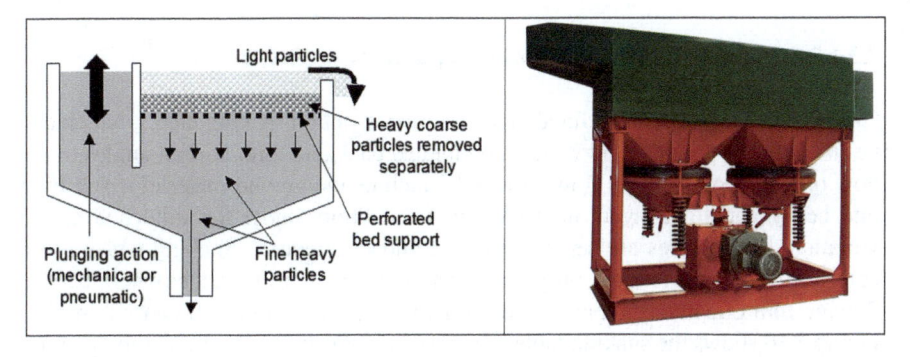

Fig. 7.9 Jig cross section and industrial-scale jigs

different speeds to the bottom, depending on their density. When this process is repeated, the different materials will stratify in relation to their density and can be recovered accordingly (Fig. 7.9). This method is very similar to the rising-current separation process; but, here, the water jet is projected through the screen in pulses. Jigging is used for separating shredded metals and particularly plastics from metals. The magnitude and direction of the water-jet force can be varied to change the separation result. A jig is generally used for concentrating relatively coarse materials down to 3 mm. When the feed is fairly similar in size, it is not difficult to achieve a good separation at low cost.

Sarvar et al. (2015) used a wet jig for +0.59 and 1.68 mm PCB particles with metal recovery between 85% and 97.5%. However, lower grades around 70% were obtained and a significant part of Au is lost at the flotation process [17].

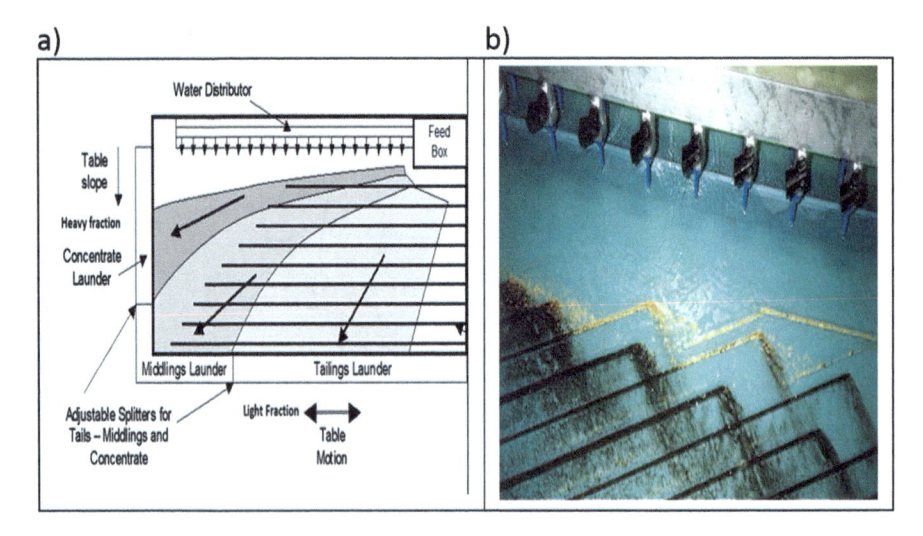

Fig. 7.10 Shaking table top view (**a**) and industrial-scale separation (**b**)

7.3.1.5 Shaking Tables and Mozley Separators

Gravity concentration on inclined planes is carried out on shaking tables, which can be smoothed or grooved and which are vibrated back and forth at right angles to the flow of water. As the pulp flows down the incline, the ground material is stratified into heavy and light layers in the water; in addition, under the influence of the vibration, the particles are separated in the impact direction. Shaking tables use a controlled vibration and the property of particle density to separate materials. The flowing film effectively separates coarse light particles from small dense particles. Figure 7.10 shows the shaking table top view and industrial-scale separation picture. Heavy fraction of material goes to the concentrate launder on the left side, and light fraction goes to the tailing launder on the right side. Mixed particle goes to the middle launder of the table. Holman Wilfley wet shaking tables can be used for recovery of PMs, Cu-wire, synthetic diamonds, chromite, heavy mineral sands, and Au. The different models process feed streams of between 5 and 2500 kg h^{-1}. Holman models are available for all fine mineral concentration (e.g., mineral sands, Sn, W, Cr, Au). Wilfley model 7000 is available for metal recycling and reprocessing of WEEE materials. 20–25% solid by weight can be used. Particle size can change from 100–150 μm to 5 mm at capacities from 0.5 to 12.5 t h^{-1} [18]. Capacity can be increased using double or triple deck tables.

Veit et al. (2014) investigated the utilization of Mozley concentrator for preextraction of metals from WPCBs. The Mozley concentrator consists of a flat tray and separation V-shaped tray for fine and coarse particles, respectively. The water flow rate and tilt tray angle were considered as parameters to be optimized. It is reported that the material size fraction of − 1 + 0.25 mm is used in the gravity process, after taking the loss of materials and interference of fine particles in the gravity process into consideration. It appeared that it was possible to pre-concentrate 85% Cu, 95% Sn, 96% Ni, and 98% Ag, while Al and Au could not be recovered due to its density and lamellar form, respectively [19].

7.3.2 Dry/Air Gravity Separation

Dry WPCB recycling equipment with air-separation type is applicable to recycle various waste and bare PCBs. The mixture of metal and nonmetal materials gained from crushing-pulverizing-classifying of WPCB raw materials is fed into the material hopper of air separator and then into the separating zone of the separator to separate metal, fiber, and resin mixtures. Since the separator is connected with a dust removal system, a horizontal air current is formed which moves the materials in horizontal direction. Meanwhile, with the action of the gravity, the materials move downward. Due to the different specific gravities of materials, nonmetal materials, such as fine dust and grains with lower gravity, are taken away by the dust removal system when the mixture passes the separating plate, leaving the metal materials with larger gravity into finished product recovery zone.

Habib et al. (2013) successfully used vertical vibration generated by a pair of connected loudspeakers to separate the metallic and nonmetallic fractions of PCBs. When WPCBs were comminuted to less than 1 mm in size, metallic grades as high as 95% (measured by heavy liquid analysis) could be achieved in the recovered products under dry conditions [20]. Dry treatment in the absence of water is limiting dewatering and sludge disposal costs. But, successive stage crushing and milling are required for higher metal liberation.

7.3.2.1 Zigzag Separators

The separation of metal from nonmetal materials is achieved. Zigzag (three-way) air classificators remove the light part of heterogeneous materials by means of an air flow in countercurrent inside a zigzag pipe. They clean residues and separate light parts in WPCB recycling. Figure 7.11 shows normal and zigzag air classificator details and dry gravity separation plant picture. Industrial-scale plant capacity is about 0.2–0.3 t h^{-1}. Low noise, no need for water, no dust pollution, power and labor savings, and no waste and above 95% metal and nonmetal recoveries are main advantages. Floor space requirement is 7∗5∗5 m and load weight is more than 10 t. Dust pollution is prevented by two-in-one dust removal device: cyclone collector and pulse bag dust collector [15]. Air classification technique is used at Argonne, Galloo, MBA Polymers, Salyp Process, Stena, R-plus (WESA-SLF), and WW Sicon for ELV recycling plants [1].

Eswaraiah et al. (2008) used air classifier to separate WPCB and recover 96.7% Cu in the sink and 98% plastics and glass fibers in the float. However, there is still 70.9% of other materials in the sink, which makes the purity of Cu only 26.8%

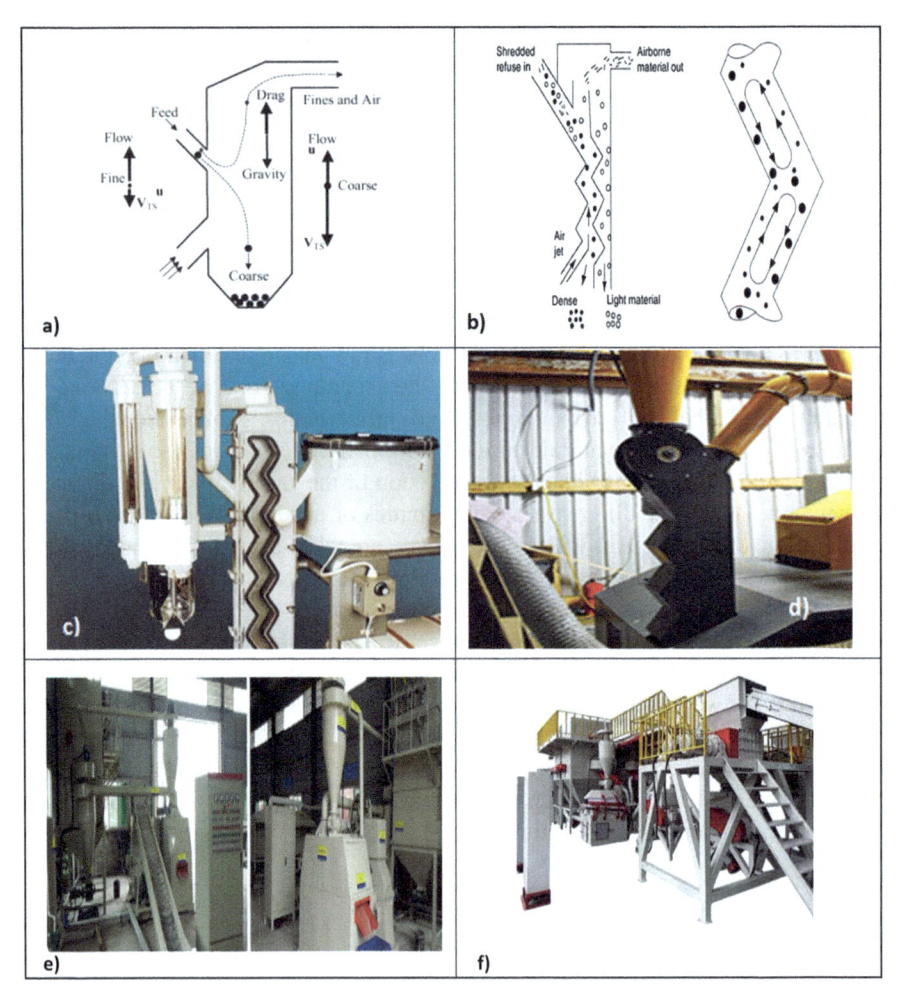

Fig. 7.11 Zigzag air classification principles (**a**, **b**), details and pilot/industrial-scale dry gravity separation equipment (**c**, **d**)/plant layout (**e**, **f**)

[21]. Habib et al. (2013) also use vertical vibration to separate MF and NMF from WPCB feedstock, and they obtained a MF fraction as a combination of Cu (~50%), Fe (~10%), Sn (~10%), Zn (~8%), and Pb (~8%), as well as some NMF [20]. Also, the NMF fraction includes 65% NMF and 25% Cu as well as a certain amount of metals. The advantages of density-based separation are the simplicity of its equipment and low energy consumption, while the common problem is low product purity, adding workload and difficulty to further refine [22].

7.3 Gravity/Density Separation

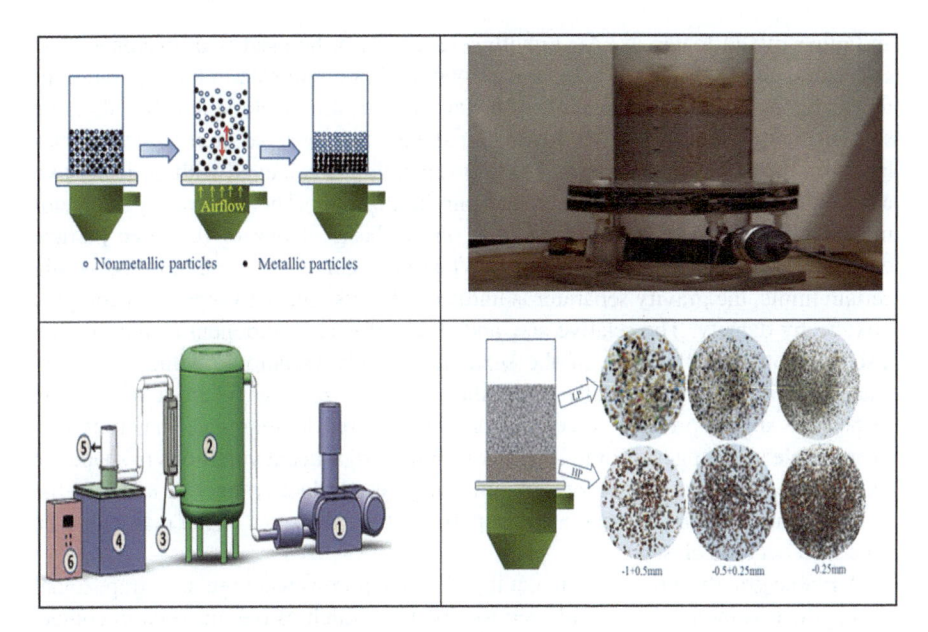

Fig. 7.12 Segregation process of metallic and nonmetallic particles in a vibrated gas-solid fluidized bed separation system: ① roots blower, ② 19 pressure tank, ③ rotameter, ④ vibrating table, ⑤ gas-solid fluidized vessel, ⑥ control unit. (Compiled from [23])

7.3.2.2 Fluidized Bed Separators (FBS)

Zhang et al. (2017) used vibrating gas-solid fluidized bed separator (FBS) for recovering residual metals from nonmetallic fraction (Fig. 7.12) [23]. Vibrated fluidized bed column segregates metals, plastic organics, and glass fibers at different size classes according to particle size, shape, and density at different superficial air flow rates (0.1–0.2 m s^{-1}), fluidization frequencies (20–30 Hz), and fluidization times (120–150 s). Compared with coarse fraction, the segregation of fine particles has a need for lower vibration frequency and superficial air velocity but longer fluidizing time. The metal recovery of − 1 + 0.5 mm, − 0.5 + 0.25 mm, and −0.25 mm size fractions are 86.4%, 82.2%, and 76.6%, respectively, meaning the bed segregation efficiency of FBS for fine particles [23].

7.3.2.3 Air/Pneumatic Gravity Separators (Air Tables)

Air gravity separator is a separation equipment that can separate mixture of powder materials into light and heavy two parts through air suspension. Air gravity separator lifts the material by vacuum over an inclined vibrating screen covered deck. This results in the material being suspended in air, while heavier particles are left behind on the deck and are discharged from outlet. Air gravity separator is suitable for all kinds of metal particles, nonmetal particles, powder materials, granular materials, and other mixed materials. Air gravity separators can be used with cyclones, dust collectors, and draft fans. Air separators play an important role in the separation of recyclable materials. The principle of the air classification technique is based on the suspension of the particles in a flowing air stream and the separation of the particles based on their density difference. Density separation is achieved with basically two components, mechanical vibration and air fluidization. The air gravity separator makes a highly sensitive dry separation on the basis of one of the three particle characteristics: density, size, or shape. When size and shape are controlled within certain limits, the gravity separator is unmatched in its ability to separate a complex mixture by density. The relative size and shape of each component of the mixture also bear on the efficiency of the separation. Wide variations in these material characteristics can dramatically affect the separation results. Where a wide range of particle size is present, screening may be required to segregate materials into manageable size ranges prior to separation. Where significant variations in shape are found to be detrimental to separation efficiency, size reduction may be added to the process. These factors become more important as the densities of the material to be separated become closer.

Air table gravity separators are equipped with a porous rectangular or trapezoidal deck, which is inclined and subjected to vibration which causes material in contact with the deck surface to convey up the inclined surface of the deck. Low-pressure air is forced through the deck to fluidize the dry mixture so that the lighter materials are lifted from the deck surface and allowed to float down the inclination of the deck. The final result, presented at the discharge face of the deck, is a continuous, graduated progression from the least to densest, smallest to largest, and least to most aerodynamic.

Air tables move back and forward shaking at 200–300 rpm. Air tables have an upward tilt. The feed side is lower and concentrate side is higher. Narrow size range feed (0.25–3.2 mm) is introduced to the air table [24]. The riffles are always taller on the feed side of the table and decrease in height as they progress toward the tailing side of the table. Material is fed perpendicular to the riffles, and the high density material remains behind the riffles and low density material flows above and over the riffles. Air tables are used for separation of metallic and plastic components from a feed of −7 mm in size after ferromagnetic separation. 76%, 83% and 91% recovery rates for Cu, Au, and Ag, respectively, achieved from low-grade WPCB or general e-scrap. Air table techniques can be utilized for the separation of particulate fractions in the 5–10 mm, 2–5 mm, and <2 mm ranges, respectively [24]. Triple/S Dynamics, Inc. model T-20 air separator has a trapezoidal deck with 1.86 m^2 surface area and separates Cu and Al granules from electric wire (i.e., PVC) and cable recycling systems. Metal particles can be separated from nonmetal particles. Extraction rate is between 60% and 80%. Exhaust hood for dust-free operations and for high-temperature applications is available. Deck power is 1.5 HP, fan power is 10–20 HP, and total weight is 2.1 t. Closed loop air supply recycles 85–90% of the air used. Dust emission is less than 40 mg m^{-3} and noise is lower than 75 dB.

Originally developed for seed separation, pneumatic or air tables have an important use in the treatment of heavy and light materials. Pneumatic tables use a throwing motion to move the feed along a riffled deck and blow air continuously up through a porous bed. On the air table, both particle size and density decrease from top down which is different from wet tables. Dry air tables use a shaking motion similar to that of wet shaking tables, but instead of water, air is used to separate heavy materials. The table deck is covered with porous material, and air is blown up through the deck from a chamber underneath. The chamber equalizes the pressure from the compressor and thus ensures an even flow of air over the entire deck surface. Generally, air tables consist of a ripped top deck mounted over a base that contains a compressor. The deck is tiltable and riffles are tapered, much like a wet shaking table. The attached motor powers the system. The dry feed is introduced at one corner of the deck. The deck is shaken laterally, and air pressure is regulated to keep lighter materials move down slope along the shortest route. Heavier particles move upslope due to the movement of table. Splitters allow an adjustable middling fraction to be collected (Fig. 7.13).

PVC (density of 1.4 g ml^{-1}) and PP (density of 0.9 g ml^{-1}) were separated by air table with an air flow velocity of 1.6 m s^{-1}, deck frequency of 11.95 s^{-1}, end slope of 4.4°, and side slope of 2.5° at particle size of 2.38–3.36 mm. PVC was settled on the bed and PP was floated on the top of the bed and was separated [16]. Figure 7.14 shows a laboratory-scale air table for plastic separation.

7.4 Electrostatic Separation (ES)

Electrostatic separation (e-sorting) adopts high-voltage electrostatic processing principle, separating the conductive metal (MF) and nonconductive nonmetal/nonferrous (NMF) materials according to their conductivity/resistivity differences. Conductors have valence electrons from a sea of electrons between positive ion cores, and insulators have valence electrons tightly bound to nucleus. Figure 7.15 shows the conductor and nonconductor (insulator/dielectric) properties of WPCB components. Conductors have high conductivity and nonconductors do not have conductivity. Semiconductors have an average (low) conductivity for electricity. In conductor metals, valence electrons form a sea of free electrons, while in insulators, valence electrons are tightly bound. Si, Ga, Ge, As, Se, In, Sn, Sb, and Te have semiconductor properties.

Electrostatic separation system is the most innovative technology for an efficient final stream granulometric separation in recycling processes. Exploiting the electric fields generated by the more dielectric conductive fraction (i.e., metallic) can be separated from the dielectric lower conductive fraction (for instance, paper, plastic, rubber). The charged nonconductors are attracted to an oppositely charged electrode and collected.

160   7 Sorting and Separation of WPCBs

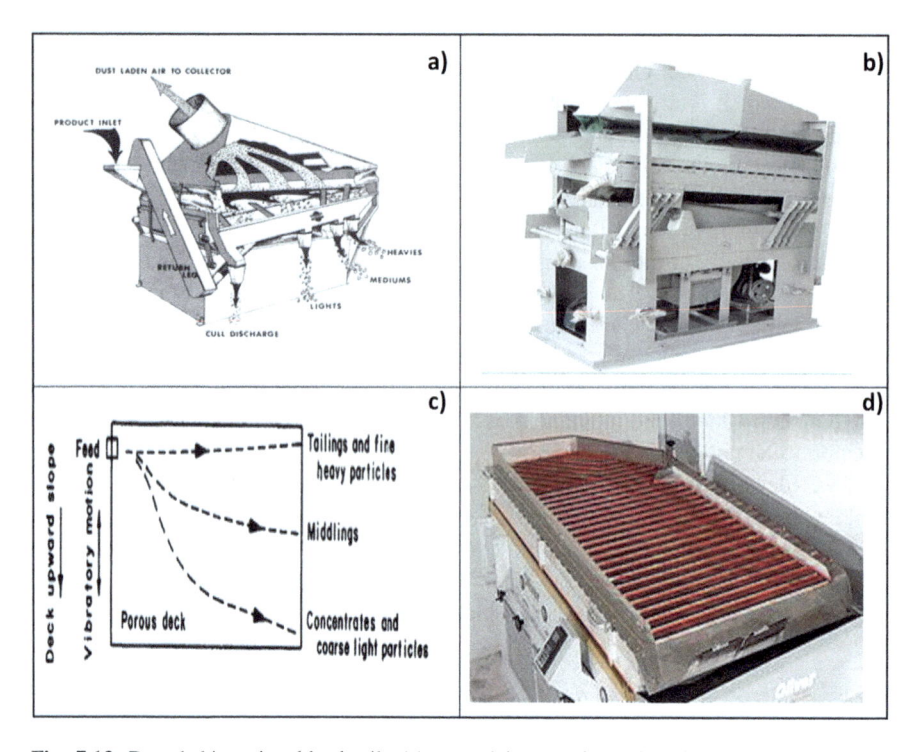

Fig. 7.13 Dry shaking air table details (**a**), material separation principles (**b**), industrial-scale equipment (**c**), and table deck surface (**d**)

Fig. 7.14 Laboratory-scale air table

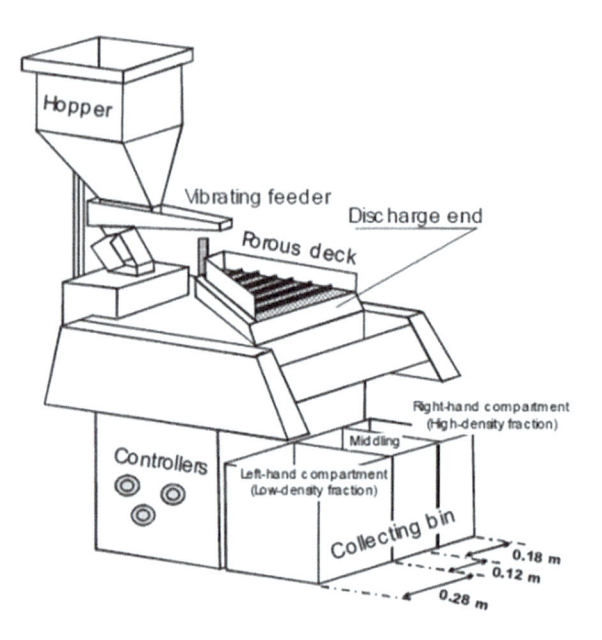

7.4 Electrostatic Separation (ES)

Conductor	Nonconductor/Insulator/Dielectric
Ag, Cu, Au, Fe, Al, Mg, Zn, Cd, Ni, Sn, graphite, C Conductivity decrease	Ceramic, glass, Plexiglas, wood, paper, plastic, Teflon, paraffin, resin

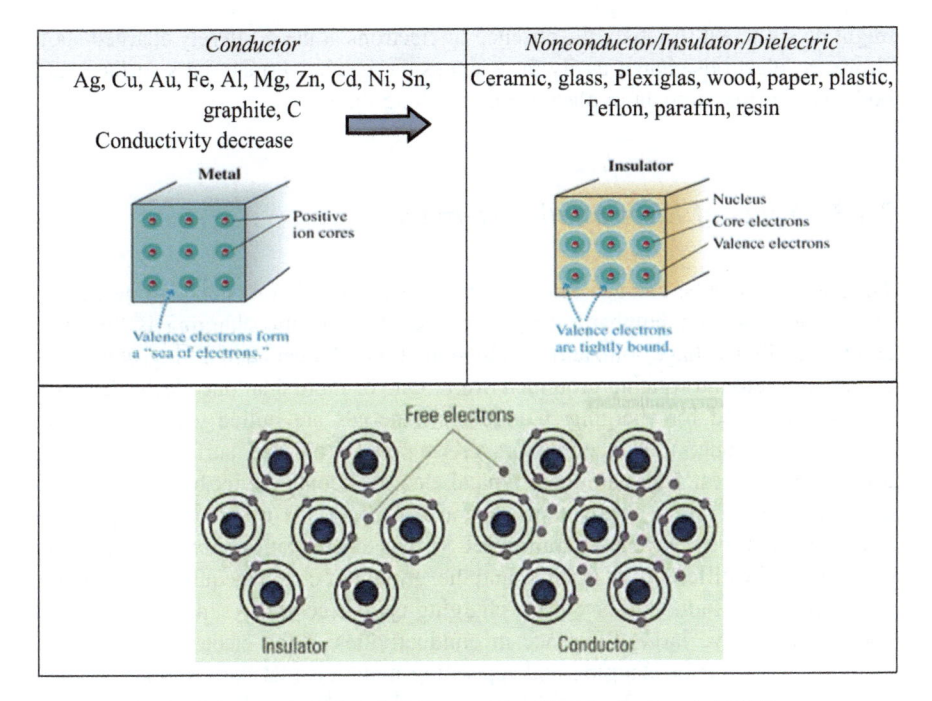

Fig. 7.15 Conductor and insulator/dielectric materials with free electrons in WPCBs

7.4.1 Conductive Materials

A conductive material, which has low electrical resistance, allows electrons to flow easily across its surface or through its volume. When a conductive material becomes charged, the charge (i.e., the deficiency or excess of electrons) will be uniformly distributed across the surface of the material. If the charged conductive material makes contact with another conductive material, the electrons will transfer between the materials quite easily. If the second conductor is attached to an earth grounding point, the electrons will flow to ground and the excess charge on the conductor will be "neutralized."

7.4.2 Insulators/Nonconductors

A material that prevents or limits the flow of electrons across its surface or through its volume is called an insulator. Insulators have an extremely high electrical resistance. A considerable amount of charge can be generated on the surface of an insulator. Because an insulative material does not readily allow the flow of electrons, both positive and negative charges can reside on insulative surface at the same time, although at different locations. The excess electrons at the negatively charged spot might be sufficient to satisfy the absence of electrons at the positively charged spot. However, electrons cannot easily flow across the insulative material's surface, and both charges may remain in place for a very long time.

7.4.3 Types of Electrostatic Separators

The particles may be charged through contact electrification, conductive induction, or high tension (ion bombardment). However, triboelectric charging is the most common. High-voltage ionization, induction, tribo-charger, and lifting electrodes can be used around rotating grounded drum. One or more than one electrode series can be employed for charging. Conductor materials are pulled toward positive-charging electrode(s). Nonconductors are scraped from the surface of rotating drum with brushes. There are three typical electrical conductivity-based separation techniques: (1) corona electrostatic separation (CES), (2) triboelectric separation (TES), and (3) eddy current separation (ECS). The electrostatic separation capability depends on the difference in polarity and the amount of charge acquired by particles to be separated. Induction or corona charging can successfully separate the mixed particles that have large difference in conductivities. Tribo-electricity or contact charging is useful for charging and separating materials that have similar conductivities. The principle of ECS is that in separation zone, gravitational, centrifugal, frictional, and magnetic deflection forces influence the falling particles; but, only magnetic force deflects the ferrous particles to a higher degree. To separate ferrous particles, the magnetic deflection force acting on the ferrous particles has to be greater than all competing forces [13]. Some examples of the use of high-voltage electrostatic separators:

- Recovery of light fractions of nonferrous metals (Cu and Al) (NFM) from the plastic fraction (NMF) of the recycling operations of cables and electrical wiring
- Recovery of light fractions from PMs (Au, Ag, and Pd) in the recycling of WPCBs, e-scrap, and WEEE

Simplicity, low energy consumption, and no wastewater discharge are some advantages of ES. High-voltage level and high roll speed are the main disadvantages of the ES process. In general, ECS is used for particles coarser than 5 mm, CES is used for particle sizes between 0.1 and 0.5 mm, and TES is used for particles finer than 5 mm.

In general, mixtures containing conductors (Cu), semiconductors (extrinsic silicon, Ge, and Ga arsenide), and nonconductors (woven glass-reinforced resin) can be separated by electrostatic separation. The results of binary mixture separation show that the separation of conductor and nonconductor and semiconductor and nonconductor need a higher-voltage level, while the separation of conductor and semiconductor needs a higher roll speed. Furthermore, the semiconductor separation efficiency is more sensitive to the high-voltage level and the roll speed than the conductor separation efficiency. An integrated process was proposed for the multiple mixture separations by Xue et al. (2012) [25]. The separation efficiency of conductors and semiconductors can reach 82.5% and 88%, respectively. ES's efficiency is influenced by electrical, mechanical, material, and environmental factors. At present, ES already has been proved to be efficient for recycling conductors and nonconductors and has been applied in industrial scale with remarkable economic and environmental benefits [25].

7.4.3.1 Corona Electrostatic Separation (CES)

The rotor type corona-electrostatic method is perhaps the most effective separation technology for the conductive (metallic) (MF) and nonconductive (nonmetallic) (NMF) fractions at present. The method has the advantage of being environment-friendly, producing no wastewater and no gaseous emissions. The feed sample in the separator will be bombarded by the high-voltage electrostatic field generated by a corona electrode and electrostatic electrode. Now, the MF will be neutralized quickly as they contact the earthed electrode and leave the rotating roller while the charged NMF are pinned by the electric image force to the rotating roller and move with the rotating roller, finally falling into the holding tanks. In the CES, electrode system, rotor speed, moisture content, and particle size have the greatest effect in determining the separation results. CES can be used for recovery of Cu or Al from chopped electrical wires and cables and for the e-sorting, downstream of the WPCB system, separates metals and allows to recover the finer fractions of PMs. The WPCBs with the metallic components removed must be reduced to very small particles which can be achieved by accelerating them at high speed to impact on a hardened plate. Then, the small particles, typically less than 0.6 mm, are passed along a vibratory feeder to a rotating roll to which is applied a high-voltage electrostatic field using a corona and an electrostatic electrode [26, 27]. The nonmetallic particles become charged and attached to the drum eventually falling off into storage bins, whereas the metallic particles discharge rapidly in the direction of an earthed electrode. Figure 7.16 shows the principles of separation operation in corona-electrostatic separator and laboratory-scale CES equipment.

There are two series of CES machines: double-roller and multi-roller electrostatic separators. Roller length changes from 1000 to 2000 mm, roller quantity 1–4, power 1.5 to 4 KW, speed 20–300 rpm, and outlet motor 0.75–2.2 KW at a production capacity between 300 and 500 kg h^{-1}. Separation efficiency is as high as 99%. Electrostatic separators feature a high precision separation with easy operation and simple maintenance. They cover a small area and easily move. It has been found that particle sizes of 0.6–1.2 mm are the most suitable size for separation in industrial applications. Therefore, a two-step crushing process has been proposed to achieve this particle size.

Li et al. (2007 and 2008) used CES with the pretreatment of pulverization of WPCB feedstock to successfully separate the MF and NMF of WPCB and achieved more than 90% recovery rate along with high capacity (0.5–1.0 t h^{-1}) with no obvious side effects compared to fluidized bed separation, which has wastewater discharge or air-current separation with a dust-releasing issue [28, 29]. The MF has

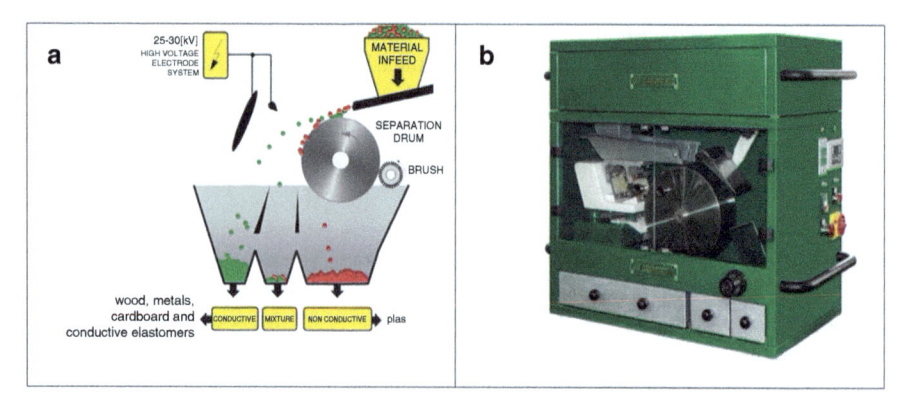

Fig. 7.16 Principle of separation operation in corona-electrostatic separator (**a**) and laboratory-scale CES equipment (**b**)

very high purity and can be sent to a smelting plant directly or with minimal refining. Li et al. (2008a) also found that as the angle of the static electrode reduced and the corona electrode angle was increased, the separation efficiency was enhanced [30] It was reported that applied voltage of 20–30 kV, center distance of 21 cm, static electrode radius of 1.9 cm, corona wire radius of 11.4 cm, static electrode angle of 20°, and corona electrode angle of 60° were the optimum operating parameters influencing the separation efficiency. Considerable work is continuing in this area with particular focus on the electrostatic behavior of the system and the field intensity [27].

Jiang et al. (2009) have designed a new two-roll electrostatic separator that takes advantage of the force of gravity to pass the mixture to the second step for recycling of metals and nonmetals from WPCB [31]. The production capacity was significantly increased for maximum 50% with 45% reduction of middling products. However, mechanical processes for recycling WPCB normally needs to undergo at least two steps, including coarse crushing and fine pulverizing, and especially during fine pulverizing, the temperature will increase up to almost 300 °C [32]. Dust and ash are common issues for mechanical separation, and although it can be avoided by using personnel protective equipment, it still adds to the cost of the process, which makes the economic feasibility of this energy-intensive process even worse. Therefore, an upgrading of NMF to make value-added products is necessary to remedy the expenditure of energy and equipment cost. Also, noble metals will be lost in the crushing process since they will attach on the surface of nonmetallic fraction and cannot be recycled [33].

Xue et al. (2012) used CES (50 kV/20 mA) for recycling conductors (Cu), semiconductors (silicon), and nonconductors (woven glass-reinforced resin) from e-waste [25]. Binary and multiple synthetic mixtures were fed onto the electric feeder that ensures a monolayer of material on the surface rotating roll. Voltage level, roll speed, and grain size were selected variables for the separation tests. Two- and multiple-stage ES flowsheets were proposed for Cu, silicon, and woven glass-

7.4 Electrostatic Separation (ES)

reinforced resin. In the first stage, woven glass was separated from mixed particles. In the second stage, Cu was separated from silicon on conductors with size − 0.45 + 0.3 mm.

CES methods are now capable of producing two streams from WPCB comprising a MF and a NMF portion with little cross-contamination; the method is dry at room temperature and as such is almost zero polluting depending on the quality of the dust extraction system. The residual Cu metal in NMF, which was separated from WPCBs without ECs by using the CES, showed that most of the residual Cu is in fine size range (7–9 µm) [34]. In order to solve the limitations of the one-roll CES, a six-roll CES was designed and investigated to satisfy the requirements of industrial applications. As a consequence, above 95% with purity of about 97.5% of metals could be recovered through this automatic line [35].

Industrial-scale electrostatic separators (Fig. 7.17) adopt physical high voltage to separate conductor metals from nonconductor nonmetals with a separation efficiency of 95–99% purity. Electrostatic separator power changes from 1.5 to 3 KW, rotational speed from 20 to 300 rpm, and static electricity from 50 to 150 KV adjustable. Considerable work is continuing in this area with particular focus on the electrostatic behavior of the system and the field intensity.

7.4.3.2 Triboelectric Separators (TES)

Static electricity is defined as an electrical charge caused by an imbalance of electrons on the surface of a material. Electrostatic charge is most commonly created by the contact and separation of two similar or dissimilar materials. Creating electrostatic charge by contact and separation of materials is known as "triboelectric charging." It involves the transfer of electrons between materials. The atoms of a material with no static charge have an equal number of positive (+) protons in their nucleus and negative (−) electrons orbiting the nucleus. In Fig. 7.18a, Material "A" consists of atoms with equal numbers of protons and electrons. Material B also

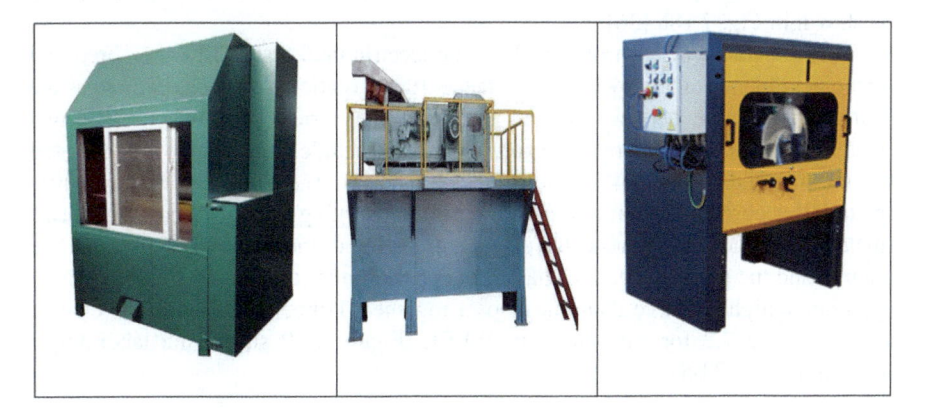

Fig. 7.17 Industrial-scale electrostatic separators

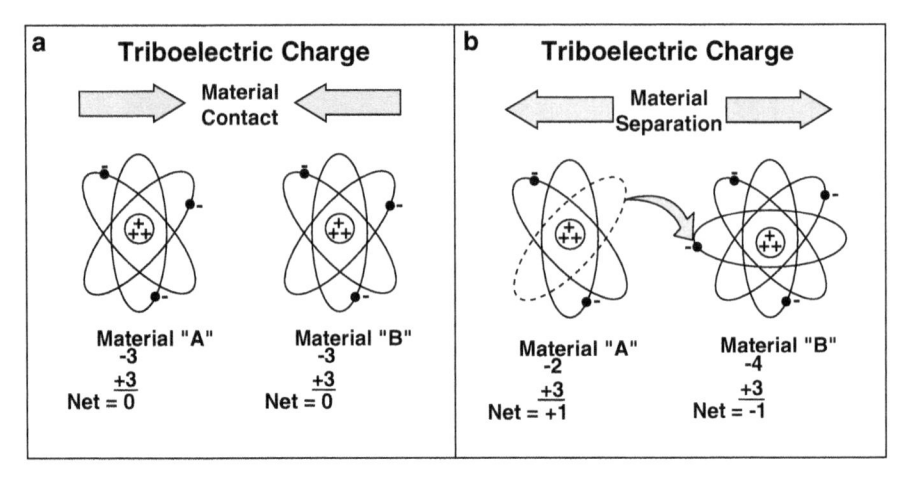

Fig. 7.18 (**a**, **b**) The triboelectric charge. Materials make intimate contact and the triboelectric charge – separation

consists of atoms with equal (though perhaps different) numbers of protons and electrons. Both materials are electrically neutral. This process of material contact, electron transfer, and separation is really a more complex mechanism than described here. The amount of charge created by triboelectric charging is affected by the area of contact, speed of separation, relative humidity, and other factors. Once the charge is created on a material, it becomes an "electrostatic" charge (if it remains on the material). This charge may be transferred from the material, creating an electrostatic discharge (ESD) event [36].

When the two materials are placed in contact and then separated, negatively charged electrons are transferred from the surface of one material to the surface of the other material. Which material loses electrons and which gains electrons will depend on the nature of the two materials. The material that loses electrons becomes positively charged, while the material that gains electrons is negatively charged. This is shown in Fig. 7.18b [36].

This imbalance of electrons produces an electric field that can be measured and that can influence other objects at a distance. Electrostatic discharge is defined as the transfer of charge between bodies at different electrical potentials. When two materials contact and separate, the polarity and magnitude of the charge are indicated by the materials' positions in the triboelectric series. The triboelectric simply lists materials according to their relative triboelectric charging characteristics. When two materials contact and separate, the one nearer the top of the series takes on a positive charge and the other a negative charge. Materials further apart on the table typically generate a higher charge than ones closer together. Triboelectric series can change like in Table 7.2 for substances in WEEE. Figure 7.19 shows the laboratory-/industrial-scale TESs.

7.4 Electrostatic Separation (ES)

Table 7.2 Triboelectric series

Positive (+++)	
Glass Nylon Pb Al Paper Steel Wood Rubber Ni, Cu	Brass, Ag Au, Pt Polyester ABS PET PS PU PE PP PVC Teflon
	Negative (- - -)

7.4.3.3 Eddy Current Separators (ECSs)

An ECS uses a powerful magnetic field to separate NFMs from waste after all ferrous metals have been removed previously by some arrangement of magnets. The device makes use of eddy currents to effect separation. ECSs are not designed to sort FMs, which become hot inside the eddy current field. This can lead to damage of ECS unit belt. ECS is applied to a conveyor belt carrying a thin layer of e-waste. NFMs are thrown forward from the belt into a product bin, while nonmetals simply fall off the belt due to gravity. ECSs use a rotating drum with permanent/electro- magnet. Figure 7.20 shows the separation principles of ECS for nonmetallic-nonferrous materials and Cu and Al separation. ECSs produce high-frequency alternating magnetic field on magnetic roller; if conductive NMF goes through the magnetic field, it will produce induced current, and if this induced current will produce magnetic field opposite with original magnetic field, then the NMF (i.e., Al, Cu, etc.) will fly ahead according to its transporting direction by repulsive force of the magnetic field; hence, NFM is separated from other NMF materials (plastics).

Not all NMF can be separated out through ECS because only material with a high σ/ρ can be separated out, where ρ is the density of the material and σ is the electrical conductivity of material. Table 7.3 shows the materials that can be separated by an ECS, from which it is obvious that Al and Cu are the most easily separated materials while stainless-steel, plastic, and glass have a zero value for the conductivity-to-density ratio, meaning it is not applicable to separate them by an ECS [37].

ECSs are mainly used for recycling Cu, Al, and other NFMs from industrial waste, and living garbage can be widely used in garbage disposal, recycling of WEEE, other environmental industry, and nonferrous metal material processing industry. The main criterion to distinguish is the ratio of material conductivity and

168   7 Sorting and Separation of WPCBs

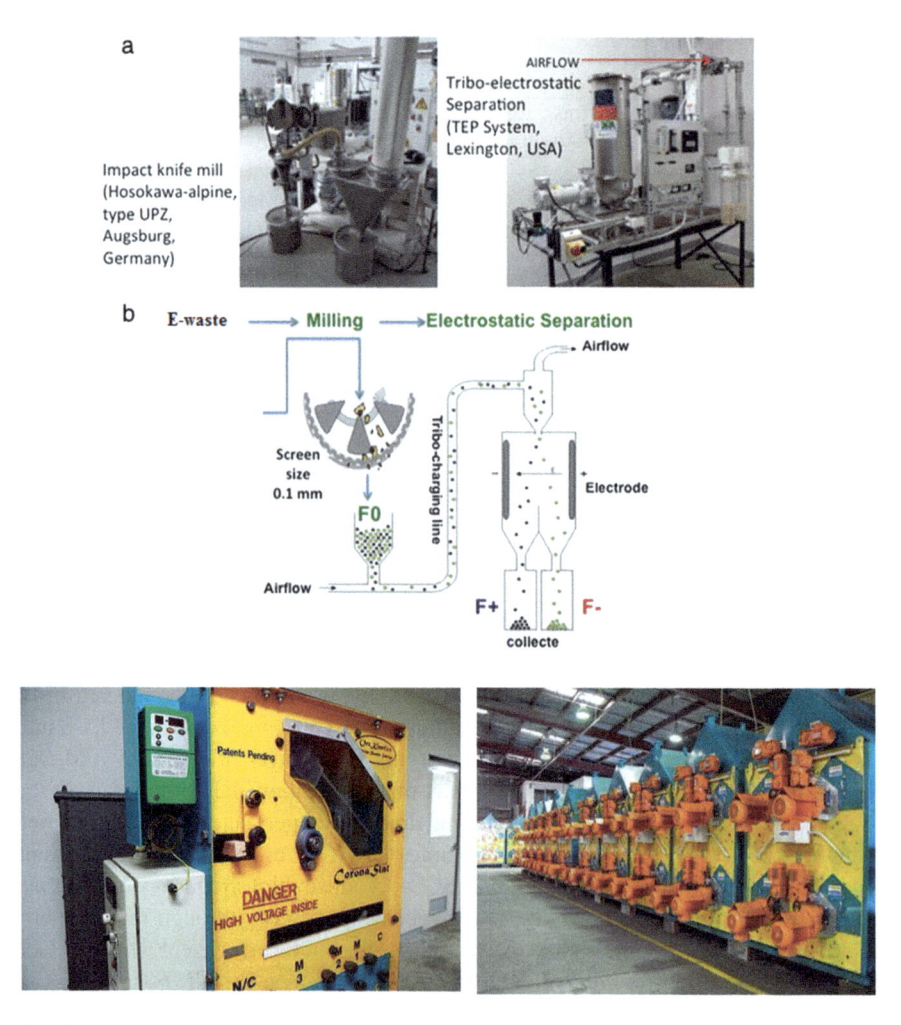

Fig. 7.19 Laboratory (**a**, **b**)- and industrial (**c**, **d**)-scale TES setups

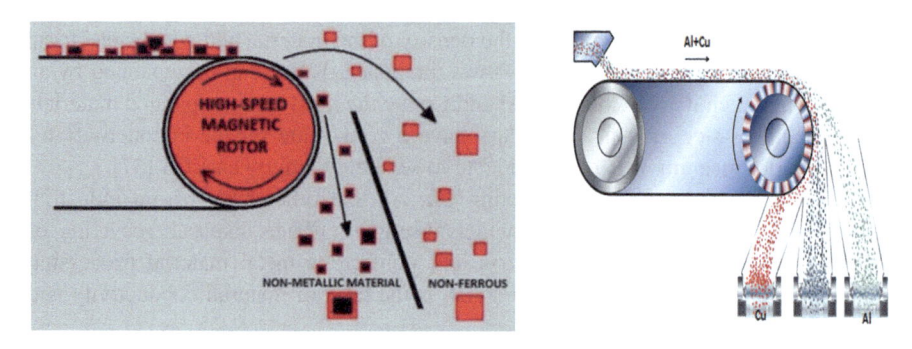

Fig. 7.20 Principals of ECS separation for nonmetallic-nonferrous (**a**) and Cu-Al separation (**b**)

7.4 Electrostatic Separation (ES)

Table 7.3 Materials that can be separated by an eddy current separator and their properties (σ electrical conductivity, ρ density, σ/ρ ratio of electrical conductivity to density)

Materials	$\sigma(10^{-8}/\Omega.m)$	$\rho(10^3 \text{ kg/m}^3)$	$\sigma/\rho(10^3 \text{ m}^2/\Omega.\text{kg})$
Aluminum (Al)	0.35	2.7	13.1
Copper (Cu)	0.59	8.9	6.6
Silver (Ag)	1.63	10.5	6.0
Zinc (Zn)	0.17	7.1	2.4
Brass (Cu + Zn)	0.14	8.5	1.7
Lead (Pb)	0.05	11.3	0.4
Plastic/glass/Fe			0.0

density values; the higher ratio value is more likely to separate. Typical particulate sizes processed tend to be in the 3–150 mm size range. High-frequency ECSs, where the magnetic field changes very rapidly, are needed for separation of smaller particles [24].

ECSs are composed of a separator and an electric controlling cabinet. Main body includes separating assembly, motor, frame, cover, etc. Separating assembly includes permanent magnet roller (FeBNd) and transporting system (includes transporting belt, driving roller, and reducer). They have a good separating result for multiple NFMs, strong adaptability, reliable structure, and strong adjustable repulsion with high separating efficiency. Magnet roller diameter is generally 300 mm. Magnet roller revolution changes up to 3000 rpm. Belt width ranges from 450 to 1250 mm and speed up to 1.0 m s^{-1}. Handling capacity changes from 2 to 2.6 t [38].

Zhang et al. (1998) used an ECS to recovery Al from e-waste and achieved more than 90% Al recovery rate with a purity around 85% for a single pass [39]. Yoo et al. (2009) used a two-step magnetic separation to recover MF from WPCB, and from the first magnetic separation at 700 Gauss, 83% of the Ni and Fe was recovered in the magnetic fraction. The second magnetic separation at 3000 Gauss increased the total amount of Ni recovery but caused a drop of the Ni purity from 76 to 56% [40].

Turn keys 200–300 kg h^{-1}, 300–500 kg h^{-1}, and 1000–1500 kg h^{-1} WPCB recycling lines with electrostatic separation are available today. Power requirement changes from 43 KW to 172 KW and area requirement from 80 to 200 m^2. Lowest capacity plants have one shredder, classifier, cyclone, dust catcher, vibrating screen, and electrostatic separator along with two bucket elevators. Highest capacity plants have one double-shaft shredder, hammer mill, bucket elevator, and storage bin; two-belt conveyor, specialized crusher, classifier, cyclone, four-level cyclone, three-in-one dust catcher, and vibrating screen; and four electrostatic separators and six bucket elevators. Figure 7.21 shows WPCB recycling flowsheet with ECSs, industrial equipment, and feed products obtained. ECSs are industrially used at Argonne, Galloo, MBA Plastics, Salyp Process, Stena, and VW-Sicon for ELV recycling plants [1].

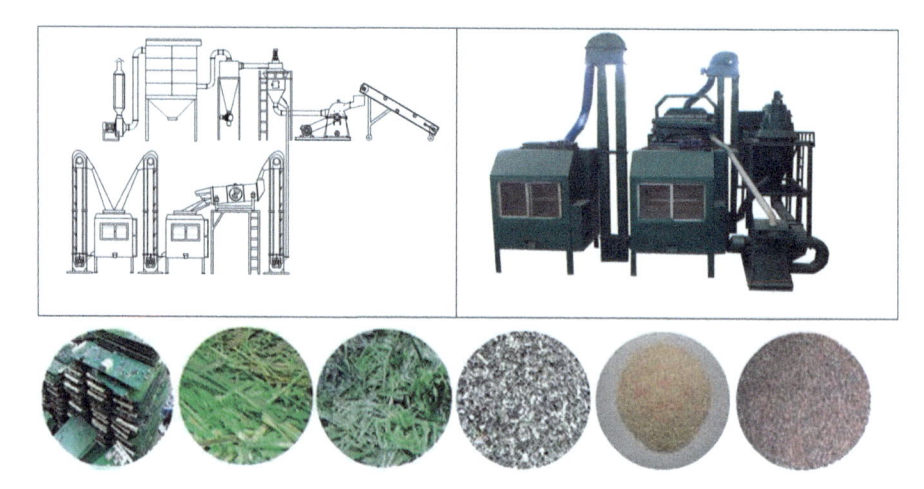

Fig. 7.21 PCB recycling flowsheet with ECSs, industrial equipment pictures, and feed products obtained

7.5 Magnetic Separation (MS)

By creating an environment comprising a magnetic force (F_m), a gravitational force (F_g) (determined by particle size and density), and a drag force (F_d) (for wet magnetic separators determined by particle diameter, shape, liquid viscosity, and velocity and for dry magnetic separators, determined by particle size, density, and air velocity), magnetic particles can be separated from nonmagnetic particles by MS. Magnetic separators exploit the difference in magnetic properties between particles. All materials are affected in some way when placed in a magnetic field.

$$\text{Magnetic attraction force } (F_m) = V * X * H * \text{grad } H \tag{7.1}$$

Where

V: Particle volume (determined by process)
X: Magnetic susceptibility
H: Magnetic field (created by the magnet system design) in mT
Grad H: Magnetic field gradient (created by the magnet system design) in mT
(mT: milli Tesla, 1kGauss = 100 mT = 0.1 T)

Materials are classified into two broad groups according to whether they are attracted or repelled by a magnet. Non-/diamagnetics are repelled from and ferromagnetics/para-magnetics are attracted to a magnet. Ferromagnetic substances are strongly magnetic and have a large and positive magnetism. The magnetic moments in ferromagnetic material are ordered and of the same magnitude in the absence of an applied magnetic field. Paramagnetic substances are weakly magnetic and have a small and positive magnetism. The magnetic moments in a paramagnetic

7.5 Magnetic Separation (MS)

material are disordered in the absence of an applied magnetic field and ordered in the presence of an applied magnetic field. In diamagnetic materials, magnetic field is opposite to the applied field. Magnetisms are small and negative. Nonmagnetic material has zero magnetism. These materials are not attracted toward magnet [41].

Ferromagnetism is the basic mechanism by which certain materials (such as Fe) form permanent magnets or are attracted to magnets. Ferromagnetism (including ferrimagnetism) is the strongest type. Ferromagnetic materials can be separated by Low-Intensity Magnetic Separator (LIMS) at lower than 2 T magnetic intensity. Paramagnetic materials can be separated by dry or wet High-Intensity Magnetic Separators (HIMS) at 10–20 T magnetic intensities. Diamagnetic materials create an induced magnetic field in a direction opposite to an externally applied magnetic field and are repelled by the applied magnetic field. Nonmagnetic substances have little reaction to magnetic fields and show net zero magnetic moment due to random alignment of magnetic field of individual atoms [41].

Strongly magnetic materials can be recovered by a magnetic separator with the use of relatively weak magnetic induction, up to 0.15 T (1,500 Gauss). Weakly magnetic materials can be recovered by a HIMS generating induction up to 0.8 T (8,000 Gauss) with modest values of the gradient of the magnetic field. Induced roll separators, field intensities up to 2.2 T, and Permroll separators can be used for coarse and dry materials (>75 μm). Fine materials reduce the separation efficiency due to particle-rotor and particle-particle adhesion/ agglomeration. For wet HIMS, Gill and Jones separators are used at a maximum field of 1.4 and 1.5 T, respectively, at −150 μm size [41].

Dry LIMSs are used for coarse and strongly magnetic substances. Magnetic field gradient in separation zone (approximately 50 mm from drum surface) changes 0.1–0.3 T. Below 0.5 cm, dry separation tends to be replaced by wet methods (wet LIMS). Concurrent and countercurrent drum separators have nonmagnetic drum containing three to six stationary magnets of alternating polarity. Separation depends on pick-up principles. Magnetic particles are lifted by magnets and pinned to the drum and are conveyed out of the field. Field intensities up to 0.7 T at the pole surfaces can be used. Coarse particles up to 6 mm–0.5 mm can be tolerated. Drum diameter is 1200 mm and length 600; 1200; 1800; 2400; 3000; and 3600 mm. Concurrent operation is normally used as *primary separation (cobber)* for large capacities and coarse feeds. Countercurrent operation is used as rougher and finisher for multistage concentration.

Moderately magnetic dry substances on a conveyor/belt can be collected by overhead, cross-belt, or disc separators using magnetic field intensities between 0.8 and 1.5 T. Very weakly paramagnetic substances can only be removed if field intensities are greater than 2.0 T. At 5–200 mm size fraction, overhead permanent magnets are used to remove ferromagnetics. Magnetic separators, such as dry low-intensity drum types, are widely used for the recovery of ferromagnetic materials from nonferrous metals (Al and Cu) and other nonmagnetic materials (plastic and glass) at −5 mm in size. The magnetic field may be generated by permanent magnets or electromagnets. There have been many advances in the design and operation of HIMS due mainly to the introduction of rare-earth alloy permanent magnets with the capability of providing high field strengths and gradients. There are some problems associated with this method. One of the major issues is the agglomeration of the particles which results in the attraction of some nonferrous fraction attached to the ferrous fraction [41]. This will lead to the low efficiency of this method. Through the process of magnetic separation, it is possible to obtain two fractions: MF, which includes Fe, steel, Ni, etc.; and NMF, which includes Cu [42]. For WEEE, magnetic separation systems utilize ferrite, rare-earth, or electromagnets, with high-intensity electromagnet systems being used extensively.

Veit et al. (2005) employed a magnetic field of 0.6–0.65 T to separate the ferromagnetic elements, such as Fe and Ni [43]. The chemical concentration of the MF was 43% Fe and 15.2% Ni on average. However, there was a considerable amount of Cu impurity in the MF as well. Yoo et al. (2009) used stamp mill, size classification, gravity separation, and a two-stage magnetic separation to PCBs [40]. The milled WPCBs of particle size >5.0 mm and the heavy fraction separated from the <5.0 mm WPCB particles were concentrated by gravity separation. In the first stage, a low magnetic field of 0.07 T was applied which led to the separation of 83% of Ni and Fe in the magnetic fraction and 92% of Cu in the NMF. The second magnetic separation stage was conducted at 0.3 T which resulted in a reduction in the grade of the Ni-Fe concentrate and an increase in the Cu concentrate grade. Hanafi et al. (2012) had an agglomeration problem of nonmetals which was pulled with ferrous materials [11]. Magnetic separation is industrially used at Argonne, Galloo, MBA Polymers, Salyp Process, Stena, R-plus (WE-SA-SLF), and VW-Sicon for ELV recycling plants [1].

7.6 Froth Flotation

Froth flotation is a process for selectively separating hydrophobic materials from hydrophilic. This is used in mineral processing, paper/waste recycling, and wastewater treatment industries more than a century. Historically, this was first used in the mining industry, where it was one of the great enabling technologies of the twentieth century. Contact angles of some hydrophobic plastics with water are PE 96.8°, PP 95°, ABS 87.3°, PVC 86.4°, PS 86.3°, and PET 76.5°. MF in WPCBs can be separated from NMFs by reverse flotation without reagents. Froth flotation methodology is a promising technique for rejecting hydrophobic plastics from the comminution product of WPCBs to recover MFs. It has been found that nearly reagent-free flotation of relatively coarse size (−1.0 mm) pulverized e-waste is feasible with a reasonably good product at a high yield and excellent recovery. The liberation studies accomplish that liberation of metal value from nonmetallic constituents at −1.0 mm size is excellent, and the particulate system is significantly rich in metal value, containing around 23% metal. Single-stage flotation enhances metal content from 23% to over 37%, contributing a mass yield of around 75% with recovery of nearly 95% metal values, suffering nominal loss of around 4% metal value only while effectively rejecting 32% of the materials in feed through float fraction. The interdependence of kinetics and process variables has been discussed, and it has been concluded that a high rotor speed aids efficient rejection of the plastics. However, addition of frother is essential to help stabilize the froth and enhance the kinetics, while efficient preconcentration is facilitated through a combination of moderate air flow with low pulp density. Generation of preconcentration through flotation route from the entire −1.0 mm comminution product stands accomplished [44]. Industrially, Argonne uses flotation for ELV recycling.

Ogunniyi and Varmaek (2009) recovered Au and Pd with 64% recovery at enrichment ratio of 3:1 [45]. Mäkinen et al. (2015) showed that even though flotation, without reagents, could produce the concentrated metal products, a relatively large amount of Cu, Ni, Pb, and Sb were found in the froth, which contributed to severe consequences of disposal and loss of metals [46]. Moreover, Vidyadhar and Das. (2013) reported that under the conditions of a stirrer speed of 1198 rpm, frother dosage of 0.61 kg/ton, and pulp density of 9.02%, as well as air flow of 5.00 lph, 37% metal content with 76% mass yield was obtained, which meant that nearly 95% metal value was recovered [44].

Gallegos-Avecedo et al. (2014) used conventional laboratory flotation cell to separate MF and NMFs from WPCBs with particle nonconventional size [47]. 0.5 ∗ 0.5 cm WPCB fragments are subjected to chemical treatment to remove resin that holds MF and NMFs. Chemical-treated samples have a variety of shapes, sizes, and materials and contain about 70% of nonmetallic and 30% metallic material. Dodecylamine is used as a cationic collector, MIBC as a synthetic frother (5–30 ppm), and NaOH as a pH regulator. Mixing speed changed from 1200 to 1600 rpms and solid content varied from 1% to 3%. Conditioning time was 20 min. and flotation time was 1 min. at a slurry temperature of 30 °C. Nonmetallic glass fiber material was floated (i.e., reverse flotation) and metallic material was sunk. At 5 ppm MIBC, 1200 rpm, and 3% solid content, 99.49% metal and 99.5% nonmetal recoveries were obtained.

Estrada-Ruiz et al. (2016) efficiently separated the metallic and nonmetallic fractions from arcade PCBs at 1.25% solid content and −250 μm particle size in a continuous laboratory column flotation [48]. Green reverse flotation technology with only water and air in the absence of reactive was used at pH: 7. Optimum superficial air velocity was of 0.4 cm s^{-1}. Hydrophobic plastics (reinforced fibers) were floated and went to the concentrate, and hydrophilic metals (mainly Sn and Pb from solder) went to the tails.

7.7 Pyrometallurgy+Supergravity+Hydrometallurgy and/or Electrolysis Separation

WPCBs contain low (Pb and Sn) and high (Cu, Zn, Au, Ag, and Pd) melting point metals. Pb and Sn can be melted at a temperature of 410 °C and Cu at 1300 °C. After size reduction of WPCBs to 1 mm, nonmetallic part is removed. Heated centrifugal separation apparatus with a gravity coefficient 1000 can be used to remove Pb-Sn alloy in 5 min. WPCB residues from this separation are reheated to 1300 °C to melt Cu for 30 min. Melted blister Cu is separated by again supergravity separation and quenched in water. Refined Cu can be produced from this blister Cu by electrolysis. The residue from the second heat goes to leaching to recovery Au, Ag, Pd, Fe, and Mn. The total recovery of Cu, Zn, Pb, and Sn over the whole separation process was 97.80%, 95.59%, 98.29%, and 97.69%, respectively. Compared with the amounts of PMs present in the original WPCBs, the contents of Ag, Au, and Pd in the Cu alloy increased by 5.16, 2, and 1.85 times, respectively, while those in the final residues increased by 2.92, 1.59, and 1.54 times, respectively. Upon combination of the appropriate hydrometallurgical process and supergravity separation of metals or alloys, this clean and efficient process provides a new way to recycle valuable metals and effectively prevent environmental pollution from WPCBs [49].

References

1. http://wedocs.unep.org/bitstream/handle/20.500.11822/8423/-Metal%20Recycling%20Opportunities%2c%20Limits%2c%20Infrastructure-2013Metal_recycling.pdf?sequence=3&isAllowed=y
2. http://eco3e.eu/en/base/design-for-dismantling/
3. Kaya M (2018) Current WEEE recycling solutions, Chap. 3. In: Veglio F, Birloaga I (eds) Waste electrical and electronic equipment recycling, aqueous recovery methods, pp 33–93. https://doi.org/10.1016/B978-0-08-102057-9.00003-2
4. Kaya M (2018) Waste printed circuit board (WPCB) recycling: conventional and emerging technology approach. In: Reference module in materials science and materials engineering/encyclopedia of renewable and sustainable materials volume. Elsevier. https://doi.org/10.1016/B978-0-12-803581-8.11246-9
5. Rocchetti L, Veglio F, Kopacek B, Beolchin F (2013) Environmental impact assessment of hydrometallurgical processes for metal recovery from WEEE residues using a portable prototype plant. Environ Sci Technol ACS 47:1581–1588. https://doi.org/10.1021/es302192t
6. Weeden GS, Nicholas H, Wang NHL (2015) Method for efficient recovery of high purity polycarbonates from electronic waste. Environ Sci Tech 49:2425–2433. https://doi.org/10.1021/es5055786
7. Razi KMHA (2016) Resourceful recycling process of waste desktop computers: a review study. Resour Conserv Recycl 110:30–47. https://doi.org/10.1016/j.resconrec.2016.03.017
8. Baxter J, Hanssen IG (2016) Environmental message framing; enhancing consumer recycling of mobile phones. Resour Conserv Recycl 109:96–101. https://doi.org/10.1016/j.resconrec.2016.02.012
9. Yoshida A, Terazono A, Ballesteros FC, Nguyen DQ, Sukandar S, Kojima M, Sakata S (2016) E-Waste recycling process in Indonesia, the Philippines, and Vietnam: a case study of cathode ray tube TVs and monitors. Resour Conserv Recycl 106:48–58. https://doi.org/10.1016/j.resconrec.2015.10.020
10. Wang JB, Xu ZM (2015) Disposing and recycling waste printed circuit boards: disconnecting, resource recovery and pollution control. Environ Sci Technol 49:721–733. https://doi.org/10.1021/es504833y
11. Hanafi J, Jobiliong E, Christiani A, Soenarta DC (2012) Material recovery and characterization of PCB from electronic waste. Procedia Soc Behav Sci 57:331–338. https://doi.org/10.1016/j.sbspro.2012.09.1194

References

12. Zhang S, Forssberg E (1997) Mechanical separation oriented characterization of electronic scrap. Resour Conserv Recycl 21:247–269. https://doi.org/10.1016/S0921-3449(97)00039-6
13. Li J, Shrivastava P, Gao Z, Zhang HC (2004) Printed circuit board recycling: a state-of-the art survey. IEEE Trans Electron Packag Manuf 27(1):33–42. https://doi.org/10.1109/TEPM.2004.830501
14. Duan C, Wen X, Shi C, Zhao Y, Wen B, He Y (2009) Recovery of metals from waste printed circuit boards by a mechanical method using a water medium. J Hazard Mater 166:478–482. https://doi.org/10.1016/j.jhazmat.2008.11.060
15. http://en.jxmingxin.com/circuit-board/30.html
16. https://www.slideshare.net/sophea79/plastic-recycling-54885628
17. Sarvar M, Salarirad MM, Shabani MA (2015) Characterization and mechanical separation of metals from computer Printed Circuit Boards (PCBs) based on mineral processing methods. Waste Manag 45:246–257. https://doi.org/10.1016/j.wasman.2015.06.020
18. Wills BA (1988) Mineral processing technology, 4th edn. Pergamon Press, England
19. Veit HM, Juchneski NCF, Scherer J (2014) Use of gravity separation in metals concentration from printed circuit board scraps. REM: Rev Esc Minas 67:73–79. https://doi.org/10.1590/S0370-44672014000100011
20. Habib M, Miles NJ, Hall P (2013) Recovering metallic fractions from waste electrical and electronic equipment by a novel vibration system. Waste Manag 33:722–729. https://doi.org/10.1016/j.wasman.2012.11.017
21. Eswaraiah C, Kavitha T, Vidyasagar S, Narayanan SS (2008) Classification of metals and plastics from printed circuit boards (PCB) using air classifier. Chem Eng Process Process Intensif 47:565–576. https://doi.org/10.1016/j.cep.2006.11.010
22. Kumar V, Lee J, Jeong J, Jha MK, Kim BS, Singh R (2015) Recycling of printed circuit boards (PCBs) to generate enriched rare metal concentrate. J Ind Eng Chem 21:805–813. https://doi.org/10.1016/j.jiec.2014.04.016
23. Zhang G, He Y, Zhang T, Wang T, Wang H, Yang X (2017b) Novel technology for recovering residual metals from nonmetallic fraction of waste printed circuit boards. Waste Manag 64:228–235. https://doi.org/10.1016/j.wasman.2017.03.030
24. Kellner D (2009) Recycling and recovery. In: Hester RE, Harrison RM (eds) Electronic waste management, design, analysis and application. RSC Publishing, Cambridge, pp 91–110
25. Xue M, Yan G, Li J, Xu Z (2012) Electrostatic separation for recycling conductors, semiconductors and nonconductors from electronic waste. ACS Environ Sci Technol 46:10556–10563. https://doi.org/10.1021/es301830v
26. Lu H, Li J, Guo JJ, Xu Z (2008) Movement behavior in electrostatic separation: recycling of metal materials from waste printed circuit board. J Mater Process Technol 197:101–108. https://doi.org/10.1016/j.jmatprotec.2007.06.004
27. Hadi P, Xu M, Lin CSK, Hui C-W, McKay G (2015) Waste printed circuit board recycling techniques and product utilization. J Hazard Mater 283:234–243. https://doi.org/10.1016/j.jhazmat.2014.09.032
28. Li J, Lu H, Guo J, Xu Z, Zhou Y (2007) Recycle technology for recovering resources and products from waste printed circuit boards. Environ Sci Technol 41:1995–2000. https://doi.org/10.1021/es0618245
29. Li J, Lu H, Liu S, Xu Z (2008) Optimizing the operating parameters of corona electrostatic separation for recycling waste scraped printed circuit boards by computer simulation of electric field. J Hazard Mater 153:269–275. https://doi.org/10.1016/j.jhazmat.2007.08.047
30. Li ZL, Zhi H, Pan XY, Liu HL, Wang L (2008) The equipment of dismantling for electronic components from printed circuit boards. Chinese patent; 2008103057561
31. Jiang W, Jia L, Zhen-Ming X (2009) A new two-roll electrostatic separator for recycling of metals and nonmetals from waste printed circuit board. J Hazard Mater 161:257–262. https://doi.org/10.1016/j.jhazmat.2008.03.088
32. Zhao M, Li J, Yu K, Zhu F, Wen X (2006) Measurement of pyrolysis contamination during crushing of waste printed circuit boards. J Tsinghua Univ Technol 46:1995–1998
33. Zhou Y, Qiu K (2010) A new technology for recycling materials from waste printed circuit boards. J Hazard Mater 175(1–3):823–828. https://doi.org/10.1016/j.jhazmat.2009.10.083

34. Gao P, Xiang D, Yang J, Cheng Y, Duan G, Ding X (2008) Optimization of PCB disassembly heating parameters based on genetic algorithm. Mod Manuf Eng 8:92–95, In Chinese
35. Li J, Xu ZM (2010) Environmental friendly automatic line for recovering metal from waste printed circuit boards. Environ Sci Technol 44(4):1418–1423. https://doi.org/10.1021/es903242t
36. https://physics.appstate.edu/laboratory/quick-guides/electrostatics-0
37. Williamsburg VA (1990) Practical applications of eddy current separators in the scrap recycling industry. Second Int. Symp. Met. Eng. Mater
38. http://www.sxrecycle.com/en/productInfo/2015111310536742.html
39. Zhang S, Forssberg E, Arvidson B, Moss W (1998) Aluminum recovery from electronic scrap by high-force eddy-current separators. Resour Conserv Recycl 23:225–241. https://doi.org/10.1016/S0921-3449(98)00022-6
40. Yoo JM, Jeong J, Yoo K, Lee JC, Kim W (2009) Enrichment of the metallic components from waste printed circuit boards by a mechanical separation process using a stamp mill. Waste Manag 29:1132–1137. https://doi.org/10.1016/j.wasman.2008.06.035
41. Bentli İ, Erdoğan N, Elmas N, Kaya M (2017) Magnesite concentration technology and caustic - calcined product from Turkish magnesite middlings by calcination and magnetic separation. Sep Sci Technol 52(6):1129–1142. https://doi.org/10.1080/01496395.2017.1281307
42. Yamane LH, Moraes VT, Espinosa DCR (2011) Recycling of WEEE: characterization of spent printed circuit boards from mobile phones and computers. Waste Manag 31:2553–2558. https://doi.org/10.1016/j.wasman.2011.07.006
43. Veit HM, Diehl TR, Salami AP, Rodrigues JS, Bernardes AM, Tenório JAS (2005) Utilization of magnetic and electrostatic separation in the recycling of printed circuit boards scrap. Waste Manag 25:67–74. https://doi.org/10.1016/j.wasman.2004.09.009
44. Vidyadhar A, Das A (2013) Enrichment implications of froth flotation kinetics in the separation and recovery of metal values from PCBs. Sep Purif Technol 118:305–312. https://doi.org/10.1016/j.seppur.2013.07.027
45. Ogunniyi IO, Vermaak MKG, Groot DR (2009) Chemical composition and liberation characterization of printed circuit board comminution fines for beneficiation investigations. Waste Manag 29:2140–2146. https://doi.org/10.1016/j.wasman.2009.03.004
46. Mäkinen J, Baché J, Kaartinen T, Wahlström M, Salminen J (2015) The effect of flotation and parameters for bioleaching of printed circuit boards. Miner Eng 75:26–31. https://doi.org/10.1016/j.mineng.2015.01.009
47. Gallegos-Acevedo PM, Espinoza-Cuadra J, Olivera-Ponce JH (2014) Conventional flotation techniques to separate metallic and nonmetallic fractions from waste printed circuit boards with particles nonconventional size. J Min Sci 50(5):974–981. ISSN 1062-7391
48. Estrada-Ruiz RH, Flores-Campos RF, Gamez-Altamirano HA, Velarde-Sanchez EJ (2016) Separation of the metallic and non-metallic fraction from PCBs employing green technology. J Hazard Mater 311:91–99. https://doi.org/10.1016/j.jhazmat.2016.02.061
49. Meng L, Zhong Y, Wang Z, Chen K, Qui X, Cheng H, Guo Z (2018) Supergravity separation for Cu and precious metal concentration from waste printed circuit boards. ACS Sustain Chem Eng 6:186–192. https://doi.org/10.1021/acssuschemeng.7b02204

Chapter 8
Industrial-Scale E-Waste/WPCB Recycling Lines

> *"Natural resources are like air – of no great importance until you are not getting any"*
> Anonymous

Abstract This chapter introduces industrial-scale state-of-the-art e-waste/WPCB recycling plant applications in the world. Simple WPCB recycling flowsheet adapts physical recycling methods to recover Cu, fiber, and resin powder. Today, integration of electronic component disassembly, solder removal, fine pulverization, dry gravity, and electrostatic separation to obtain Cu-rich mixed powder and glass fiber +resin powder are very important in most of the current plants (with 1.0–1.5 t h^{-1} e-waste processing capacity) in the world. Umicore's integrated smelter-refinery has the biggest e-waste recycling capacity. MGG, Elden, Daimler Benz, NEC, Dowa, Sepro, Shanghai Xinjinqiao, SwissRTec, WEEE Metallica, Hellatron, Aurubis, Attero, Noranda, Rönnskar, and Taiwan e-waste/WPCB recycling practices, aims, capacities, and application flowsheets are presented and discussed in detail. It seems that pyrometallurgical treatment methods are most widely used than hydrometallurgical methods in the world. But, aqueous recovery methods for e-waste recycling are gaining more importance. Lastly, academic and industrial research and practices for e-waste recycling are compared.

Keywords Recycling lines · Recycling chain · Integrated smelter-refinery · Pyrometallurgical processing

8.1 Unpopulated/Populated WPCB Recycling Lines

WPCB recycling line adapts a physical recycling method to get Cu, fiber, and resin powder. In the recycling of populated boards, fiber and resin powder mix with EC material. Therefore, unpopulated/bare WPCB recycling after EC removal is much more beneficial due to ease of special and precious metal separation which do not

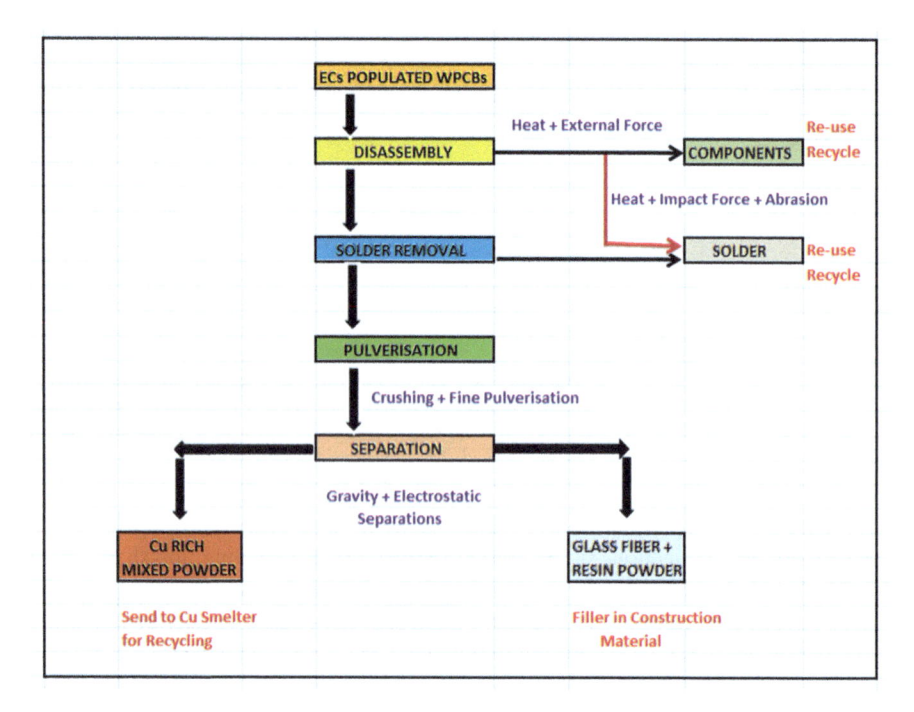

Fig. 8.1 Integration of disassembly, solder removal, and pulverizing with separation processes for WPCB recycling

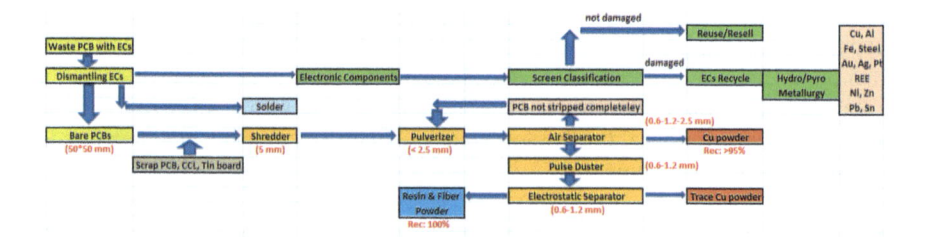

Fig. 8.2 Waste/scrap WPCB recycling flowsheet after dismantling ECs

mix with board fiber and resin. Dismantled ECs are separately and efficiently recycled according to their metal contents. After dismantling the ECs from WPCBs, bare PCBs are shredded and pulverized to obtain mixed metal and resin powders. Then, through the air separator and electrostatic separator, Cu metal powder (Cu, Sn, Pb) is separated from resin powder (Fig. 8.1). To prevent dust pollution in the production process, pulse dusters (i.e., impulse precipitators) are used. Typical WPCB recycling flowsheet for industrial application is given in Fig. 8.2. The recovery rate is high and the purity of metal is as high as 98%. Later, ECs can be recycled by chemical methods (i.e., leaching) through the Au recovery

8.1 Unpopulated/Populated WPCB Recycling Lines

Table 8.1 The five stages of WPCB recycling chain

Stage	Size	Objective		Process
1	5 * 5 cm 1 * 2 mm	Size reduction Liberation		Shredder Hammer mill (Pulverizer)
2		Nonmetal-metal separation Nonmetal-metal separation Ferrous-nonferrous separation	Plastic/ceramic floats Dense metal sinks Denser nonmetals/ceramics Metals	Settling chamber with water Electrostatic separator Magnetic separator
3		Metals dissolution		Leaching (H_2SO_4)
4		Cu Au Ag Pd		Electro winning (AR)
5		Acid neutralized		Waste treatment
		Nonsoluble	Membrane separation	

equipment to get pure Au and other PMs [1]. These types of flowsheets are not very common now but gaining importance recently.

The pulse air separator together with high-voltage electrostatic separator, which is different from the common vibrating separator and electrostatic separator, has a larger capacity and high rate of separation (99.8%) without pollution. More than 95% of metals can be recovered. Technical features of 1 t h^{-1} capacity WPCB recycling line dimensions are 20,000 * 15,000 * 4100 mm (l * w * h), total power demand is 160 KW, and space need is 300 m^2. A low energy consumption, low noise, high automation, and high efficiency and land occupation are the main characteristics of the recycle. Total power requirement changes from 55 to 120 KW and area requirement from 90 to 200 m^2 [2].

Table 8.1 shows the five stages of WPCB recycling chain. The objectives of the processes are given along with suitable particle sizes. The first stage is size reduction for liberation using shredders and pulverizes. The second stage contains physico-mechanical separation methods for MF and NMF and ferrous and nonferrous separations. The third stage is hydrometallurgical leaching for BM and PM dissolutions. The fourth stage includes metal refining with electrowinning, and the last stage comprises waste treatment for neutralization.

Four-shaft shredder, crusher, sorter, vibrating screen classifier, cyclone dust collector, pulse dust collector, blower and exhaust gas purification tower for clean air discharge flowsheet, layout, and plant pictures are shown in Fig. 8.3. This system operates in a totally sealed negative pressure and prevents dust emission and secondary pollution.

Due to the complex nature of feed material, sophisticated flowsheets and sufficient economies of scale are crucial for integrated smelting and refining (ISR). Such facilities exist in Belgium, Canada, Germany, Japan, and Sweden. These ISRs have

Fig. 8.3 Typical WPCB recycling flowsheet, layout, and industrial application pictures [3]

8.1 Unpopulated/Populated WPCB Recycling Lines

Table 8.2a 1000–1500 kg h^{-1} capacity plant details

Name	Model	Power (KW)	Quantity (PCS)	Dimension (m)
Double roller crusher	SX-600	22	1	1.8 * 1.8 * 2
Hammer mill	SX-400	18.5 + 3	1	1.8 * 1.8 * 2
Bucket elevator	SX-220	2.2	1	0.8 * 0.8 * 5.5
Storage bin	SX-1000 * 3000		1	3 * 1.2 * 4
Conveyor belt	SX-600 * 4 m	1.1	2	4 * 1 * 2
Specialized crusher	SX-700	55	2	2 * 2 * 1.7
Classifier	SX-800	1.1	2	1.2 * 1.2 * 4
Cyclone	SX-800		2	1 * 1 * 4
Four level cyclone	SX-1600		2	1.8 * 1.6 * 4
Three-in-one dust catcher	SX-1500		2	1.5 * 1.5 * 5.5
Bucket elevator	SX-180	1.1	6	0.6 * 0.6 * 5.5
Vibrating screen	SX-1000	0.37	2	3 * 1.5 * 3.5
Electrostatic separator	SX-1000-2	1.1 + 2.2	4	2 * 1.8 * 2.3
Outlet	1000–1500 kg h^{-1}			
Total power	172 KW			
Area	200 m^2			

Table 8.2b 300–500 kg h^{-1} plant

Name	Model	Power (KW)	Quantity (PCS)	Dimension (m)
Crusher	SX-700	37	1	2 * 2 * 1.7
Classifier	SX-800	1.1	1	1.2 * 1.2 * 4
Cyclone	SX-800		1	1 * 1 * 4
Deduster	SX-1500		1	1.5 * 1.5 * 5.5
Bucket elevator	SX-180	1.1	3	0.6 * 0.6 * 5.5
Vibrating screen	SX-1000	0.37	1	3 * 1.5 * 3.5
Electrostatic separator	SX-1000-2	1.1 + 2.2	2	2 * 1.8 * 2.3
Outlet	300–500 kg h^{-1}			
Total power	49 KW			
Total area	90 m^2			

been recovered and supplied back to the market in 20 metals since 1998. About 25% of the annual production of Ag and Au and 65% of Pd and Pt originate from EoL recyclables (e-waste materials plus catalysts). The e-waste fractions are mixed with other complex PM-bearing materials such as automotive and petrochemical catalysts, industrial wastes, and by-products from the nonferrous industries.

Tables 8.2a, 8.2b, and 8.2c shows equipment details and technical parameters of 1.0–1.5 t h^{-1}, 0.3–0.5 t h^{-1}, and 0.2–0.3 t h^{-1} capacity plants, respectively. 1.0–1.5 t h^{-1} plant requires 28 pieces of equipment, 200 m^2 area, and 172 KW energy. 0.3–0.5 t h^{-1} capacity plant requires ten pieces of equipment, 90 m^2 area, and 49 KW energy. 0.2–0.3 t h^{-1} capacity plant requires eight pieces of equipment, 80 m^2 area, and 43 KW energy. Figure 8.4 shows both industrial-scale Cu powder, resin-glass fiber production plant layout, and equipment picture.

Table 8.2c 200–300 kg h^{-1} plant

Name	Model	Power (KW)	Quantity (PCS)	Dimension (m)
Specialized crusher	SX-700	37	1	2 * 2 * 1.7
Classifier	SX-800	1.1	1	1.2 * 1.2 * 4
Cyclone	SX-800		1	1 * 1 * 4
Dust catcher	SX-1500		1	1.5 * 1.5 * 5.5
Bucket elevator	SX-180	1.1	2	0.6 * 0.6 * 5.5
Vibrating screen	SX-1000	0.37	1	3 * 1.5 * 3.5
Electrostatic separator	SX-1000-2	1.1 + 2.2	1	2 * 1.8 * 2.3
Output	200–300 kg h^{-1}			
Power	43 KW			
Area	80 m^2			

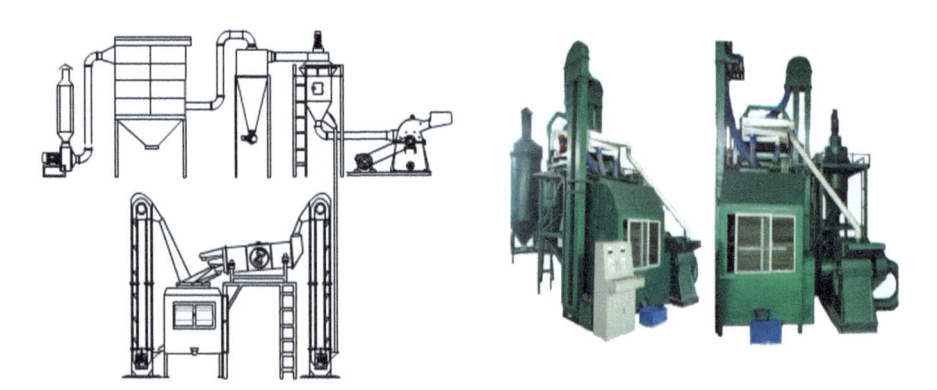

Fig. 8.4 General industrial Cu powder, resin-glass fiber production plant flowsheet, and pictures

8.2 Umicore's Integrated Smelters-Refineries (ISRs), Hoboken, Antwerp

The unique processes are based on complex Pb/Cu/Ni metallurgy using BMs as collectors for PMs and SMs. ISR has two main routes: the precious metal operations (PMO) and the base metal operations (BMO) (Fig. 8.5a). Umicore ISR recovers 20 different metals (Au, Ag, Pd, Pt, Rh, Ir, Ru, Cu, Pb, Zn, Ni, Sn, Bi, In, Se, Te, Sb, As, Co, REE) from WEEE with modern technology and with world-class environmental and quality standards (Fig. 8.5b) [4]. At Umicore in a total of over 200 different types of raw materials, about 350,000 tons are processed each year in a highly flexible flowsheet [5]. The plant is one of the world's largest and complex PM recycling facilities with a yearly capacity of 50 tons of PGMs (i.e., 7% of world's mine production), 100 tons of Au, and 2400 tons of Ag. Industrial by-products are 86% by volume (75% by revenue), and EoL recycles are 14% by volume (25% by revenue) for Umicore. For most of the e-waste materials, the smelter is the first step. The smelter uses vertical *IsaSmelt furnace* submerged lance combustion technology

8.2 Umicore's Integrated Smelters-Refineries (ISRs), Hoboken, Antwerp

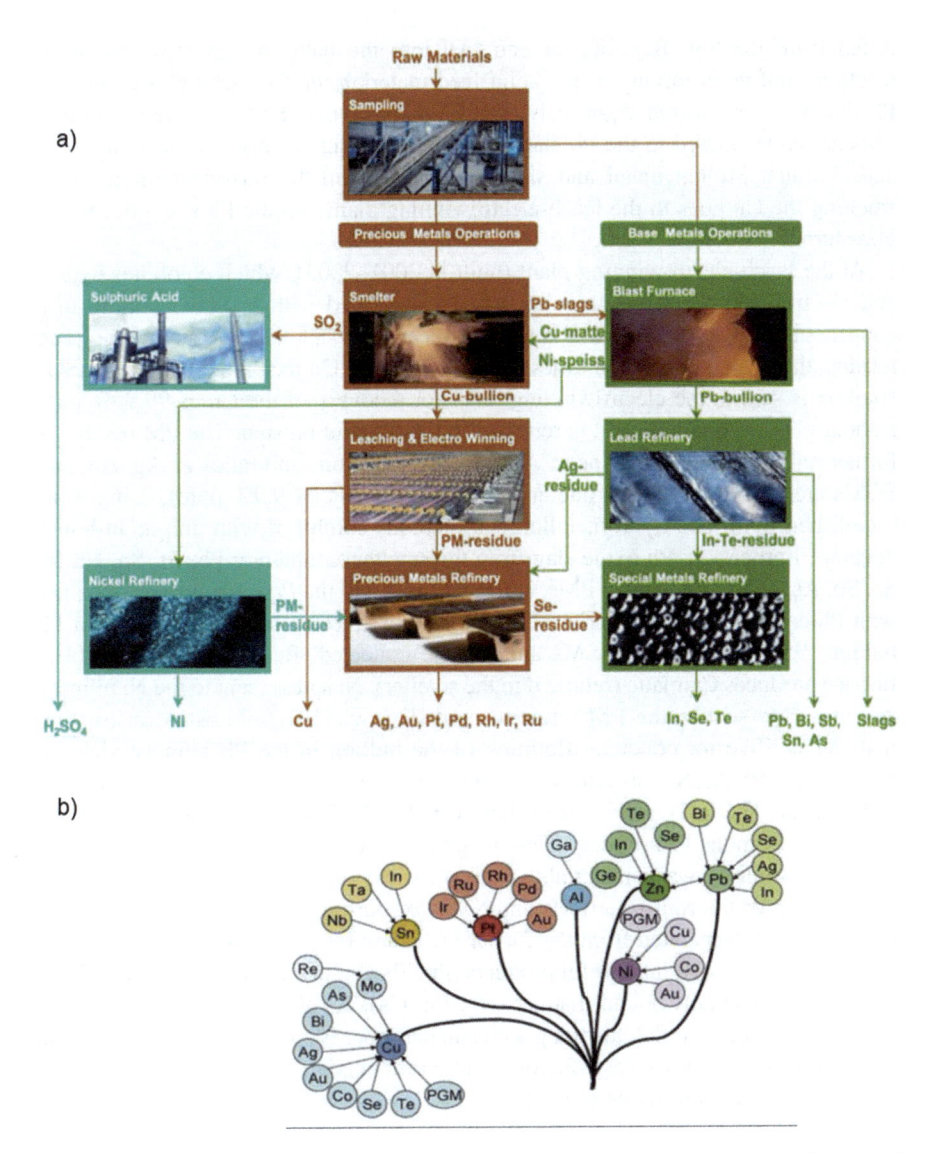

Fig. 8.5 Umicore integrated smelting and refining (ISR) operation flowsheet (**a**) and recovered metals (**b**)

using oxygen-enriched air (90%) and oil/natural gas injection and is equipped with extensive off-gas emission control installation and processes about 1000 tons of feed material per day. At about 1200 °C, enriched air and fuel are injected through a lance in a liquid bath, and coke (4.5% by mass) is added for chemical reduction of the metals. Organic components from the circuit boards function as an additional reducing agent and fuel, thus being classified as feedstock recycling. The furnace is fed from the top. Blowing air and fuel into the bath ensures rapid chemical reactions and good mixing as the solid feed material, the Cu metal phase, and the Pb slag phase are stirred vigorously. The PMs dissolve in the Cu, while most other SMs are concentrated in the Pb slag together with oxide compounds such as silica and alumina. Molten metal and slag are tapped from the furnace bottom. After smelting the Cu goes to the leach-electrowinning plant and the Pb slag goes to the blast furnace.

At the leach-electrowinning plant (built in 2001–2003), which combines hydro- and electrometallurgy, the granulated Cu is dissolved with H_2SO_4 resulting in a copper sulfate ($CuSO_4$) solution, and the PMs are concentrated residue. In the leach residue, the PM content is ten times higher than in the Cu feed material. The $CuSO_4$ solution is sent to the electrowinning plant for recovery of the Cu as 99.99% pure cathodes. The remaining acid is returned to the dissolution step. The PM residue is further refined at the PM refinery. All possible variations and ratios of Ag, Au, and PGMs are recovered one by one as high-purity metals (>99.9% pure), using well-established pyro- and hydrometallurgical methods combined with unique in-house developed processes. Pb oxide slag from the smelter, containing Pb, Bi, Sn, Ni, In, Se, Sb, As, and some Cu and PMs, is further treated in the *Pb blast furnace* together with Pb-containing raw materials. The furnace produces about 200–250 tpd of Pb bullion (95% Pb) in which SMs and Ag are collected. Besides bullion the blast furnace produces Cu matte (returned to the smelter), Ni spiess (sent to the Ni refinery and any PMs sent to the PM refinery), and slag, which is sold as a construction material/additive for concrete. Refining of the bullion in the Pb refinery yields – besides Pb – Bi, As, Sn, Sb, and two residues. The Ag residue is further treated in the PM refinery. The In-Te residue is further treated in the SM refinery together with the Se residue from the PM refinery. The integrated process achieves high PM recoveries from complex e-waste materials. A gold recovery of over 95% has been reported. Independent of the route, the PMs and SMs take through the flowsheet, and in the end they are all separated from the carrier metal and recovered, while substances of concern are converted into useful products (Pb, Sb, As) or captured and immobilized in an environmentally sound manner (Hg, Be, Cd). As also the produced slag is a certified building material, the integrated smelter process converts less than 5% of its feed materials mix into a waste fraction that is sent to controlled deposits; for e-scrap the final waste is even much lower [4].

8.2.1 Off-Gas Treatment

The Cu smelter, the Pb blast furnace, and the Pb refinery are equipped with heat recovery systems, which generate high-pressure steam for internal use. IsaSmelt process is equipped to deal with the complex mix of metals, organic resins, and plastics as it is connected to an external off-gas treatment plant. Dust is captured by electrostatic precipitator and returned to the smelter; toxic dioxin emissions are prevented; and SO_2 gases are converted to H_2SO_4 for use in plant and external sales. SO_2 and NO_x emissions are continuously monitored from stack. Dust emissions are prevented by intensive water sprinkling to the roads and stocks and covering conveyor belts to minimize the precious metal losses.

8.2.2 Resource Efficiency

Smelting of WPCBs requires 1500 kJ kg^{-1} and further refining needs 6500 kJ kg^{-1}. The energy content of WPCBs is around 9600 kJ kg^{-1}. WPCBs fulfill the necessary energy during smelting and generate excess energy for steam generation. Furthermore, the recovery of metals from WEEE materials avoids the high CO_2 emissions associated with primary metal production. The CO_2 emission to produce PMs from ore ranges between 9380 and 13,954 tons CO_2/ton PGMs compared to 3.4 tons CO_2/ton for Cu. The primary production of the metal demand (4.7 million t y^{-1}) for WEEE manufacturing accounts for an annual CO_2 emission of 23.4 million tons, almost 1/1000 of the world's CO_2 production. The CO_2 emission of the Umicore process when used to recover 75,000 tons of metal from 300,000 tons of recyclables and smelter by-products is only 3.73 tons CO_2/ton metal produced compared to 17.1 tons CO_2/ton metal when using a primary production route. This single operation can thus prevent CO_2 emissions and lower the environmental footprint of the metals substantially [4].

8.3 Austrian Müller-Guttenbrunn Group (MGG)

The Austro-American Joint Venture, MBA Polymers Austria Kunststoffverarbeitung GmbH located in Kematen/Ybbs, was founded in November of 2004 between American-based MBA Polymers Inc. and the Austrian Müller-Guttenbrunn Group. On an area of 20,000 m^2 including around 8000 m^2 of built-up area, one of the most advanced plastic recycling facilities developed with a capacity of 50,000 tpy. MGG Polymers produce high-grade plastic granulates and compounds, mainly from PP, HIPS, ABS, and PC/ABS as used in various applications for electronic and automotive components. The raw material for MBA Polymers Austria granulates mainly come from EoL plastic from WEEE.

MGG is one of the key recyclers of e-waste in seven Central and Southeast European countries with numerous facilities in the region from Austria and the countries along the Danube. The process handles collection (with an own fleet of trucks, rails connections, and harbor links along the Danube), depollution, shredding, and ferrous metal separations followed by nonferrous separations and even plastic recycling. Mixed plastic is first separated from nonplastics and then sorted by type and grade using fully automated process steps converting the material to high-value new materials to be used for many demanding applications. The produced plastics are RoHS and REACH compliant and can be used for new electronic

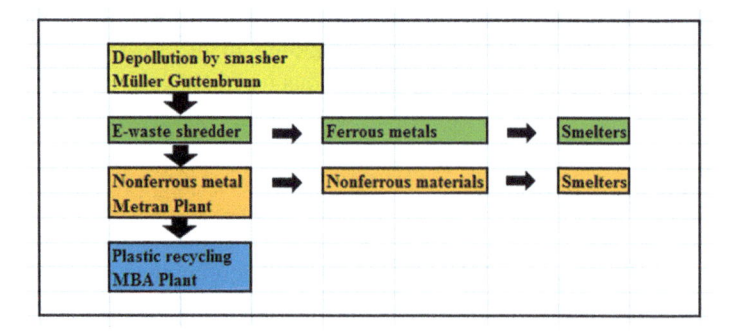

Fig. 8.6 MGG plant flowsheet

products. Depollution and shredding steps include a smasher, hand sorting, shredding, and ferrous recovery at Müller-Guttenbrunn plant. Low-energy opening of e-waste with "smasher" and picking out pollutants and valuables before shredding are the main difference of this plant from other recycling plants. Nonferrous metals are separated by sieving, eddy current (Foucault current) separation, two-step heavy media density separation, optical separation, air table separation for PM concentration, and finally smelting to generate Cu and PMs at Metran Plant. MBA Polymers process the recycled plastics to result in newly compounded and pelletized ABS, HIPS, and PP. This flowsheet provides the lowest possible loss of PMs (Fig. 8.6). Plastics are recycled to produce plastic raw materials by MBA Polymers. MGG treats yearly 80 Kt of e-waste. Recovery rate of 85% is achieved which is over EU target in 2018. MGG recovers 850,000 t of metals and contributes to a saving of over 1 million tons of CO_2 [6].

8.4 Eldan Recycling, Spain

Elden Recycling produces combi plants for WEEE and cable in Zaragoza, Spain. The WEEE is firstly grab fed into super pre-chopper SC1412 and then ring shredder S1000, with 40 ∗ 125 mm screen. Overband electromagnet removes ferrous metals. Then, two-stage ECS is used to remove nonferrous metals. Heavy granulators (FG1504) and two separating air tables (C26) separate organics and metals. Screens and shaking tables are also used to separate organics from metals. Capacity changes from 800 to 7000 kg h^{-1} (Fig. 8.7). Ring shredder has 10,000 m^3 h^{-1} filtering system. Cyclones and filters recover light materials and dust. The Eldan plants are designed for automatic processing, i.e., staff is required for surveillance only.

In the Eldan WEEE recycling plants, most of the items mentioned in the EU WEEE Directive Annex 1B can be processed. Exceptions are refrigerators/freezers, which instead are processed in Eldan refrigerator systems. The system offered by Eldan Recycling for processing of WEEE is modular and designed to process the following types of input:

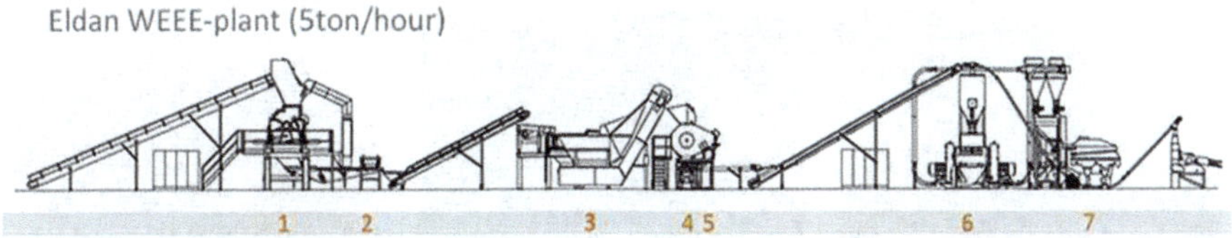

Fig. 8.7 Example of a setup of large WEEE recycling system (5 t h^{-1}) [7]. (1) Shredder. (2) Eddy current separator. (3) Tumble back feeder. (4) Multipurpose rasper. (5) Eddy current separator. (6) Heavy granulator. (7) Separation table

- Computer scrap containing the mainframe of computers, PCs, keyboards, monitors without glass tubes, printers, faxes, etc.
- Small home appliances containing videos, TV sets without glass tubes, record players, CD players, hair dryers, toasters, vacuum cleaners, coffee makers, irons, microwave ovens, etc.
- Handheld tools such as drilling machines, grinding machines, etc.
- Electrical scrap such as contactors, relays, main breakers, fuses, contact bars, switches, instruments, etc.
- Electronic and telegraphic scrap such as electromechanical switchboards/relays, computerized switchboards, printed circuit boards, etc.
- Small electrical motors up to approximately 1–1.5 HP

From an Eldan WEEE recycling plant, the following output fractions are retrieved:

- Various fraction of ferrous and stainless steel (SS)
- Various fraction of NFMs (Al, Cu, brass, and PCBs)
- Refining material containing Cu, brass, Zn, Pb, PMs, etc.
- Organic fraction with plastic, rubber, tree, textile, etc.
- Dust

There are three different capacity WEEE systems at Elden. Small-capacity system has a capacity of 1.5 tons h^{-1}, complete-capacity system 3.0 tons h^{-1}, and large-capacity system 5–7 tons h^{-1}. Figure 8.8 shows industrial-scale input WEEE materials, shredder, and output systems at Elden.

8.5 Daimler Benz in Ulm, Germany

Germans have developed a mechanical treatment approach, which has the capacity to increase metal separation efficiencies, even from fine dust residues generating post-particulate comminution, in the treatment of WPCB assemblies. They considered a purely mechanical approach to be the most cost-effective methodology, and a major objective of their work was to increase the degree of purity of the recovered metals, such that minimal pollutant emissions would be encountered during

Fig. 8.8 Input WEEE materials, shredder, and output systems at Elden

subsequent smelting process. Their process comprises the initial coarse-size reduction to 2 * 2 cm, followed by magnetic separation for ferrous materials, which is followed by a low-temperature grinding stage. The embrittlement of plastic components at temperatures <70 °C was found to enable enhanced separation from NFM components when subjected to grinding within a hammer mill. In operation, the hammer mill was fed with liquid nitrogen at −196 °C, which served both to impart brittleness to the plastic feedstock constituent and to effect process cooling; the grinding of material within such an inert atmosphere eliminated any likelihood of oxidation by-product formation from the plastics, such as dioxins and furans. Subsequent to this enhanced grinding stage, the MFs and NFMs were separated by sieving and electrostatic separation. This process was claimed to be cheaper even for low-grade WPCBs [8]. Yang (2013) developed a mechanical separation + hydrometallurgy technology with a relatively high PM recovery. This technology includes primary crushing, liquid nitrogen refrigeration, classification, electrostatic separation, and hydrometallurgical dissolving [9].

8.6 NEC Group in Japan

NEC Group combined the automatic disassembly with mechanical separation. ECs from WPCBs were removed on a conveyor with IR heating, and solder (about 4%) was taken away by belt sanding, after fine pulverization gravity and electrostatic separation techniques were used to obtain Cu-rich powder and glass fiber-resin powders. The main objective of this work was to reduce the intrinsic material loss from mechanical treatment and utilize more fully the uneven material distribution between the bare boards and ECs [10].

8.7 DOWA Group in Japan

DOWA's recycling and smelting complex at Kosaka accepts e-waste such as used WPCBs and the residue from the Zn smelting process as the main raw ingredients to produce a large number of metals by leveraging technological strengths in the highly efficient recovery of precious and rare metals [11]. The DOWA group has established its technology to collect 17 different kinds of valuable metals, including not only Cu and Zn but also a small amount of rare metals contained in mineral ores through its long experience in the mining and smelting business. In the DOWA group's recycling network centering on Kosaka Smelting and Refining and Akita Zinc, 11 recycling-related companies accept and receive various types of recycled materials for extraction of valuable metals, detoxification, and final disposal.

Kosaka Smelting and Refining accepts mobile phones and electronic substrates as material for recycling besides mineral ores and also accepts smelting residues from Akita Zinc, incineration residues from incinerators of the DOWA group companies, and low-quality materials containing valuable metals to extract Cu, Au, or Ag. Kosaka and Akita cooperate with each other in collecting valuable metals: Kosaka also supplies smelting residues to Akita Zinc, and Akita extracts metals such as Zn, which Kosaka cannot extract in its plants. In this way, the DOWA recycling network forms Japan's preeminent metal recycling complex, where each plant not only collects specific metals but also accepts residues containing metals which it cannot extract as materials to be recycled for other plants to extract to enable effective collection of a great variety of metals. Figure 8.9 shows the material recycling flowsheet for DOWA group.

Kosaka Smelting and Refining has Japan's first furnace that can process materials for recycling valuable metals from electronic substrates and mobile phones (TSL Furnace) besides mineral ores. Figure 8.10 shows smelting furnace and products obtained at Kosaka. The furnace can process Zn residues from Akita Zinc and various materials from plants of the DOWA group in addition to scraps to extract Co, Au, and Ag. Kosaka is one of the major refineries of the DOWA group together with Akita Zinc and will significantly contribute to building resource-recycling society in Japan.

In 2010, DOWA started up the first recycling model plant in China that has three functions: home appliance recycling, hydrometallurgical processing, and pyrometallurgical processing. This plant not only enabled metal recycling from a variety of types of materials but also enabled pyrometallurgical processing of CFC from home appliances and effluent from hydrometallurgical processing (Fig. 8.11) [12].

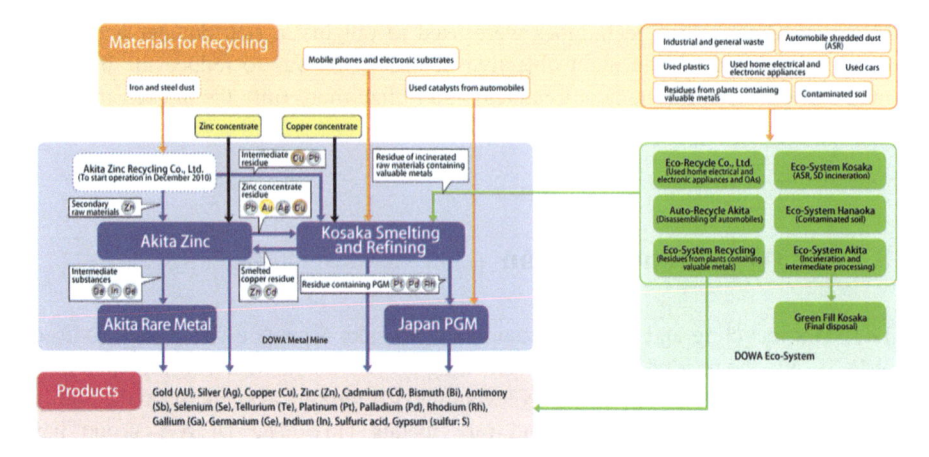

Fig. 8.9 Materials recycling along with e-waste in DOWA group smelters

Fig. 8.10 DOWA's Kosaka smelting furnace in Japan along with general view of the plant

8.8 PCB Manufacturing Waste Recycling in Taiwan

In Taiwan, the commercial recycling processes for the wastes of the PCB industry mainly focus on the recovery of Cu and PMs. Recently, the average price of Cu has risen significantly due to the imbalance of demand and supply. This is the driving force behind the successful development of the Cu recycling industry in Taiwan. Nevertheless, there are still many issues that need to be addressed. Although the recycling of these valuable materials is economically self-sustaining, there are many research activities focusing on the improvement of current technology in order to elevate the technology level and strengthen the compatibility of recycling industry.

The recycling of the nonmetal portion of WPCBs, however, is relatively small. It has been demonstrated, in a small commercial scale, that the plastic material can be

8.8 PCB Manufacturing Waste Recycling in Taiwan

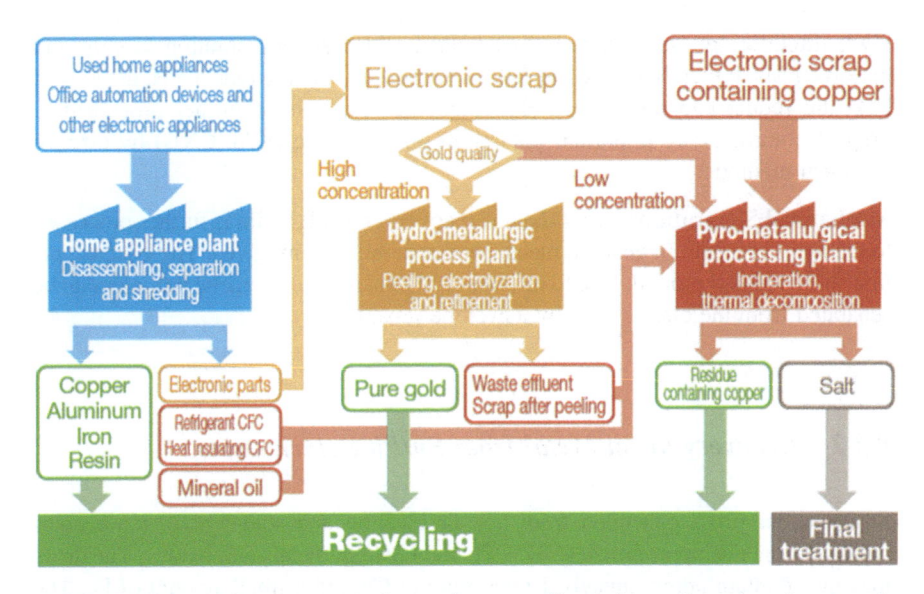

Fig. 8.11 DOWA's first e-waste hydro- and pyrometallurgical recycling plant flowsheet in China [13]

used for artwork materials, artificial wood, and construction materials. Nevertheless, the niche market is quite limited. Most of the nonmetal wastes of WPCBs are therefore treated as landfill (76–94%). In the USA, the nonmetal portions of WPCBs are currently used as raw materials for production by several industries. In plastic lumber, it gives strength to the "wood"; in concrete it adds strength, making the concrete lighter and providing an insulation value ten times higher than that of standard concrete. It is also being used in the composite industry as filler in resins to make everything from furniture to award plaques. More research on this issue is needed in the future. In view of the current commercial processes, the recycled products are not of great value. The development of more innovative recycled products will help the industry by extending the market to new terrain. In addition to the efforts by the recycling industry, the PCB industry itself should also promote and practice waste minimization. Facilities can significantly reduce waste production to minimize the secondary environmental risk of waste transportation.

8.8.1 Recovery of Cu from Edge Trim of PCBs

PCB edge trim has high Cu content ranging from 25% to 60%, as well as PM content (>3 ppm). The process for recovery of Cu and PMs from PCB edge trim is similar to that from WPCBs. In general, the edge trim is processed alone with WPCBs. The recycling process includes [14]:

Hydrometallurgy: Edge trim is first treated with stripping solution to strip and dissolve PMs, typically Au, Ag, and Pt. After adding suitable reductants, the ions of PMs are reduced to metal form. The recovered Au can be further processed to prepare commercially important potassium gold cyanide ($KAu(CN)_2$) by electrochemical methods.

Mechanical Separation: After the PM recovery, the edge trim is further processed to recover Cu metal. In general, mechanical separation is involved. The edge trim is first shredded and ground. Due to difference of density, the Cu metal particles can be separated from the plastic resin by a cyclone separator.

8.8.2 Recovery of Sn Metal from Sn/Pb Solder Dross

Sn/Pb solder dross generated from hot air leveling and solder plating processes typically contains approximately 37% Pb and 63% Sn metals and oxides. The dross may also contain approximately 10,000 ppm of Cu and a small amount of Fe. The dross is first heated in a reverberator furnace (1400–1600 °C) and reduced to metals by carbon reduction:

$$SnO_{2(s)} + C_{(s)} \rightarrow Sn + CO_{2(g)} \tag{8.1}$$

$$2MO_{(s)} + C_{(s)} \rightarrow 2M + CO_{2\ (g)} \tag{8.2}$$

During the deslagging operation, the Fe impurity is removed in order to reach the standard of 63Sn solder, of which Cu < 0.03% and the trace amount of Cu should also be removed. This can be achieved by placing the molten metal in a melting furnace with the addition of S. The S reacts with Cu to form copper monosulfide (CuS), which can be removed as slag. The tin-lead ratio is analyzed with X-ray fluorescence (XRF) and readjusted to meet the standards in Taiwan by adding high-grade Sn and Pb metals. Figure 8.12 shows the recycling process [14].

8.8.3 Recovery of Cu Oxide from Wastewater Sludge

PCB manufacturing industry wastewater sludge typically contains high amounts of Cu (>13% in dry base). The simple and straightforward recycling includes heating the sludge to 600–750 °C to remove the excess amount of water and to convert the $Cu(OH)_2$ to CuO. However, the energy consumption and environmental impacts should be evaluated in detail before industrial usage for each country.

Fig. 8.12 Sn/Pb dross recycling process in Taiwan

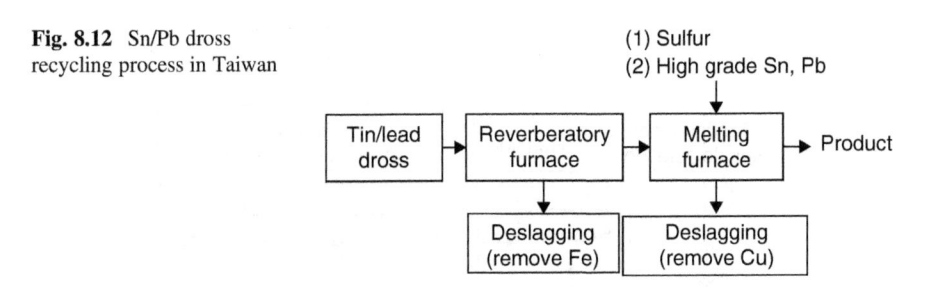

```
                                              (1) Sulfur
                                              (2) High grade Sn, Pb
                                                       ↓
  ┌──────────┐    ┌──────────────┐    ┌──────────┐
  │ Tin/lead │ →  │ Reverberatory│ →  │ Melting  │ → Product
  │  dross   │    │   furnace    │    │ furnace  │
  └──────────┘    └──────────────┘    └──────────┘
                         ↓                  ↓
                  ┌──────────────┐    ┌──────────────┐
                  │  Deslagging  │    │  Deslagging  │
                  │ (remove Fe)  │    │ (remove Cu)  │
                  └──────────────┘    └──────────────┘
```

8.8.4 Recovery of Cu from Spent Basic Etching Solution

The spent solution is generated from the etching process. Spent basic etching solution contains about 130–150 g l^{-1} of Cu. The spent solution is first adjusted to a weak acidic condition, at which most of the Cu ions are precipitated out as Cu(OH)$_2$. Cu(OH)$_2$ is filtered and further processed to recover Cu similar to that as used in sludge recycling. The Cu remaining in the filtrate (about 3 g l^{-1}) is further recovered with selective ion exchange resins. Since the filtrate is acidic, the spent solution can be used to neutralize basic etching solution at the beginning of this process. Cu(OH)$_2$ can also be further converted to Cu(SO)$_4$. Cu(OH)$_2$ is dissolved in concentrated H$_2$SO$_4$. After cooling, crystallization, filtration or centrifugation, and drying, Cu(SO)$_4$ is obtained. Figure 8.13 shows the recycling process flowsheet [14].

8.8.5 Recovery of Cu(OH)$_2$ from CuSO$_4$ Solution in PTH Process

Spent CuSO$_4$ generated from PTH manufacturing contains Cu ions at a concentration between 2 and 22 g l^{-1}. The spent solution is loaded into the reactor. The solution is agitated, while the temperature is lowered by a chiller to 10–20 °C, at which the CuSO$_4$ crystal precipitates out of solution. The CuSO$_4$ crystal is recovered by centrifugation. The pH of the effluent is further readjusted to basic condition to recover the remaining Cu as Cu(OH)$_2$, of which the recycling process is as described previously. Figure 8.14 shows the process flowsheet.

8.8.6 Recovery of Cu from the Rack Stripping Process

The stripping process is performed to remove Cu from the rack and uses HNO$_3$. The Cu in the spent HNO$_3$ is in the form of Cu ion. Therefore, the Cu ion (approximately 20 g l^{-1}) can be recovered directly by electrowinning. Under suitable electrochemical conditions, the Cu ions can be recovered as metal Cu. The other metal ions in the spent solution can also be reduced and deposited along with Cu on cathode. After the

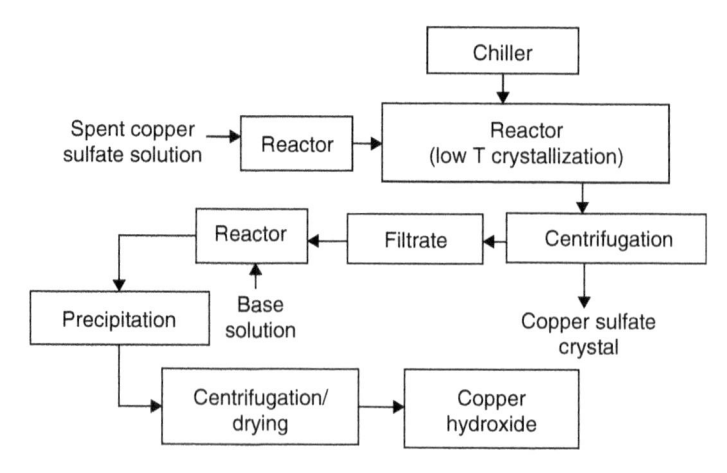

Fig. 8.13 Recovery of Cu from acidic/basic etching solution

Fig. 8.14 Recovery of Cu(OH)$_2$ from CuSO$_4$ solution in PTH process [7]

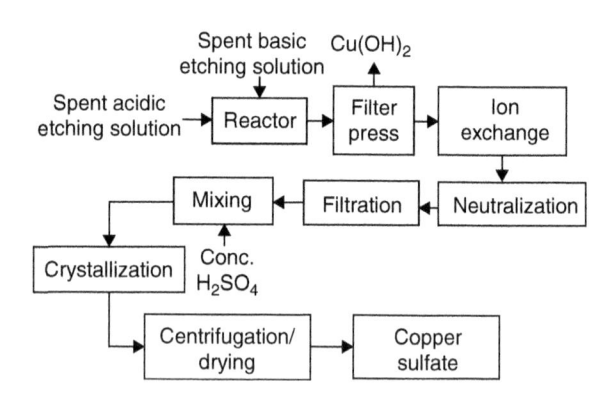

electrochemical process, the HNO$_3$ solution contains about 2 g l^{-1} of Cu and some trace amount of other metal ions. The solution can be used as nitric solution to strip the rack. The stripping efficiency is not affected by the presence of the metal ions.

8.8.7 Recovery of Cu from Spent Sn/Pb Stripping Solution in the Solder Stripping Process

After the etching process, the protective Sn/Pb solder plate should be removed to expose the Cu connections. Sn/Pb solder can be stripped by immersing PCBs in HNO$_3$ or HF stripping solution (20% H$_2$O$_2$, 12% HF). The spent solution contains 2–15 g l^{-1} Cu ion, 10–120 g l^{-1} Sn ion, and 0–55 g l^{-1} Pb ion. Cu and Pb can be recovered by an electrochemical process. During the process, Sn ion is precipitated out as oxides, which is filter pressed to recover valuable tin oxides. The filtrate is low in metal ions and can be used as Sn/Pb stripping solution after composition readjustment. The recycling process flowsheet is given in Fig. 8.15 [14].

Fig. 8.15 Recycling of Sn/Pb spent stripping solution

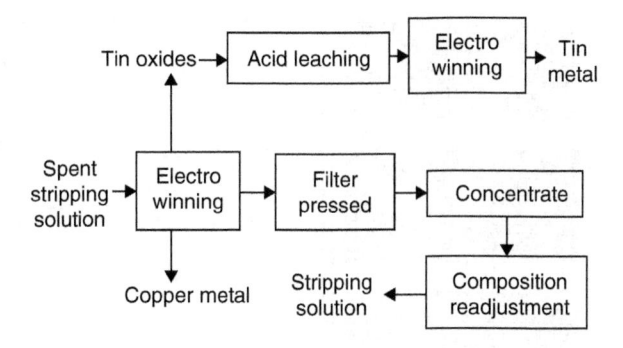

Current research activities in Taiwan are related to the metal separation and development of more innovative recycled products. Wastewater from the manufacturing of PCBs contains high level of Cu^{2+} and small amount of other metal ions (mainly Zn^{2+}). Separation of Cu ions from other metals can improve the purity of recycled Cu. A D2EHPA-modified Amberlite XAD-4 resin prepared by solvent-nonsolvent method can remove Zn ions, leaving Cu ions in the solution. Ion-exchange isotherm showed that D2EHPA-modified Amberlite XAD-4 resin has higher Zn ion selectivity than Cu ion. The selective extraction results demonstrated that D2EHPA-modified Amberlite XAD-4 resin can separate Zn/Cu mixed ion solution. After ten batches of contacts, the relative Cu ion concentration increases from 97% to more than 99.6%, while the relative Zn ion concentration decreases from 3.0% to less than 0.4% [14]. As pointed out previously, Cu in wastewater is traditionally recycled as copper oxides and sold to smelters. The other alternative is to prepare CuO particles directly from wastewater. This will significantly increase the value of recycled product. CuO particles can be used to prepare high-temperature superconductors, materials with giant magnetoresistance, magnetic storage media, catalysts, pigment, gas sensors, p-type semiconductor, and cathode materials.

8.9 Sepro Urban Metal Process in Canada

Sepro Minerals Systems Corp. has Sepro Urban Metal Process in BC, Canada. Sepro has released a simplified flowsheet and process description for its e-waste/WEEE Au and heavy metal recovery process. Combined with thermal pretreatment, the Sepro e-waste/WEEE recovery process is ideal for recovering Au, Cu, and other heavy metals from PCBs and other e-waste/WEEE materials [15]. Specifically, the process starts with chipped e-waste/WEEE and utilizes grinding, size separation, low-g gravity separation, and high-g gravity separation to generate two concentrates and a clean reject stream (Fig. 8.16). The two concentrates include a coarse Cu concentrate composed mostly of Cu foil as well as a fine heavy metal concentrate containing Au, Ag, Pt, Pd, Cu, and other heavy metals. The resulting reject stream is clean and nonhazardous for conventional landfill disposal or subsequent reuse.

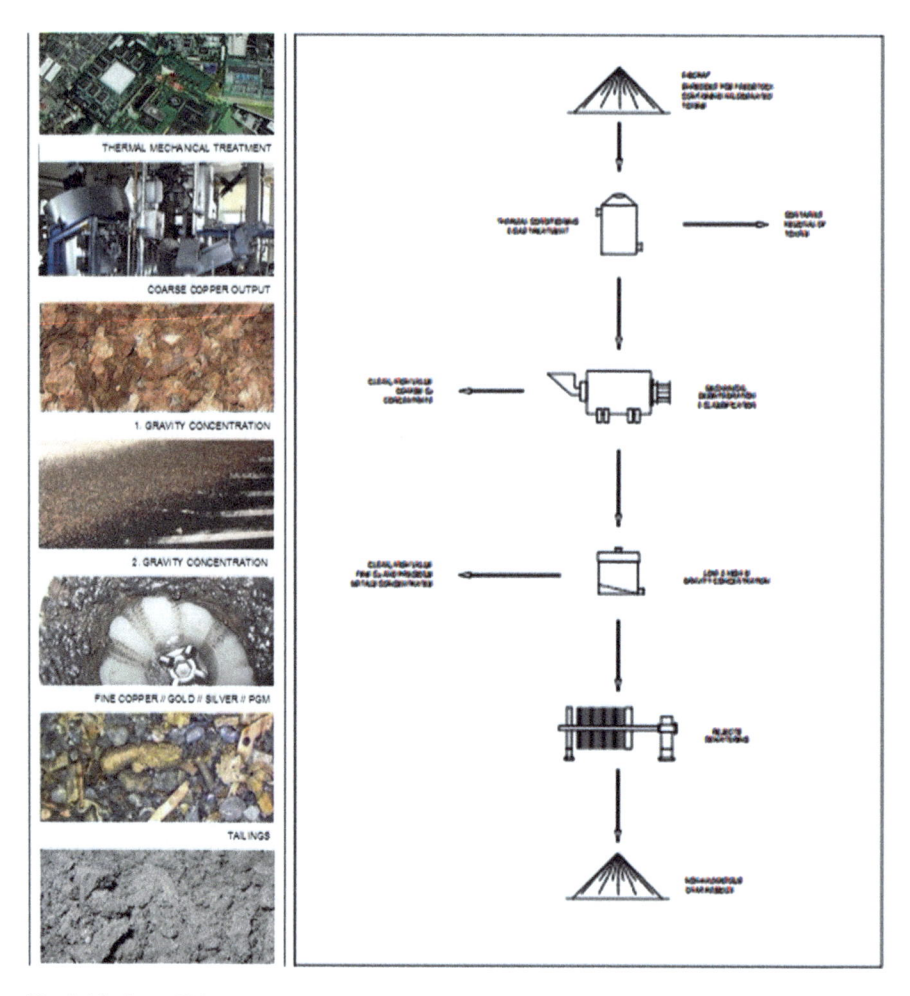

Fig. 8.16 Sepro Urban Metal Process flowsheet

The high value of PMs in WPCBs constitutes a valuable resource and at the same time a challenge from process view. The new approach reaches unmatched liberation of PMs and Cu from the complex circuit boards to allow for optimum recoveries. Toxins contained in WPCBs are safely removed and will not negatively affect the sales value of the metal concentrates. Crushed WPCBs are fed to a thermal treatment system where WPCBs become fractural and toxins are removed safely. Thereafter, the WPCBs are mechanically disintegrated in a mill, and coarse Cu is removed from the mill discharge as a clean product. The remaining fine slurry is then processed by Sepro's low-g and high-g gravity concentration stages to recover fine Cu and PMs. The gravity concentration rejects are dewatered and disposed of as nonhazardous residue [9].

Fig. 8.17 Pyrolyzed WPCB photos

8.9.1 WEEE/E-Waste Pyrolysis Process of Sepro

Pyrolysis can be used to help turn WPCBs into cash quickly and effectively. Sepro has been working closely with e-waste pyrolysis specialists to develop complete, turnkey processes for the recovery of PMs and HMs from e-waste. In simple terms, pyrolysis means heating without oxygen. In other words, it means applying heat but not burning or incinerating the material. Pyrolysis is a process that's been used since the 1950s and has gone through many improvements over the past several decades. Recently, many improvements have been made that allow for the production of clean, natural gas from recycled plastics and other waste. In fact, many countries now consider the pyrolysis of biomass to be a renewable energy source. Figure 8.17 shows pyrolyzed WPCB boards. Pyrolyzed boards look very different from the original WPCBs. When WEEE/E-waste is pyrolyzed, there are several changes that take place in the WPCBs. Firstly, the solder connections on the board melt and release many of the metal components. Secondly, the epoxies in the glass fiber breakdown, and the layers of the board start to separate. Thirdly, the glass fiber becomes brittle and can be easily broken apart. These physical changes, along with a few others, release the HMs and the PMs contained in the board. The Cu foil that is sandwiched between layers of glass fiber is released in relatively large pieces. The PMs fall away from melted solder connections and the brittle fibers.

8.10 Shanghai Xinjinqiao Environmental Co., Ltd., and Yangzhou Ningda Precious Metal Co., Ltd., in China

Two-step crushing and cyclone air separation-corona electrostatic separation (CAS-CES) are used to separate metallic and nonmetallic materials in Shanghai Xinjinqiao Environmental Co., Ltd. (established in 2000; employee: 200 full time), and Yangzhou Ningda Precious Metal Co., Ltd. (established 2014; employee: 120 full time), in China with the productivity of 5000 tons of WPWBs per year;

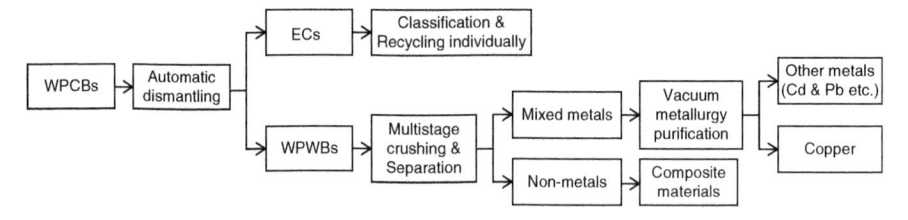

Fig. 8.18 Chinese e-waste recycling flowsheets

vacuum distillation is used to reclaim metals including Cu purification and other metal, such as Pb, Cd, Zn, Bi, etc., separation sequentially. Nonmetal components are used as fillers for other products, like nonmetal – modified asphalt, nonmetallic plate products, phenolic molding products, and wood plastic composite products [16].

Yangzhou Ningda Noble Metal Co., Ltd., dissembles e-waste and recycles and processes PMs. The company is based in China. As of August, 2014, it operates as a subsidiary of Shenzhen Green Eco-Manufacture Hi-tech Co. Ltd. This line is the first full-function demonstration line of dismantling WEEE in Jiangsu Province. It can dismantle and dispose over 2 million sets of waste appliances a year. The company's capacity is 5000 tons of WPCBs. The process includes two-step crushing and cyclone air separation-corona electrostatic separation. The main products of the company are Ge and In metals.

In China, Fig. 8.18 shows integrated process for WPCB recycling. Firstly, WPCBs are automatically dismantled with desoldering. ECs, solder, and depopulated WPCBs are separated. Removed ECs are classified according to type and recycled individually based on metal contents, while unpopulated WPCBs are size-reduced, separated, and metallurgically extracted. Metals and non-metals are separately beneficiated. Process flowsheet of separation for metals and nonmetals from WPWBs is given in Fig. 8.2. Unpopulated WPCBs are shredded to 50.0 mm; two stages of hammer milled to 2.5 mm; two stages of cycloned, vibrating screed at 1.2 and 0.6 mm; and six-roll electrostatic separator are used to recover more than 95% of metallic fraction. Cyclone overflow are sent to dust collector which catches almost all nonmetals [17].

At Shanghai Xinjinqiao Environmental Co., vacuum distillation is used to reclaim metals including Cu purification and other metal, such as Pb, Cd, Bi, etc., separation sequentially (Fig. 8.19) [17].

8.11 SwissRTec AG

SwissRTec AG provides engineering, process equipment, and plants to international recycling industry. It is based in Kreuzlingen, Switzerland. SwissRTec is a producer of shredders, composite material recycling, and recovery solutions. It is an expert in

8.11 SwissRTec AG

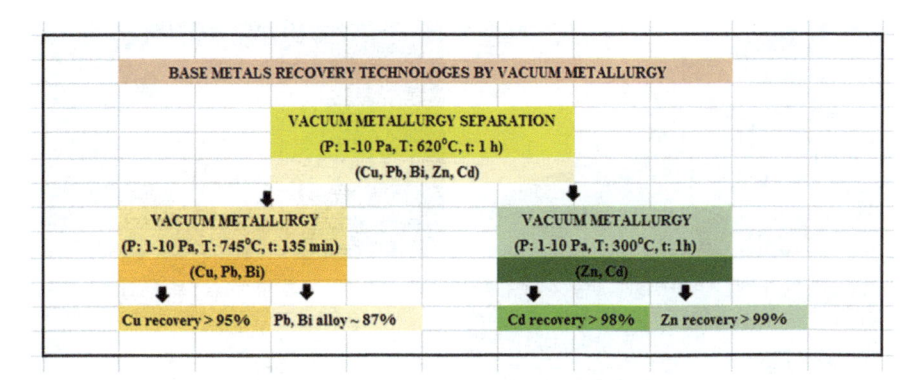

Fig. 8.19 Vacuum metallurgy separation Cu, Cd, Zn, Pb, and Bi in China

design, manufacturing, and commissioning of turnkey e-waste recycling facilities. SwissRTec offers a variety of solutions to composite waste streams including WEEE, cables, and WPCBs. Composites are made up of more than one raw material. E-waste composite contains wires, WPCBs, plugs, plastics, etc. By breaking up, disintegrating, and separating these composites valuable, high-yield raw materials such as Cu, Al, plastics, and steel are being recovered. SRT1 (shredding and separation plant) and SRT2 (delamination and separation plant) are modular processing systems that can be used individually or in combination depending on material input and operational requirements. SwissRTec systems have throughput range from 1 to 25 tons h^{-1}.

High yields, low operating cost, excellent output quality, energy and wear efficiency, high operational availability and easy maintenance, capability of processing large WEEE, automatic control, and high operating reliability are the main advantages of this system [18]. SRT1 plant has Kubota vertical shredder to liberate and downsize even bulky material in a single step with its breakers and grinders. The combination of continuous impacts, shearing, compression, and grinding delivers uniformly shredded particles, perfectly prepared for further handling or separation. Massive, high-speed rotating breakers perform the primary downsizing operation, and rotating grinders carry out the secondary grinding step. Figure 8.20 shows SRT2 plant, vertical shredder, and SRT2 products. High-quality separation of ferrous, nonferrous, plastic, and WPCB from input WEEE material can be achieved with MSs, ECSs, screens, and sensor sorters according to the following flowsheet (Fig. 8.21). For even higher yields, the nonferrous metal concentrate can be further processed and purified in the SRT2 delamination and separation plant. The heart of the SRT2 system is delamination impact milling technology where the material is exposed to thousands of collisions that cause the composite particles to break up and disintegrate into individual components. There are three types of delamination mills with a diameter from 1.0 to 2.0 m and throughput from 2 to 7.5 tons h^{-1}. A size reduction and an excellent ball shaping of metals facilitate and enhance these subsequent steps of separation. Lower-density Al metal is separated from higher-density Cu metal at high purity with air tables (Fig. 8.21).

Fig. 8.20 SRT2 plant, vertical shredder, and SRT2 product metals

In SRT plants, process type is dry mechanical and throughput is between 1 and 10 tons h^{-1}. Power requirements range from 300 to 1600 kW. Plant space requirement changes from 200 to 1000 m^2, and height changes from 6 to 12 m. Low- to medium-grade WPCBs are separated to a variety of metals (Cu, Al, steel), plastics, and some PMs, which are further processed and refined by Italimpianti refining equipment. Plant layout is shown in Fig. 8.22 [18].

8.12 WEEE Metallica, France

A French company was established in Isbergues in 2014. WEEE Metallica with a capacity of over 25,000 tons y^{-1} is a market leader in the recycling of e-waste with a specialization in the treatment of WPCBs. Using their patented WPCB pyrolysis process, WEEE Metallica extracts valuable metals from PCBs and delivers metal concentrates to smelters around the world. Shipments of WPCB arrive daily at Isbergues facility. The WPCBs are prepared for processing by using an automated sequence of shredding, Fe separation by magnetic band, and Al separation by eddy current. Multistep low-heat pyrolysis process at 500 °C is the heart of the operation. The fully automated pyrolysis furnace eliminates all the organic components (plastics and resins) and retains all the useful metals. The resulting gases from the pyrolysis are treated and neutralized in an independent post-combustion system. The gases are collected directly from the furnace by extraction and go through several stages of remediation. These include the post-combustion – which destroy organic halogens, DeNox (catalytic nitrogen oxide reduction), as well as sodium

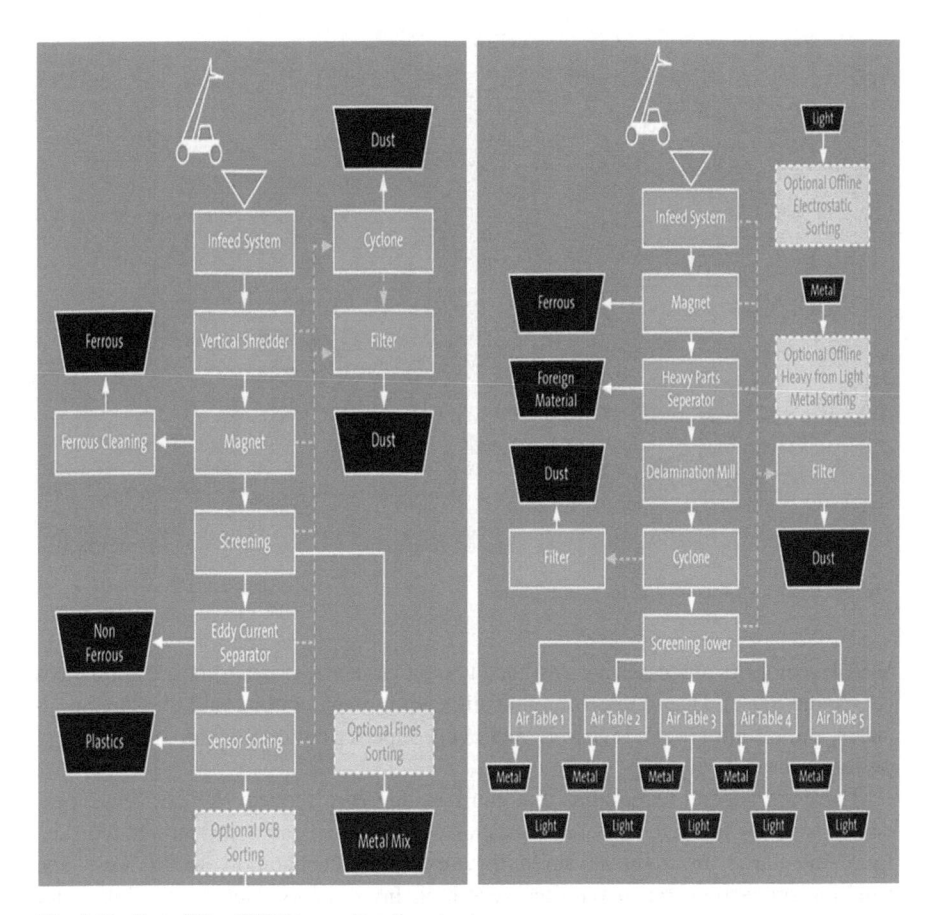

Fig. 8.21 SwissRTec WEEE recycling flowsheet

bicarbonate and activated carbon injection. During these treatments, the gases reach temperature of above 1100 °C. The resulting material becomes the main feeding ingredient for the recycling furnace. After the removal of organic and ferrous components, the remaining final product contains high-grade Cu and PMs which can be delivered to smelters around the world [19]. Figure 8.23 shows the general plant view of WEEE Metallica.

8.13 Hellatron Recycling, Italy

Hellatron Recycling division is an Italian company active on the development and distribution of innovative environment and recycling technology solutions for several waste streams such as solid urban, car batteries, solar panels, alkaline batteries, and WEEE in general. Hellatron Recycling is one of the market leaders on the

Fig. 8.22 SwissRTec WEEE recycling plant

WEEE sector. Its processes are designed to treat large and small domestic appliances (LDA and SDA), IT and telecommunication equipment, cooling and freezing appliances, and monitor and televisions, achieving superior environmental and recycling performances.

There are several techniques that can be utilized to recycle WEEE. Some processes are very manual, and some are extremely sophisticated and automated (i.e., involving large mechanical shredding, expensive separation techniques, and smelting processes), but it requires very high investments. Hellatron solution for WEEE recycling was designed to optimize the valorization of the materials coming out of the plant (i.e., ferrous, Cu, Al, and others) with a reduced plant dimension/investment, highly automated, and with low utilization of manpower (Fig. 8.24). This environmentally friendly, compact, and innovative process represents a big step toward an efficient and responsible WEEE recycling. The key advantages of Hellatron process in relation to traditional technologies to treat WEEE are:

- Hellatron WEEE process is designed to optimize:
  - Investment required: much lower than traditional plants that treat this type of material
  - Manual intervention: limited to few working stations securing correct depollution and material sorting
  - Degree of separation of the outgoing material: clean and well-separated materials – best cost vs benefit ratio regarding fractions valorization
- Flexible process, having the possibility to have several waste streams as input material (SDA/LDA/IT/PC).
- Reduced plant dimension with scalable capacity.

Fig. 8.23 WEEE Metallica plant view

- Purely pneumatic material flow inside and between machines (continuous material cooling by air).
- Innovative milling process, with no traditional cutting or grinding tools, reducing energy consumption and increasing material recovery.
- Limited maintenance costs.
- Easy to operate process [20].

8.14 Aurubis Recycling Center in Lünen (Germany)

A growing proportion of Cu production at Aurubis takes place by processing recycling materials such as Cu and Cu alloy scrap, Cu-bearing residues from foundries and semis fabricators, shredder materials, galvanic slimes, slags, WPCBs, ashes, and filter dust. A number of PM-bearing raw materials are also processed, including e-waste in various qualities and dimensions (Fig. 8.25). Aurubis fabricates products made of recycling materials at several of its sites (Lünen and Hamburg in Germany, Olen in Belgium, and Pirdop in Bulgaria).

Aurubis procures Cu and Cu alloy scrap as well as Cu-bearing residues. Cu scrap is used as input in the convertors and anode furnaces of primary and secondary smelting processes, while alloy scrap and residues are used in Aurubis AG's Kayser Recycling System (KRS). Aurubis recovers Cu and PMs from the delivered WEEEs

Fig. 8.24 Hellatron WEEE process flowsheet

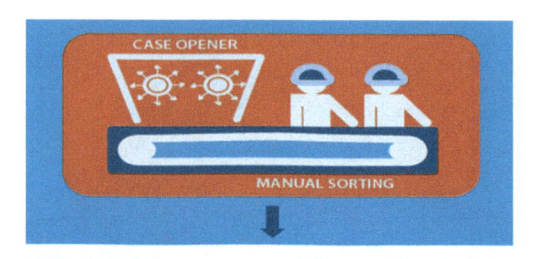

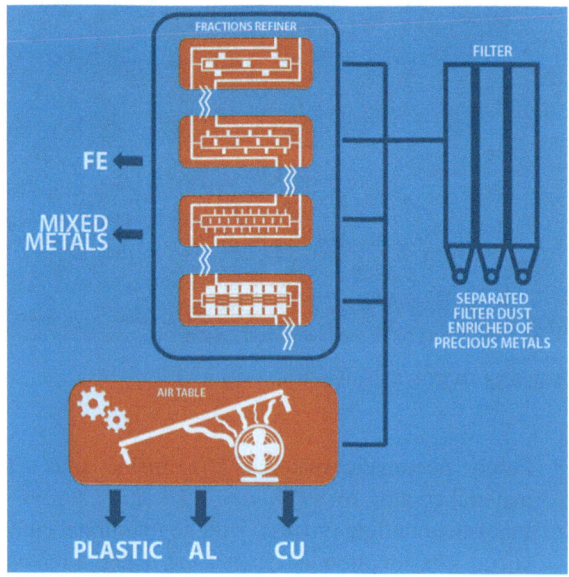

with environmentally sound methods. The KRS is well suited for utilizing recycling materials with low Cu and PM contents and very complex materials such as e-waste. During material preparation, very coarse, moist, or dusty materials can be treated. Al and plastics are separated from the materials in some cases and sold to other recycling companies.

Pyrometallurgical preparation – smelting and refining – begins in the KRS. The central operation is a submerged lance furnace which is almost 13 m high. A special feature is the use of a submerged combustion lance, which is immersed into the furnace from above and supplies the process with heating oil, oxygen, and air. The reduction process is very fast in the submerged lance furnace. Charging times are short. The iron silicate sand extracted in that step of the process has very low residual Cu contents. Cu, Ni, Sn, Pb, and the PMs contained in the raw materials are enriched in an alloy with a Cu content of about 80% (i.e., black Cu). In a top-blown rotary converter (TBRC), the Cu content is further enriched to 95%, and Sn and Pb are separated into a slag. The Sn-Pb slag is subsequently processed into a Sn-Pb alloy in the directly connected Sn-Pb furnace. During the KRS process, Zn is enriched in the KRS oxide, a flue dust (Fig. 8.26) [21].

8.14 Aurubis Recycling Center in Lünen (Germany)

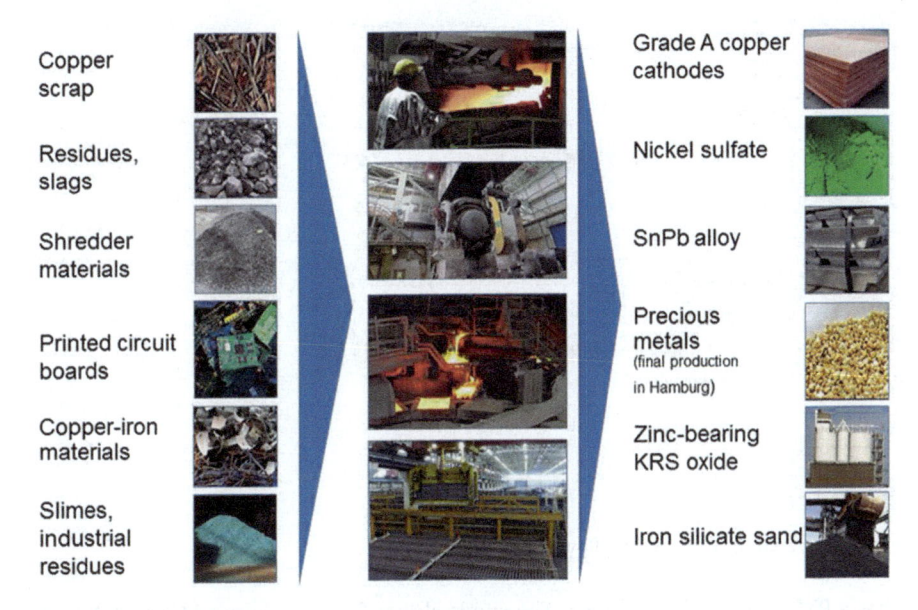

Fig. 8.25 Aurubis Recycling KRS system feed and products

Unwrought Cu is produced with an average Cu content of 95% and is further processed in the anode furnace in a molten form. It is refined with additional amounts of Cu scrap. The melt is initially oxidized with air and oxygen and then deoxidized with natural gas after the slag that has formed has been removed. At the end of the pyrometallurgical process, the meanwhile 99% pure Cu is cast into Cu anodes. This is the starting product for the final refining stage of secondary Cu production, the Cu tank house process, where high-grade Cu cathodes are produced in a quality identical to Cu cathodes from primary Cu production. Important by-products, in particular Au and Ag, are enriched in the anode slimes. Ni is extracted as crude nickel sulfate from electrolyte treatment.

Cu scrap, e-waste, and residues are used at Aurubis Hamburg (Germany). The basic material for Hamburg's secondary Cu production process consists of a variety of recycling materials rich in PMs as well as intermediate smelter products originating both from Aurubis' production plants and from external metal smelters and PM-separating plants. The focus of secondary Cu production in this case is on enriching and recovering PMs and separating various by-metals that result from the Cu production process. KRS is well suited to recycle materials with low levels of Cu and PMs.

In accordance with the requirements of the specific raw materials, these materials are processed in a modern electric furnace in various smelting campaigns. The most important target is the pyrometallurgical separation of Pb and Cu and the enrichment of PMs. By-elements still existing during Cu production, such as Pb Bi, Sb, and Te, are separated in the connected Pb refinery and sold as Pb bullion, Pb-Bi alloy, Sb concentrates, and Te concentrates. The PMs are fortified in a so-called rich Pb, which

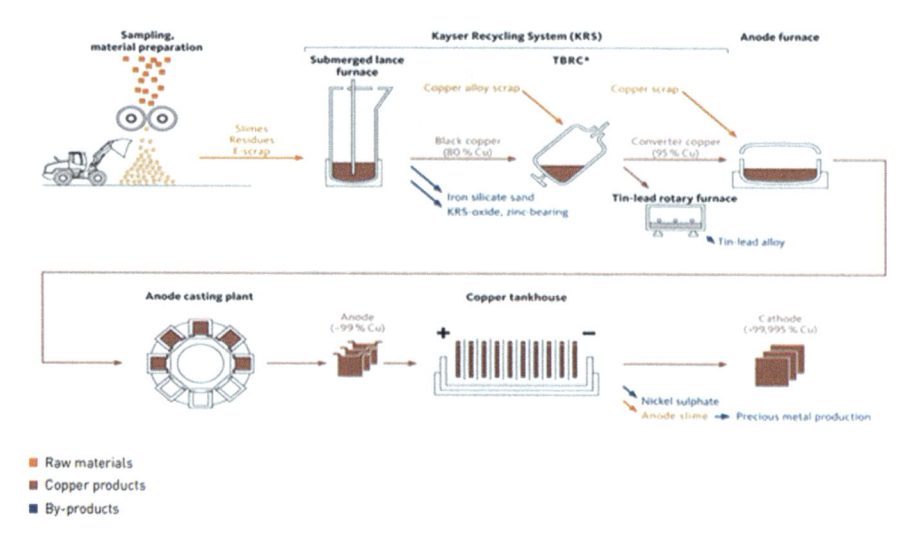

Fig. 8.26 Aurubis Cu recycling flowsheet [13]

has a PM content of about 70%. The anode slimes from the Aurubis Cu tank houses are processed together with the rich Pb from the Pb refinery in the PM production plant. With Aurubis' modern, environmentally friendly PM production facilities, the company covers a wide range of feed materials rich in PMs. Thus, Aurubis also processes anode slimes from other Cu smelters, bullion, PM, and coin scrap as well as PM-bearing sweeps and slag. In addition to fine Ag, fine Au, and concentrates of the PGMs, sales products from this process include wet Se which is processed at the Aurubis subsidiary Retorte GmbH.

8.15 Attero Recycling, Roorkee, India

With 2.7 million tons of e-waste generated, India now stands as the fifth largest e-waste generator in the world. Old mobile phones to the tune of 170 million are discarded or replaced annually contributing heavily to ever-growing e-waste. A recent report by the Centre for Science and Environment (CSE) found that Moradabad gets over 50% of India's total discarded PCB e-waste; and over 50,000 people are involved in the business of informal/illegal recycling [22]. India loses 50% of the Au during crude dismantling, according to GeSI and Solving the E-waste Problem (StEP), an international initiative that finds solutions to e-waste management problem. According to Chaturvedi, 95% of the country's e-waste is recycled by scrap dealers, but they extract not more than 15% of the PMs. India makes huge losses considering the country's projected growth for e-waste generation is 34% every year, according to reports by nonprofit Toxics Link. Mumbai, the highest e-waste-producing city in the country, throws away 19,000 tons of e-waste in a year.

Business trends show that by 2020, e-waste from mobile phones will be an astounding 18 times the current level. Computer waste will rise by 500% [23].

The biggest e-waste recycling unit in the country, Roorkee-based Attero Recycling, has the capacity to treat 12,000 tons of e-waste in a year. Attero recycles mobile phones, WPCBs, display units, batteries, and IT goods. The company claims it uses an environmentally sound technology, but technical experts differ. There are a few more such as Mumbai-based Eco Recycling Limited and Bengaluru-based E-Parisaraa Private Ltd. which have capacities of 7200 tons per annum and 1800 tons per annum, respectively. But they are still upgrading their technology.

WPCBs are complex assemblies that include numerous materials, and these materials require large quantities of energy and other materials to manufacture. They also include significant quantities of metals such as Cu, Pb, and Ni. While some of these are toxic in nature, all of them are valuable resources. The recycling WPCBs begin with the circuit boards being passed through the Component Removal Machine. This machine automatically separates all the assembled components on the circuit board. It is then pulverized and shredded. The output from these processing stages is the blank board and its components. The bare/blank board primarily contains Cu, which is smelted and electro-refined to obtain 99.9% pure Cu bars. The separated components are classified as heavy and light chips and include transistors, diodes, connectors, and miscellaneous parts such as capacitors and heat sinks among others. Magnetic separation is carried out to separate ferrous metals as well as some of the Cu alloys. This is followed by ECS to separate nonferrous metals. All of these components are smelted to obtain pure metals. Plastic components are separated through density-based separation [24].

8.16 Noranda Smelter in Quebec, Canada

It is currently owned by Xstrata and treats about 100,000 tons of e-waste annually in addition to its feed of mined Cu concentrates, and subsequently 99.1% pure Cu is produced.

8.17 Rönnskar Smelter in Sweden

The Rönnskar smelter in Sweden also recycles both, high- and low-grade Cu scrap, accounting for 100,000 tons of Cu per year via two separate furnaces (the Kaldo furnace handles the e-waste fraction). The respective flowsheets are given by Khaliq et al. (2014) [25]. Smelter system produces a mixed Cu alloy, which will be refined to extract metals, such as Cu, Ag, Au, Pd, Ni, Se, Zn, Pb, Sb, In, and Cd. Gas cleaning and handling systems are also integrated, which capture off-gases to carry out additional combustion and thermal energy recovery. Rönnskar smelter uses

Table 8.3 Comparison of academic research and industrial practices

	Academic research	Industrial practices
Target materials	Mainly metals	Metals, glass, ABS/PC
Disassembly	Manual	Manually and automatic
Size reduction	Fine particle size (<5 mm)	Larger particle size (<50 mm)
Material extraction	Mainly hydrometallurgy	Mainly pyrometallurgy
Recycling rates	Metals (80–99%) by acid; Au (15–98%) by thiosulfate; Au (39–95%) by thiourea; Au (10%) and Cu (11–72%) by microoganisms	Au (>95%) by smelting; Au (80%) by biomining; Precious metals (98%) by eVOLV

fossil fuel as both a reducing agent and fuel, and thus, shredded e-waste scrap (−30 mm) was considered as an energy recovery source in addition to the usual metal extraction process.

8.18 Comparison of Academic Research and Industrial-Scale E-Waste Recycling Practices

Based on reported information on current industrial recycling practices for recovering materials from WPCBs, generic comparisons between academic research and industrial practices are summarized in Table 8.3. Academic researches focus on mainly metals, while industry focuses on metals, glass, and plastics. Dismantling in the industry is performed both manually and automatically. Coarser sizes are used in the industry. Hydrometallurgy is the main metal extraction route for academic research, and pyrometallurgy is the main extraction route for industry.

In industrial-scale e-waste recycling, mixed type of WEEE/WPCB material is feed in a bunker. Primary shredding to approximately 150 mm via rotor shear is performed. Shredded material spreads via vibrating feeder on a conveyor for hand picking to remove contaminants (such as batteries and hazardous waste). An inclined belt is used to transfer material to the secondary shredder with noise/dust control systems. Materials are shredded approximately 25–30 mm. Cu wool is sorted via vibratory feeder. Heavy and light fractions (e.g., foils, dust) are sorted by zigzag air shifter. Light fractions are discharged into big bags. Ferrous fraction in heavy materials is separated by overband magnet during conveying. Rare-earth Nd drum magnet and eddy current separator are used to separate nonferrous fraction and remaining waste plastic. Grinding to minus 8 mm is performed via a hammer mill. Ground material is sieved into 4 and 8 mm by screening. Density shifters are used to sort Al, Cu, Ni, and brass.

References

1. Kaya M (2018) Current WEEE recycling solutions. In: Veglio F, Birloaga I (eds) Waste electrical and electronic equipment recycling, aqueous recovery methods. Elsevier Science, pp 33–93. https://doi.org/10.1016/B978-0-08-102057-9.00003-2
2. http://www.sxrecycle.com/
3. http://www.varygroup.com/product/15
4. Mesker CEM, Hagelüken C, Van Damme G (2009) Green Recycling van EEA: Special en Precious Metal EEE. In: Howard SH, Anyalebechi P, Zhang L (eds) TMS 2009 annual meeting & exhibition, San Francisco, California, USA, EPD Congress 2009 Proceedings. TMS, Warrendale, pp 1131–1136. ISBN: 978-0-87339-732-2
5. http://www.ewasteguide.info/files/UNEP_2009_eW2R.PDF
6. http://www.mgg-recycling.com/?page_id=483
7. http://eldan-recycling.com/en/electronic-waste-recyclin
8. Kellner D (2009) Recycling and recovery. In: Hester RE, Harrison RM (eds) Electronic waste management, design, analysis and application. RSC Publishing, Cambridge, pp 91–110
9. Yang M (2013) Study on the technologies and mechanism of copper and stannum extraction from waste printed circuit boards by PEG-NOx catalysis and oxidation. Ph.D. Dissertation, Donghua University
10. Yokoyama S, Iji M (1993) (Resources and Environmental Protection Research Laboratories NEC Corp., Japan), Proceedings of the IEEE/Tsukuba International Workshop on Advanced Robotics, Tsukuba, Japan, 8–9 Nov 1993
11. http://www.dowa.co.jp/en/products_service/metalmine.html
12. http://www.dowa-csr.jp/csr2010/html/english/special02.html
13. http://www.dowa-csr.jp/csr2014/html/english/csr/about_eco.html
14. https://www.epa.gov/sites/production/files/2014-05/documents/handout-10circuitboards.pdf
15. http://seprourbanmetals.com/solutions/e-waste/
16. https://www.pcworld.co.nz/slideshow/524767/pictures-old-electronics-don-t-die-they-pile-up/
17. Wang JB, Xu ZM (2015) Disposing and recycling waste printed circuit boards: disconnecting, resource recovery and pollution control. Environ Sci Technol 49:721–733. https://doi.org/10.1021/es504833y
18. http://www.swissrtec.ch
19. https://www.weeemetallica.com/our-solution
20. http://www.hellatron-recycling.it/solution/2/electric-eletronic-waste
21. https://www.aurubis.com/en/products/recycling/technology
22. http://attero.in/blogs/ewaste-free-swachh-bharat/
23. http://www.downtoearth.org.in/coverage/wasted-e-waste-40440
24. http://www.attero.in/recycling-solutions.html
25. Khaliq A, Rhamdhani MA, Brooks G, Masood S (2014) Metal extraction processes for electronic waste and existing industrial routes: a review and Australian perspective. Resources 3(1):152–179. https://doi.org/10.3390/resources3010152

Chapter 9
Recycling of NMF from WPCBs

"By recycling you can change tomorrow, today"
Anonymous

Abstract Cheap nonmetal fraction (epoxy resin, glass fiber, ceramics, etc.) accounts for 70% of e-waste/WPCB. It contains most of the hazardous brominated flame retardants. This chapter covers the direct and chemical recycling of NMF from WPCBs. Chemical recycling consists of combustion (smelting/incineration), pyrolysis, and depolymerization processes by using supercritical fluids, hydrogenolytic degradation, plasma treatment, hydrothermal methods, and gasification process. The purpose of thermal treatment of e-waste is elimination of plastic/epoxy resin components. Details of pyrolysis and hydrothermal depolymerization of WPCBs with different solvents are presented in this chapter.

Keywords Nonmetal fraction (NMF) · Chemical recycling · Pyrolysis · Depolymerization · Solvents · Subcritical fluids

9.1 Direct and Chemical Recycling of NMF from WPCBs

Despite diversity possessed by MF recycling techniques, a common drawback shared by them is the absence of NMF treatment, which will give rise to the toxic emission including BFR organic compounds, secondary particulates, and other kinds of pollutants. Therefore, there is a need for a safe NMF recycling technology. The effective separation for MF and NMF is indispensable for the purpose of more detailed and efficient recycling. At the beginning, the purpose of recycling of NMF from WPCB is simply to avoid the potential hazardous environmental impacts when they end up in landfill or incineration. The composition of NMF is a random combination of epoxy resin and glass fiber. The reuse of NMF is hindered due to understanding of the chemistry of them. Therefore, it is common to be directly used as filler, concrete, and modifier which means NMF was used without any forms of modification and is a low value-added material [1–9]. The modification of NMF to make value-added products (i.e., catalyst, absorbent, and filter support) can greatly improve the value of NMF. Thus, the value added to NMF can possibly compensate for the high cost derived from mechanical separation.

9.2 Direct Recycling of NMF from WPCBs

The direct recycling of NMF can help solve the disposal problem of NMF. However, it is still weak for them to compensate for the high operation cost of the MF and NMF separation due to the low value-added products they produced by NMF. With the requirement to better use NMF, which takes more than half of the weight of NMF, the modification of NMF followed by the upgrading was studied by many research groups.

Guo et al. (2008) has done a series of studies on the feasibility of replacing wood flour in the production of phenolic molding compound (PMC) using the NMF of WPCB [6, 7]. The addition of 20% NMF can significantly improve the impact strength and heat deflection temperature (HDT) and reduces flexural strength and Raschig fluidity of the phenolic molding. Yokoyama and Iji (1995) also investigated the use of NMF as filler in the reproduction of resin-type construction materials, and the comparison of mechanical properties of these materials with those of reference materials with silica powder was conducted [3]. The NMFs in the materials show reinforcement in mechanical strength and thermal expansion properties of the epoxy resin mold, which has better performance than talc, calcium carbonate, and silica. This is probably because of the compatibility between the NMFs and the epoxy resin matrix and also the incorporation of glass fiber.

Niu and Li (2007) used recycled WPCBs for cement solidification, which is actually a method to use the WPCBs as a raw material for concrete [9]. It is proved that the cement solidification can be significantly improved in terms of the compressive strengths (4.89 MPa for slag cement and 7.93 MPa for Portland cement) and the impact resistance of 200 (maximum) of NMF can turn it into strong monoliths. The leaching test shows that the leaching of Pb in the raw material can be effectively prevented (<5 mg l^{-1}) even under an acidic environment.

Zheng et al. (2009) used NMFs as reinforcing fillers in PP composites [8]. The NMFs are modified by a silane coupling agent KH-550, and PP powder S1003 is applied as the matrix polymer. NMF and PP powder are premixed and the products are obtained by extruding. The mechanical characterization shows that the tensile strength, tensile modulus, flexural strength, and flexural modulus are significantly improved 28.4%, 62.9%, 87.8%, and 133.0%, respectively, by the adding of NMF. The optimum content of the NMF added is 30 wt.% based on technical, environmental, and economic consideration.

However, Lu et al. (2000) found that WPCB can be liberated effectively with a size between 0.5 and 1.2 mm, while for a better recovery ratio, the over-pulverized phenomenon is very serious. This limits the further use of the glass fibers that are the support material of PCB (about 50 wt.%) [10]. The mechanical properties of NMF were sacrificed in order to achieve a high MF recovery rate. Therefore, the direct reuse of NMF is seriously influenced due to this phenomenon. A more advanced technology of NMF modification is indispensable.

9.3 Chemical Recycling of NMF from WPCBs by Degradation/Modification/Depolymerization of Thermoset Organic Polymers by Solvents

Other than landfill or combustion, the NMF can be reused in different fields, such as building materials, additives, etc. Chemical NMF recycling refers to decomposition of the waste organic polymers into monomers or useful chemicals by means of chemical reactions at high temperatures. Chemical recycling consists of combustion (smelting/incineration), pyrolysis, and depolymerization process by using supercritical fluids, hydrogenolytic degradation, plasma treatment, hydrothermal method, and gasification process. The purpose of thermal treatment of e-waste is elimination of plastic/epoxy resin components. Additionally, the gaseous product of the process after cleaning may be used for energy recovery or as syngas. The refining of the products (gases and oils) is included in the chemical recycling process and can be done with conventional refining methods in chemical plants. Metal fractions in the ash can be treated by pyrometallurgical and hydrometallurgical approaches. Biotechnological processes are being still in their infancy.

One thought for the modification is based on the thought that NMF is a carbon (C) source for potential reuse considering the high content of C in its composition due to the addition of epoxy resin or other polymers. Normally, the C content in NMF is between 30% and 40%, which varies with the source of WPCB [5]. Currently, studies on thermochemical methods to recycle WPCBs are being investigated to convert the resin fraction into monomers and to recover the metals. Pyrolysis processing has been investigated where the polymer can be thermally degraded to produce oils and the metals collected in the char after the reaction [11]. However, the metals are often recovered together with the char, so that further processing via separation processes becomes necessary, along with Br containing emission gas. In addition, for resins containing fire retardant bromine compounds, the liquid phase contains Br, which thereby contaminates the product oil. Gasification has been investigated as a thermochemical treatment method which converts the organic resin into gas products, mostly CO and H_2. However, high temperatures are used to gasify all of the polymer fractions [12].

9.4 Pyrolysis

Pyrolysis of polymers leads to the formation of gases, oils, and chars which can be used as chemical feedstock or fuels. Pyrolysis degrades the organic part of the unpopulated PBC wastes, making the process of separating the organic, metallic, and glass fiber fractions of WPCBs much easier and recycling of each fraction more viable. Additionally, if the temperature is high enough, the pyrolysis process will melt the solder remaining on the WPCBs. The combination of the removal and recovery of the organic fraction of WPCBs and the removal of the solder aids the separation of metal components.

Zhou and Qui (2010) conducted vacuum pyrolysis of WPCBs at 240 °C. Pyrolysis solid yield was between 70% and 76%, liquid yield was between 20% and 28%, and gas yield was between 2.7% and 4.3%. Gaseous product contains CO, CO_2, CH_4, and H_2. The residue of pyrolysis contained various metals (e.g., Cu, Fe, Al, Au, Ag, Pt, Pd, etc.) that could be further recovered by leaching [13]. In addition, pyrolysis significantly reduces the volume of the waste which is beneficial for further treatment. The combination of mechanical separation, pyrolysis, and chemical processing could increase efficiency of e-waste recovery.

Thermal behavior of epoxy resins, the most common polymer matrix in WPCB, has been widely investigated as a basis for pyrolytic recycling. In thermogravimetry brominated epoxy resins are less thermally stable than the corresponding unbrominated ones. They exhibit a steep weight loss stage at 300–380 °C depending on the hardener, those hardened by aromatic amines and anhydrides decomposing at higher temperature [14, 15]. Mostly brominated and unbrominated phenols and bisphenols are found in the pyrolysis oil; however, the balance between phenols/bisphenols and brominated/unbrominated species depends on the temperature and residence time in the reactor. Higher temperatures and longer times make debromination more extensive [16, 17]. WPCB particle size also affects the decomposition temperature and degradation is postponed when particles are larger than 1 cm^2 due to heat transfer limitation [18].

Ke et al. (2013) pyrolyzed NMF of WPCB in the temperature range of 500–800 °C and used physical or chemical activation to obtain activated C [19]. Physical activation with H_2O as an activation reagent produced granular activated C with a surface area as high as 1019 m^2 g^{-1} and pore volume of 1.1 cm^3 g^{-1}, while chemical activation with KOH as activation reagent obtained the same product with higher surface area of 3112 m^2 g^{-1} and a pore volume of 1.13 cm^3 g^{-1}.

Rajagopal et al. (2016) used the activated C prepared by physical activation with CO_2 subsequent to pyrolysis of NMF of WPCB and apply it in supercapacitor [20]. The NMF activated at 850 °C for 5 h shows the highest surface area of 700 m^2 g^{-1} as well as 0.022 cm^3 g^{-1} pore volume. Electrochemical characterization of the activated C prepared under optimum conditions shows a specific capacitance of 220 F g^{-1} at the current density of 30 mV s^{-1} and 156 F g^{-1} at the current density of 100 mV s^{-1}, which is comparable to activated C prepared by other methods. Also,

9.4 Pyrolysis

the activated C has an energy density of 15.84 Wh kg^{-1} at 850 W kg^{-1}, which is very high compared to commercially available supercapacitors based on activated carbons that have an energy density ranging from 4 to 5 Wh kg^{-1} with power density values of 1 to 2 KW kg^{-1} [21]. Moreover, the retention value of the activated C was studied to be 98% for over 1000 cycles.

Hadi et al. (2003) have developed a thermal-alkaline activation process to functionalize NMF to produce an aluminosilicate adsorbent for HM uptake from wastewater [22]. NMF was mixed with KOH solution and activated in 300 °C to develop the porosity of NMF which is a nonporous material. After activation, equilibrium isothermal adsorption tests show that the modified novel material called ANMF had a high uptake capacity for Cu (2.9 mmole g^{-1}), Pb (3.4 mmole g^{-1}), and Zn (2.0 mmole g^{-1}), which is much higher than commercial adsorbents used in industry. Xu et al. (2014) studied the factors affecting the adsorption capacity including contact time, initial Cd ion concentration, pH, and adsorbent dosage when this material was used for Cd uptake [23]. The results showed that pH had an important effect on the Cd uptake capacity, and the maximum uptake capacity for Cd was 2.1 mmole g^{-1} obtained when pH: 4. This provides a possibility for the porous structure tuning for the NMF of WPCB, which will significantly enlarge the potential applications of NMF as a catalyst, adsorbent, and filter support.

Disposal of WPCBs is regarded as a potential major environmental problem due to their heavy metal contents. Therefore, recycling WPCBs represent an opportunity to recover the high-value resin chemicals and the high-value metals that are present. The resin is predominantly a thermosetting polymer such as phenolic- and cresol-based epoxy, bisphenol A, epoxy resin, or cyanate esters and polyamides. Hydrometallurgical treatment by acid/alkali dissolution destructs nonrenewable organic resin resources. The nonmetallic fraction, which is the resin fraction of the board, can find application areas as a filler for thermosetting resin composites [24], a reinforcing filler for thermoplastic resin composite materials [8], as a raw material for concrete [9, 25, 26] or as a modifier for viscoelastic materials [3].

Other than the conventional physical processes mentioned above, attention to supercritical water, whose temperature and pressure are over its critical point (374 °C, 218 atm), has been drawn as a technique to pretreat WPCBs before chemical processes. Compared with ambient liquid water, several unique properties, such as lower dielectric constant, lower energy of hydrogen bond, and high solubility of organic compound, were possessed by water near its critical point. Reactions could occur in a homogeneous supercritical phase because of the high miscibility of organic compounds in supercritical water, which contributes to a fact that there are no interphase mass transport limitations to lower reaction kinetics. Moreover, high pressure and temperature of supercritical water provide high enough velocity and temperature rise to destroy particle structure and make particles more porous. These further result in the diffusion of supercritical water through particle structures and the creation of big hole structures [27, 28]. Typically, the supercritical techniques applied to recycling of WPCBs contain supercritical water oxidation (SCWO), which is in the presence of oxygen gas, and supercritical water depolymerization (SCWD) under reducing atmosphere [27].

Xiu et al. (2013) investigated both SCWO and SCWD with regard to the separation of organic substances from metals and ceramics [27]. The effects of temperature, retention time, and initial pressure were studied. It appeared that as temperature increased, the weight loss of solid phase increased for both methods. However, the degradation rate of organic matter on behalf of SCWO was higher than that under the SCWD process, which could be attributed to the combination of hydrolysis and oxidation in the case of SCWO process, compared with SCWD where only hydrolysis occurred. In addition, polymers with high molecular weight were eventually decomposed into low molecular compounds along with SCWD; however, in the process of SCWO, most organic matter was eventually oxidized into CO_2 and H_2O, resulting in high pressure generated in the reactor. The higher final pressure in the process of SCWO also provided an explanation that SCWO was a superior method. Xiu et al. (2013) also found HCl could be applied to acid leaching of BMs after the pretreatment of SCWO. That could be explained by the fact that SCWO converted Cu metal to copper oxides that could be soluble in HCl, even though HCl is a nonoxidizing acid [27]. However, since water not only has high critical temperature and pressure but also unique properties, such as ion product and dielectric point, which could contribute to high requirement of reactors, several alternatives are undertaken.

Xiu and Zhang (2010) used advanced degradation of brominated epoxy resin and simultaneous transformation of glass fiber from WPCBs by SCWO and supercritical methanol (SCMO) oxidation process [28]. The reasons for utilizing methanol as a supercritical fluid are that the critical point of methanol is 240 °C, 8.09 MPa, which is lower than that of water. Additionally, the boiling point of methanol is lower than that of water. The results indicated that temperature, pressure, reaction time, and the solid/liquid ratio affected the performance of supercritical methanol. The highest conversion happened either at 380 °C with 120 min or 420 °C with 60 min. The difference with the two conditions was that the oil recovery at higher temperature with shorter reaction time was less than that at low temperature with longer reaction time. Based on the elemental analysis of ICP-OES, it turned out that most metals were concentrated after the supercritical methanol process. Particularly, the content of Cu after the treatment was approximately three times of that in the original material, and Ag were significantly concentrated up to 7902 ppm [29]. For SCMO recycling, most of the heavy metals are converted to metal oxides with complete conversion, except for Mn and Ni (50%). The Cu extracted from WPCB will undergo an electrokinetic process and can be synthesized to nanoparticles, which functions as a photocatalyst.

Xing and Zhang (2012) also used SCWO to detoxify bromorganic compounds inside WPCB and achieved maximum debromination rate of 97.8% [30]. The bromide was concentrated in water in the form of HBr. Also, the bromide free oil with main components consisting of phenol (58.5%) and 4-(1-methylethyl)-phenol (21.7%) was collected. Moreover, Cu was recovered in the purities of near 95% with a recovery rate as high as 98.1%. However, supercritical technologies need high temperature (>400 °C) and high pressure (>20 MPa). Thus, this process needs high-quality equipment, which increases the cost significantly.

9.4 Pyrolysis

Table 9.1 The hydrothermal depolymerization of WPCBs with different solvents

Solvent	Temperature (°C)	Time (h)	Resin removal (%)	Gas produced (g/g WPCB)
Ethanol (C_2H_5OH/ C_2H_6O) [structure shown] Critical point: 241 °C, 6.14 Mpa	300 400	3 3	56 50	0.24 2.84
Acetone (C_3H_6O) [structure shown] Critical point: 235 °C, 4.71 Mpa	300	3	37	–
Water (H_2O) Acetic acid/water KOH/water NaOH/water Critical point: 373 °C, 23.3 Mpa	400 400 400 400	0 0 0 0	81.0 81.9 93.6 94.1	0.28 0.34 0.30
Tetralin ($C_{10}H_{12}$)	340	5	99	

Utilizing sub- and supercritical fluids to hydrothermal depolymerize the resin fraction of the WPCB represents a potentially cleaner and cheaper technology for recycling the waste, as it enables the recovery of metals in addition to converting the resin into valuable chemicals feedstock and/or fuels. The cleaned mixture of metals obtained would also become easier to process or separate. There are various studies with different solvents which have been investigated for the treatment via depolymerization of plastic wastes reported in the literature. The most common solvents are water and alcohols, and the process has been mainly applied to pure polymers such as PET rather than WPCBs. The solvothermal depolymerization of WPCBs can be carried out using water, ethanol, acetone, tetraline, etc. between 300 and 400 °C. When water is used, alkalis (NaOH and KOH) and acetic acid can be used as additives to promote the removal of the resin fraction of the WPCBs (Table 9.1) [31]. At 400 °C, 81% of resin removal was achieved when water alone was used, and 94% of resin removal was achieved in the presence of NaOH and KOH in water as the solvent. Addition of alkalis to the water solvent increased the phenol yield up to 62.4%, and the residues were recovered in a clean state, ready for metal separation. Gas products consisted of CO_2, H_2, CO, CH_4, and C_2H_2 hydrocarbons. The organic content of the liquid obtained from depolymerization contains phenol and phenolic compounds which were the precursors of the original thermosetting resin. Brominated plastics ended up mostly in the aqueous phase, which results in oil recovery with near-zero Br content, due to dissolution in the water medium. The clean residue could be further processed for the recovery of valuable metals, such as Cu, Ag, Au, Pt, etc.

Today, there are very few possibilities for the recycling of cross-linked polymers (such as epoxy resins reinforced with glass or C fibers). This is because these materials are insoluble and infusible. It is very difficult to cleave cross-linked polymers into soluble compounds. Epoxy resins can be liquefied by transfer of hydrogenation with various hydrogen donors such as tetraline (1,2,3,4-tetrahydronaphthaline) ($C_{10}H_{12}$), 9,10-dihydroanthracene ($C_{14}H_{12}$), or indoline (C_8H_9N). Degradation liquid products analyzed by GC/MS were bisphenol A and its fragments phenol and p-isopropylphenol A as well as phthalic anhydride and its fragments benzoic acid and benzene [32]. Reinforced epoxy resin circuit boards covered with Cu foil on both sides and reinforced with 59.5% glass fiber mats was liquefied in 1 * 5 (W * L) cm in size in an autoclave. At 340 °C without grinding WPCB pieces, 99% of the resin was liquefied for feedstock recycling by hydrogenolysis using tetraline and 9,10-dihydroanthracene (Table 9.1). After cleaning with C_4H_8O, the unharmed glass fiber mats and Cu foil were recovered with very little contamination [32]. When adding an amine like ethanolamine to the reaction mixture, cross-linked epoxy resins can be liquefied at 280 °C in 24 h.

Simultaneous recovery of high-purity Cu and PVC from electric cables by diisononyl phthalate (DINP) plasticizer extraction using diethyl ether, methanol, and hexane and ball milling is achieved by Xu et al. (2018) [33]. Maximum separation rate of 77% and highest Cu purity at 150–600 μm of 99% were achieved using samples of which 100% of the plasticizer had been extracted from PVC resin at 5 h by diethyl ether, because the PVC become brittle after plasticizer removal. Huang et al. (2009) used DC arc plasma to treat WPCBs, which decomposed the WPCB in molten bath at 1400–1800 °C in a DC arc furnace [34]. The product from this process is homogeneous and vitreous slag and small molecular gases (i.e., HCl, H_2S, NO_x, and SO_2). Although the safe disposal of solid residue is achieved, it produces a certain amount of air pollution, and the energy consumption is huge due to the high temperature and long duration time.

References

1. Guo J, Cao B, Guo J, Xu Z (2008) A plate produced by nonmetallic materials of pulverized waste printed circuit boards. Environ Sci Technol 42:5267–5271. https://doi.org/10.1021/es800825u
2. Guo J, Guo J, Cao B, Tang Y, Xu Z (2009) Manufacturing process of reproduction plate by nonmetallic materials reclaimed from pulverized printed circuit boards. J Hazard Mater 163:1019–1025. https://doi.org/10.1016/j.jhazmat.2008.07.099
3. Yokoyama S, Iji M (1995) Recycling of thermosetting plastic waste from electronic component production processes. In: Proceedings of the 1995 IEEE International Symposium on Electronics and the Environment, ISEE, pp 132–137

4. Mou P, Xiang DDG (2007) Products made from nonmetallic materials reclaimed from waste printed circuit boards. Tsinghua Sci Technol 12:276–283. https://doi.org/10.1016/S1007-0214(07)70041-X
5. Ban BC, Song JY, Lim JY, Wang SK, An KG, Kim DS (2005) Studies on the reuse of waste printed circuit board as an additive for cement mortar. J Environ Sci Health A Tox Hazard Subst Environ Eng 40:645–656
6. Guo J, Li J, Rao Q, Xu Z (2008b) Phenolic molding compound filled with nonmetals of waste PCBs. Environ Sci Technol 42:624–628. https://doi.org/10.1021/es0712930
7. Guo J, Rao Q, Xu Z (2008c) Application of glass-nonmetals of waste printed circuit boards to produce phenolic molding compound. J Hazard Mater 153:728–734. https://doi.org/10.1016/j.jhazmat.2007.09.029
8. Zheng Y, Shen Z, Cai C, Ma S, Xing Y (2009) The reuse of nonmetals recycled from waste printed circuit boards as reinforcing fillers in the polypropylene composites. J Hazard Mater 163:600–606. https://doi.org/10.1016/j.jhazmat.2008.07.008
9. Niu X, Li Y (2007) Treatment of waste printed wire boards in electronic waste for safe disposal. J Hazard Mater 145:410–416. https://doi.org/10.1016/j.jhazmat.2006.11.039
10. Lu MX, Zhou CH, Liu WL (2000) The study of recovering waste printed circuit board by mechanical method. Tech Equip Environ Pollut Control 10:30–35
11. Pimenta S, Pinho ST (2011) Recycling carbon fiber reinforced polymers for structural applications: technology review and market outlook. Waste Manag 31:378–392. https://doi.org/10.1016/j.wasman.2010.09.019
12. Hall WJ, Williams PT (2007) Separation and recovery of materials from scrap printed circuit boards. Resour Conserv Recycl 51(3):691–709. https://doi.org/10.1016/j.resconrec.2006.11.010
13. Zhou Y, Qiu K (2010) A new technology for recycling materials from waste printed circuit boards. J Hazard Mater 175(1–3):823–828. https://doi.org/10.1016/j.jhazmat.2009.10.083.15
14. Yamawaki T (2003) The gasification recycling technology of plastics WEEE containing brominated flame retardants. Fire Mater 27(6):315–319. https://doi.org/10.1002/fam.833
15. Luda MP (2011) Recycling of printed circuit boards. In: Kumar S (ed) Integrated waste management, vol II. InTech, Rijeka. https://cdn.intechopen.com/pdfs-wm/18491.pdf
16. Luda MP, Balabanovich AI, Zanetti M, Guaratto D (2007) Thermal decomposition of fire retardant brominated epoxy resins cured with different nitrogen containing hardeners. Polym Degrad Stab 92(6):1088–1100. ISSN: 01413910
17. Luda MP, Balabanovich AI, Zanetti M (2010) Pyrolysis of fire retardant anhydride cured epoxy resins. J Anal Appl Pyrolysis 88(1):39–52. https://doi.org/10.1016/j.jaap.2010.02.008
18. Quan C, Li A, Gao N (2009) Thermogravimetric analysis and kinetic study on large particles of printed circuit board wastes. Waste Manag 29:2253–2360. https://doi.org/10.1016/j.wasman.2009.03.020
19. Ke YH, Yang ET, Liu X, Liu CL, Dong WS (2013) Preparation of porous carbons from nonmetallic fractions of waste printed circuit boards by chemical and physical activation. Xinxing Tan Cailiao/New Carbon Mater 28:108–114. https://doi.org/10.1016/S1872-5805(13)60069-4
20. Rajagopal RR, Aravinda LS, Rajarao R, Bhat BR, Sahajwalla V (2016) Activated carbon derived from non-metallic printed circuit board waste for supercapacitor application. Electrochim Acta 211:488–498. https://doi.org/10.1016/j.electacta.2016.06.077
21. Burke A (2007) R&D considerations for the performance and application of electrochemical capacitors. Electrochim Acta 53:1083–1091. https://doi.org/10.1016/j.electacta.2007.01.011
22. Hadi P, Barford J, McKay G (2013) Toxic heavy metal capture using a novel electronic waste-based material-mechanism, modeling and comparison. Environ Sci Technol 47:8248–8255. https://doi.org/10.1021/es4001664
23. Xu M, Hadi P, Chen G, McKay G (2014) Removal of cadmium ions from wastewater using innovative electronic waste-derived material. J Hazard Mater 273:118–123. https://doi.org/10.1016/j.jhazmat.2014.03.037

24. Tohka A, Lehto H (2005) Mechanical and thermal recycling of waste from electric and electronic equipment, Helsinki University of Technology, Department of Mechanical Engineering. Energy Engineering and Environmental Protection Publications, Espoo
25. Siddique R, Khatib J, Kaur T (2008) Use of recycled plastic in concrete: a review. Waste Manag 28(10):1835–1852. https://doi.org/10.1016/j.wasman.2007.09.011
26. Panyakapo P, Panyakapo M (2008) Reuse of thermosetting plastic waste for lightweight concrete. Waste Manag 28(9):1581–1588. https://doi.org/10.1016/j.wasman.2007.08.006
27. Xiu FR, Qi Y, Zhang FS (2013) Recovery of metals from waste printed circuit boards by supercritical water pre-treatment combined with acid leaching process. Waste Manag 33:1251–1257. https://doi.org/10.1016/j.wasman.2013.01.023
28. Xiu FR, Zhang FS (2010) Materials recovery from waste printed circuit boards by supercritical methanol. J Hazard Mater 178:628–634. https://doi.org/10.1016/j.jhazmat.2010.01.131
29. Cui H, Anderson CG (2016) Literature review of hydrometallurgical recycling of printed circuit boards (PCBs). J Adv Chem Eng 6(1):142. https://doi.org/10.4172/2090-4568.1000142
30. Xing M, Zhang F (2012) A novel process for detoxification of BERs in waste PCBs. Procedia Environ Sci 16:491–494. https://doi.org/10.1016/j.proenv.2012.10.067
31. Yıldırır E, Onwudili JA, Williams PT (2015) Chemical recycling of PCB waste by depolimerization in sub and supercritical solvents. Waste Biomass Valorization 6:959–965. https://doi.org/10.1007/s12649-015-9426-8
32. Braun D, von Gentzkow W, Rudolf AP (2001) Hydrogenolytic degradation of thermosets. Polym Degrad Stab 74:25–32. https://doi.org/10.1016/S0141-3910(01)00035-0
33. Xu J, Tazawa N et al (2018) Simultaneous recovery of high-purity copper and polyvinyl chloride from thin electric cables by plasticizer extraction and ball milling. RSC Adv 8:6893–6903. https://doi.org/10.1039/C8RA00301G
34. Huang K, Guo J, Xu Z (2009) Recycling of waste printed circuit boards: a review of current technologies and treatment status in China. J Hazard Mater 164:399–406. https://doi.org/10.1016/j.jhazmat.2008.08.051

Chapter 10
Hydrometallurgical/Aqueous Recovery of Metals

> *"The Earth does not belong to us...*
> *We belong to the Earth"*
>
> Chief Seattle

Abstract This chapter basically covers solder strip leaching, base metal leaching, and precious metal leaching techniques. Previously performed laboratory and industrial leach test results, conditions, reagents, extraction ratios, leach reactions, and flowsheets are reviewed and discussed in detail. Alternative leaching reagents are compared from chemistry, research level, and commercial extend points of view. Analytic hierarchy process (AHP) is used to compare various leaching processes. Brominated epoxy resin leaching, purification of metals from leachates by solvent extraction, and industrial scale refining solutions are also covered in this chapter. Occupational, health, and safety hazardous characteristics determination tests were briefly mentioned.

Keywords Hydrometallurgy · Solder strip leach · Base metal leach · Precious metal leach · BFR leach · Purification · Refining

Leaching is the initial step in a hydrometallurgical process; it is also the most important key during both BM and PM recoveries from WPCBs. Therefore, discussion and summary on hydrometallurgical technology progress especially on progress of leaching process have been provided in this book. Research on recovery of metals from WPCBs by hydrometallurgical processing began in the 1970s; the initial goal of that is exactly the recovery of such PMs as Au and/or Ag.

In the last few decades, many studies have been carried out in order to develop environmentally benign production processes which allow the parallel recovery of metals and the efficient treatment of the toxic compounds. The most promising results have been obtained in the case of hydrometallurgical processes which have the distinct advantage of offering an eco-friendly and selective separation of the metals from the nonmetallic parts of the WPCBs [1]. However, in some hydrometallurgical processes, the use of high volumes of leaching solutions leads to the formation of large quantities of waste. Therefore, it is important to find reliable and cost-effective heavy metal recovery techniques that do not produce any secondary pollution threats to the environment and human health during the processing of WPCBs [2, 3].

Before leaching several components should be manually removed from WPCBs (Li batteries, Al heat sinks, Cr/Ni plated bronze screws from the peripheral interfaces (RS232, VGA, LPT, etc.), and the cylindrical Al electrolytic capacitors) in mechanical pretreatment step to avoid unwanted problems:

- Risk of explosion when the extremely reactive inner of the Li batteries comes in contact with the leaching solution.
- Risk of leaching solution contamination with extremely toxic polychlorinated biphenyls presented in some cylindrical Al electrolytic capacitors.
- Increased total time of leaching due to the high thickness of the mentioned screws.
- Significant and unjustified consumption of leaching agent for the Al dissolution due to the small commercial value of Al and the difficulty of Al recovery from the resulting solution. Moreover, the reaction between Al and the aqueous solutions of Br generates high amount of gaseous H and is extremely vigorous and exothermic, increasing the risk of explosion.

10.1 Solder Stripping Leach

Additionally, recycling of other valuable BMs, including Pb, Sn, Ni, and Zn, is also of significance in terms of economic and environmental perspectives. Studies on recycling of Pb and Sn from solder are reported using alkaline [4, 5], HNO_3 [5–7], HCl [5], or fluoroboric (HBF_4) acids [8] as leaching reagents. Ranitović et al. (2015) reported that NaOH was not a reliable leaching reagent to extract Pb and Sn because of metal hydroxide dissolution and precipitation. HNO_3 is capable of extracting Pb; yet, Sn could not be extracted resulting from the fact that Sn is oxidized to insoluble form [5]. On the contrary, it is also shown that 90% of Sn could be extracted using HCl at high temperature; however, the precipitation of AgCl resulted in an unacceptable loss of Ag [5]. In the investigation of Pb leaching from solder, Jha et al. (2012) reported that 99.99% Pb could be leached out from the swelling liberated solder under the conditions of 90 °C, 0.2 M HNO_3, and S/L radio of 1:20 (g/mL) in 2 h [7]. Meanwhile, 98.7% Sn could also be recovered at 90 °C with 3.5 M HCl for 2 h. Mecucci et al. (2002) proposed a flowsheet to selectively separate Cu, Pb, and Sn [8]. The flowsheet contains shredding, HNO_3 acid leaching, and electrodeposition, as well as electrolyte regeneration. It was found that Cu and Pb could be completely dissolved in the HNO_3 acid of 6 M, whereas at the high concentration of HNO_3, the formation of metastannic acid resulted in the fact that Sn was able to be removed as a precipitate (metastannic acid). The reaction was shown in Eq. (10.1):

10.1 Solder Stripping Leach

$$Sn + 4HNO_3 = H_2SnO_3 \downarrow + 4NO_2 + H_2O \quad (10.1)$$

The results indicated the feasibility of simultaneous recovery of Pb and Cu through electrodeposition, and Sn could be recovered by electrodeposition after the dissolution of metastannic acid in the presence of HCl acid. Park and Fray (2009) studied the separation of Zn and Ni ions in a diluted AR using TBP, Cyanex 272, and Cyanex 301. It appeared that over 99% zinc could be recovered using Cyanex 301 at low pH (pH < 6), while 20% Ni could be extracted. The separation factor was approximately 21.700 at pH 6 [9].

Other several regenerable leaching systems were proposed and tested, e.g., electro-generated Cl_2 in HCl solution [10], $FeSO_4/H_2SO_4$ [11], $Fe_2(SO_4)_3/H_2SO_4$ [12] $FeCl_3$/HCl [13–15], and $SnCl_4$/HCl (for solder stripping) [16]. Unfortunately, the electrochemical regeneration of chloride-based leaching agents presents the risk of chlorine evolution, requiring well-sealed equipment. Also, the presence of sulfate induces a low rate of the solder alloy dissolution if large amounts of Pb are present. The Br_2-based lixiviants can be also used, but some authors suggest that these are unattractive due to the high vapor pressure of Br_2 (28 kPa at 35 °C) [17]. Contrarily, other researchers indicate that the use of the adequate complexing agents like bromide or organic ammonium perhalides can resolve this problem [18].

Dorneanu (2017) used bromine-bromide (Br_2/KBr) media for preliminary leaching and dismantling of WPCBs. In this electrochemical recycling system, all metallic parts were removed simultaneously with the lixiviant regeneration and the partial electrodeposition of the dissolved metals (Fig. 10.1) [19]. For dismantling, specific energy consumption was 0.65 kWh kg^{-1} of treated WPCB. After leaching tests, the remaining undissolved parts (fiberglass boards, ECs, plastics, etc.) preserve their original shape and structure, allowing an easier consequent separation/classification and a more efficient recycling.

Jeon et al. (2017) separated Sn, Bi, and Cu from Pb-free solder paste by swelling and ammonia leaching with cupric ion (Cu^{2+}) followed by HCl acid leaching with stannic ion (Sn^{4+}) [20]. Cupric ions act as an oxidant. They leached Cu with 5 M NH_3 (25% in water) solution, 1 M (($NH_4)_2CO_3$), 0.1 M $CuCO_3$ at 50 °C, 400 rpm, and 1% pulp density. Sn and Bi were leached with 0.5 M HCl, 10,000 mg L^{-1} Sn^{4+} ($SnCl_4$) at 50 °C, and 1% pulp density. Waste solder paste can be swelled by methyl ethyl ketone (MEK) at 30 °C, 200 rpm, and 5% pulp density. Resin can be precipitated by distilled water addition.

Cu can be leached by ammonia with cupric ion (Cu^{2+}), $CuCO_3$ and $(NH_4)_2CO_3$, and $CuSO_4$ and $(NH_4)_2SO_4$ or $CuCl_2$ and NH_4Cl. Cupric ammonia complex ion ($Cu(NH_3)_4^{2+}$) could oxidize Cu metal as follows [21–23]:

$$Cu(NH_3)_4^{2+} + Cu = 2Cu(NH_3)_2^+ \quad (10.2)$$

100% Cu recovery can be obtained within 15 min. in carbonate and sulfate media. Cu is recovered from the ammonia-based solution through an electrochemical process. Cu electrodeposition is a conventional process to recover Cu from a solution. It is widely used under acidic conditions, which is also the current industrial

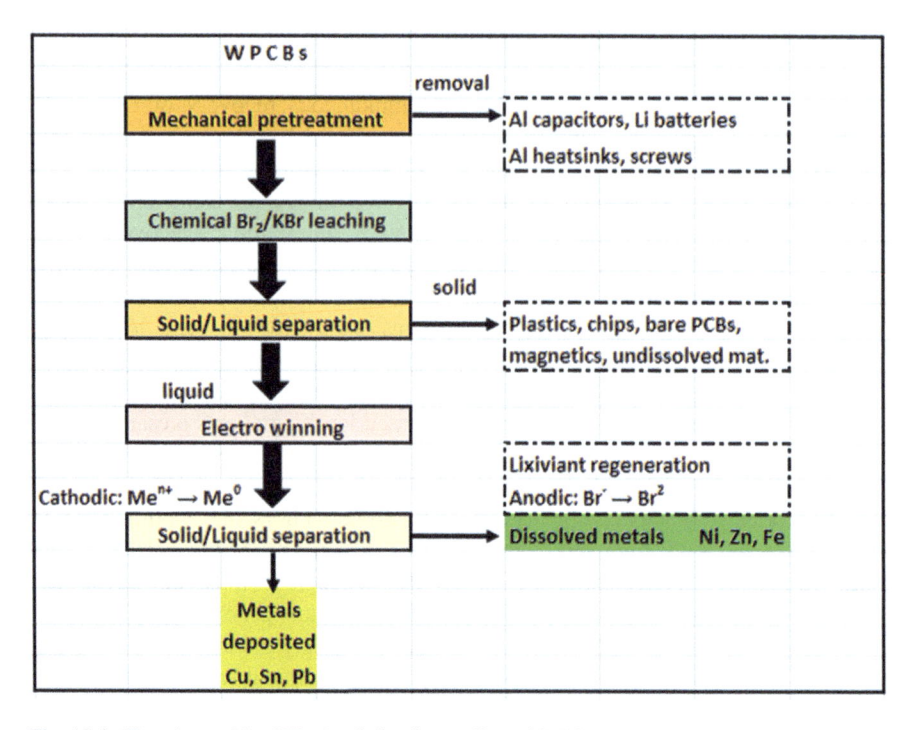

[Figure 10.1: Flowsheet diagram showing WPCBs processing]

Fig. 10.1 Flowsheet of Br$_2$/KBr leach for dismantling of WPCBs

practice [10]. But, current efficiency is quite low in electrodeposition as compared to traditional electrowinning from acidic solutions. Current electrodeposition efficiency was increased using cylindrical electrodes instead of plate electrodes. At 4-h deposition time, 99.8% pure Cu was obtained [24].

Kim et al. (2016) reported that stannic ions (Sn^{4+}) and tin metal in HCl solutions gave the following reaction [25]:

$$SnCl_4 + Sn = 2SnCl_2 \quad (10.3)$$

Jian-guang et al. (2016) also observed that a new generated thin layer fine Cu particle was covered on leach residue [16]. The reason for some Cu dissolution by SnCl$_4$ is given by

$$Cu + SnCl_4 = CuCl_2 + SnCl_2 \quad (10.4)$$

$$CuCl_2 + Sn = Cu + SnCl_2 \quad (10.5)$$

Yang et al. (2016) leached 99% of Sn from WPCBs using SnCl$_4$ and HCl at a temperature of 60–90 °C, and then Sn was recovered from the purified solution by electrodeposition. They reported that with a S/L ratio of 1:4, applying 1.1 times of stoichiometric SnCl$_4$ dosage and HCl concentration of 3.5–4.0 mol L^{-1} at a

10.1 Solder Stripping Leach

temperature of 60–90 °C, 99% of Sn can be leached from the metal components of WPCBs. The suitable purification conditions were obtained in the temperature range of 30–45 °C with the addition of 1.3–1.4 times of the stoichiometric quantity of Sn metal and stirring for a period of 1–2 h, followed by adding 1.3 times of the stoichiometric quantity of Na$_2$S for sulfide precipitation about 20–30 min at room temperature. The purified solution was subjected to membrane electrowinning for Sn electrodeposition. Under the condition of catholyte Sn^{2+} 60 g L^{-1}, HCl 3 mol/L and NaCl 20 g L^{-1}, current density 200 A m^{-2}, and temperature 35 °C, a compact and smooth cathode Sn layer can be obtained. The obtained cathode Sn purity exceeded 99% and the electric consumption was less than 1200 kWh t^{-1}. The resultant SnCl$_4$ solution generated in anode compartment can be reused as leaching agent for leaching Sn again [26]. Yang et al. (2017) leached approximately 87% of Sn-Pb solder successfully with spent tin stripping (TSS) solution at 2 h [27]. Sun et al. (2015) used a two-stage leaching design for Cu and PMs [24]. They used ammonia-ammonium carbonate (99.0%)-based leaching solutions for Cu and H$_2$SO$_4$ for PMs. More than 95% Cu extraction was achieved. In the downstream to recovery of Cu from leached solution, direct electrowinning from ammonia-based leaching solutions usually ends with low current efficiency.

Zhang et al. (2017) leached Pb-Sn solder with methanesulfonic acid (MSA) containing H$_2$O$_2$ for effective dismantling of WPCB assemblies. 3.5 M MSA and 0.5 M H$_2$O$_2$ with a reaction time of 45 min were found to be optimum conditions for the selective desoldering separation of ECs from WPCBs [28]. Although these techniques were conducted at lower temperatures, new chemical reagents (such as SnCl$_4$, HCl, HBF$_4$, H$_2$O$_2$) were consumed.

Pb, Sn, and In were successfully recovered from alloy wire e-waste through acid/alkali leaching [29]. The scrap material was leached using a 5 M HCl-HNO$_3$ solution at 80 °C for 1.45 h, and a NaOH solution was added to the leach liquor for metal precipitation. Under these experimental conditions, the metal recoveries were 94.7% for Pb, 99.5% for Sn, and 99.7% for In. Barakat (1998) also investigated metal recovery from Zn solder dross used in WPCBs by means of leaching using a 3% H$_2$SO$_4$ solution at 45 °C for 1 h [29]. Zn and Al entered the solution, whereas Pb and Sn remained within the residue. Al was selectively precipitated as a Ca-Al-CO$_3$ by treating the sulfate leachate with CaCO$_3$ at pH 4.8. The ZnSO$_4$ solution was either evaporated to obtain zinc sulfate crystals or precipitated as a basic carbonate at pH 6.8. The undissolved Pb and Sn were leached by using a hot 5.0 M HCl solution. The majority of the PbCl$_2$ (73%) was separated by cooling the leached products down to room temperature. From the soluble fraction, Sn was recovered as tin oxide hydrated by means of alkalization with NaOH at pH 2.4. The recovery efficiency of the metal salts was 99.1% for Zn, 99.4% for Al, 99.6% for Pb, and 99.5% for Sn [30].

Lee et al. (2003) investigated the recovery of valuable metals and the regeneration of the expended WPCB HNO$_3$ etching solutions [31]. HNO$_3$ was selectively extracted from the expended etching solution using tributyl phosphate (TBP), whereas a pure HNO$_3$ solution was extracted using distilled water. After HNO$_3$ extraction, pure Cu metal was obtained through electrowinning, and Sn ions were precipitated by adjusting the pH of the solution with Pb(OH)$_2$. Pb, with a purity of 99%, was obtained by cementation with an Fe dust.

Menetti and Tenório (1996a) depicted the steps for Au, Ag, Cu, Fe, Al, Sn, and Zn recovery from three types of e-waste using physical separation methods. Special attention was given to comminution, electrostatic, and magnetic concentration procedures [32]. Menetti and Tenório (1996) also investigated Au and Ag recovery from the physical treatment of e-waste to produce metallic concentrates [33]. Au and Ag present in a sample of e-waste were leached with 1.0 N HNO$_3$ solution. The leached pulp was treated with a 1.0 N H$_2$SO$_4$ solution to cause AuSO$_4$ precipitation. The remaining solid residue was melted at 1085 °C in the presence of Na$_2$CO$_3$, producing small pellets of Au.

Veit et al. (2005, 2006) used dismantling, comminution, mechanical separation (size, magnetic and electrostatic separation), hydrometallurgy, and electrowinning to recover BMs (mainly Cu, Pb, and Sn) from nonmetals (polymers and ceramics) [34, 35]. Cu content reached a mass of more than 50% in most of the conductive fractions. A significant content of Pb (8%) and Sn (24%) could also be observed. Cu and Sn separation cannot be achieved with mechanical separation. Thus, hydrometallurgical method was necessary for Cu and Sn separation. In the second stage, the fraction concentrated in metals was dissolved with AR or H$_2$SO$_4$ and treated in an electrochemical process to recover the metals separately, especially Cu. The results demonstrate the technical viability of recovering Cu by means of mechanical processing followed by an electrometallurgical technique. Cu content in the solution decayed quickly in all the experiments, and the Cu obtained by electrowinning was found to be above 98% in most of the tests.

Castro and Martins (2009) studied the recovery of Sn and Cu by recycling WPCBs from EoL computers with mineral acid leaching and NaOH precipitation [30]. Firstly, they were taken apart the WPCBs by dismantling; WPCB powder was obtained by grinding. Nonmetal parts were removed by washing followed by drying. Then, the metallic parts at 0.208 mm in size were leached with different mineral acids (H$_2$SO$_4$, HNO$_3$, or HCl) or acid mixtures for Cu and Sn metal extraction. Sn-Cu-bearing precipitates were obtained from leach liquor at different initial pH values using NaOH. 2.18 N H$_2$SO$_4$, 3.0 N HCl, 2.18 N H$_2$SO$_4$ + 3.0 N HCl, and 3.0 N HCl + 1.0 N HNO$_3$ acid solutions were used at 60 °C and 1/10 S/L ratio in leaching tests to determine the best leach system for maximum Sn and Cu extraction. The best acid for leaching both Sn and Cu was 3.0 N HCl + 1.0 N HNO$_3$ in 2 h. Sn and Cu extractions were 98.1% and 93.2%, respectively. H$_2$SO$_4$ exhibited the poorest results (2.7% Sn and <0.01% Cu). Two-stage 2.18 N H$_2$SO$_4$ + 3.0 N HCl mixture leaching has not exhibited better Sn and Cu extraction than only 3.0 N HCl leaching. It is possible to conclude from this study that the 3.0 N HCl and 3.0 N HCl + 1.0 N HNO$_3$ leach systems were the most appropriate to recover Sn and Cu simultaneously. The precipitate obtained at the initial pH of the neutralization of the leach liquor from the 3.0 N HCl system presented the most significant mass of Sn recovered (1.2 g out of 50 g WPCB) among all precipitates produced in this experimental work. The precipitate containing the largest mass of recovered Cu (0.26 g) was obtained at an initial pH 4.2 of the neutralization of the leach liquor from the 3.0 N HCl + 1.0 N HNO$_3$ system [36].

HNO$_3$ leaching of Pb from solder material of WPCBs was studied by Jha et al. (2012) [7]. Solder material on the dismantled and shredded WPCBs was swelled using organic *n*-methyl-2-pyrrolidone. Cu metal sheets were separated from epoxy resin containing solder. After washing, drying, and cutting, small pieces of epoxy resin were leached with 0.2 M HNO$_3$ at 90 °C for 2 h for Pb extraction. Then, Sn-containing epoxy residue was leached with 3.5 M HCl at 90 °C for 2 h for Sn extraction. Almost complete Pb dissolution and 98.7% Sn dissolution were obtained. Pb- and Sn-free resin can be directly disposed to the environment. HBF$_4$ containing H$_2$O$_2$ was used to dissolve tin solder by Zhang et al. (2015), and almost 100% of the solder was dissolved [37].

Jiang et al. (2012) have developed a novel process based on green chemistry and green engineering methodologies for reclaiming valuable materials from waste electronics. They have demonstrated that they can recover metals and valuable components from EoL products using cost-effective, sustainable, and scalable methods (e.g., systems that are closed-loop, energy efficient, and environmentally benign). This process is developed by ATMI Inc. and includes both chemical desoldering and PM reclaim from WPBCs and ICs near room temperature with all metals recovered and resold. In this process no shredding or grinding is required at this point which may lead to loss of up to 40% of PMs and/or the formation of dangerous metal fines, dust containing BFRs, and dioxins. After rinsing the WPCBs to remove surface dirt, dust, etc., all chips and components (capacitors, resistors, heat sinks, connectors, etc.) are chemically desoldered. This step requires less than 20 min in an acidic solution at a temperature between 30 and 40 °C. The solder is collected as a salt and resold, while the resulting chips are still fully functional. Note that the Au connectors remain intact on the board demonstrating the selectivity of the desoldering chemistry. The chemistry is highly selective toward both Pb/Sn and Sn/Ag solders leaving Cu, Au, etc. on the board and BM components intact. Pb, Sn, and Ag salts (typically oxides) are recovered in high purity (>95% recovery) and are resold. The bath chemistry can be loaded to over 250 g L^{-1}, and both this chemistry and the rinse water are recycled multiple times. Tests of the desoldered chips show that they are still operational and thus may be reused. Alternatively, ICs may be ground and the trace PMs extracted using leaching at 30 °C for approximately 5–10 min (depending on composition and thickness of Au surface). BM components (primarily steel and aluminum) are also collected and resold. Au, Ag, and Pd metals are collected with >99% efficiency and >99.9% purity via chemical reduction or electrowinning and are sold. After rinsing, the bare WPCB may be shredded and the Cu collected and sold. >99% Cu recovery with >99.5% purity was demonstrated. Because no solder or BMs are present, the value of these boards is far higher than with competing technologies. The remaining chopped fiberglass may be used as filler in, for example, cement [38].

Figure 10.2 shows ATMI automated desoldering, Au leaching, and Cu leaching recycling pilot line photo and flowsheet. ATMI process does not require comminution cost, there is no need for hazardous lixiviants like cyanide or AR, low volumes of chemicals are used, and room temperature is sufficient.

Zhu et al. (2012) reported that [EMIM$^+$][BF4$^-$] could completely separate ECs and solders from PCBs at 240 °C due to the hydrogen bond of [EMIM$^+$][BF4$^-$]

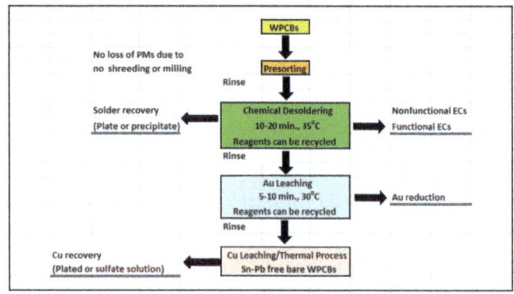

Fig. 10.2 ATMI automated WPCB recycling pilot line (**a**) and flowsheet (**b**) [38]

[39]. Zeng et al. (2013) investigated the utilization of water-soluble ionic liquid [BMIm]BF$_4$ in dismantling ECs and tin solder from WPCBs as functions of temperature, retention time, and turbulence [40]. The results indicated that approximately 90% of the ECs could be separated at 250 °C, 12 min, and 45 rpm. This study also reported that due to high efficiency and reusability of ionic liquids, [BMIm]BF$_4$ dismantling is more favorable than mechanical dismantling, when the capacity of the plant is over 3000 tpd.

10.2 Base Metal Leaching

A large amount of BMs, such as Cu, Pb, Zn, Cd, etc., can be found in WPCBs. Generally, leaching, as the first step of hydrometallurgical treatment, is to dissolve the constituents of the e-scrap to form a pregnant solution by a suitable lixiviant. Acid leaching with an oxidant reagent is widely used for the first-stage leaching of BMs from WPCBs. The BM leaching, particularly, Cu, was generally conducted by using acids such as H_2SO_4, HNO_3, AR, and HClO with various oxidants including H_2O_2, O_2, Fe^{3+}, and Cl_2.

Silvas et al. (2015) recycled WPCBs with physical processing and selective sequential stage leaching for Cu extraction: first leaching on H_2SO_4 media (1 M, 1:10 S/L ratio at 75 °C for 4 h) and second leaching on oxidant under the same conditions with 240 mL of H_2O_2. Here, H_2O_2 worked as reductant or oxidant agent [41]. During all tests, the pH was between 0 and 0.5 and ORP was between 220 and 280 mV. In alkaline media, H_2O_2 is reductor and at acid pH, it has oxidant behavior (Eqs. (10.6) and (10.7), respectively) [41]:

$$H_2O_2 + 2OH^- \rightarrow O_2 + H_2O + 2e^- \quad (-0.15\,\text{V}) \tag{10.6}$$

$$H_2O_2 + 2H^+ + 2e^- \rightarrow 2H_2O \quad (+1.77\,\text{V}) \tag{10.7}$$

In the presence of H_2O_2 and H_2SO_4, peroxysulfuric acid (H_2SO_5) acts as a strong oxidant for Cu:

10.2 Base Metal Leaching

$$H_2SO_4 + H_2O_2 \rightarrow H_2SO_5 + 2H_2O \qquad (10.8)$$

$$Cu + H_2SO_5 \rightarrow Cu^{2+} + SO_4^{2-} + H_2O \qquad (10.9)$$

Molecular oxygen obtained from the increased rate of decomposition of H_2O_2 is adsorbed onto Cu surface. The presence of oxygen on Cu surface improves the contact with H_2SO_4 [42].

It is well known that three inorganic acids can chemically attack metallic Cu in the presence of dissolved oxygen (Eqs. 10.10 and 10.11). High acid concentrations provide an adequate amount of hydronium ions, and continuous stirring of the solid-liquid mixture can improve the incorporation and dissolution of oxygen gas in the liquid:

$$2Cu + O_2 + 4H^+ \rightarrow 2H_2O + 2Cu^{2+} \qquad (10.10)$$

For HCl, the following reaction can also take place:

$$4Cu + 8Cl^- + O_2 + 4H^+ \rightarrow 2H_2O + 4CuCl_2^- \qquad (10.11)$$

Reactions among BMs, H_2SO_4, and H_2O_2 are given below along with Gibbs free energies:

$$Cu + H_2SO_4 + H_2O_2 \rightarrow CuSO_4 + 2H_2O \quad \Delta G = -77.941\,kcal/mol \qquad (10.12)$$

$$Zn + H_2SO_4 + H_2O_2 \rightarrow ZnSO_4 + 2H_2O \quad \Delta G = -127.965\,kcal/mol \qquad (10.13)$$

$$Sn + H_2SO_4 + H_2O_2 \rightarrow SnSO_4 + 2H_2O \quad \Delta G = -136.895\,kcal/mol \qquad (10.14)$$

$$Fe + H_2SO_4 + H_2O_2 \rightarrow FeSO_4 + 2H_2O \quad \Delta G = -115.847\,kcal/mol \qquad (10.15)$$

$$Ni + H_2SO_4 + H_2O_2 \rightarrow NiSO_4 + 2H_2O \quad \Delta G = -101.244\,kcal/mol \qquad (10.16)$$

In addition, H_2O_2 is environmentally friendly, since the final decomposition products are oxygen and water. It is easy to handle and is found in several industrial applications. Compared to many other oxidant agents such as hypochlorite and permanganate, H_2O_2 does not carry any additional or interfering substance except H_2O for the reaction system. The leaching in oxidant media aimed the Cu dissolution since it presents high solubility in oxidant H_2SO_4 media according with the Pourbaix diagram. To understand and to analyze the results, Pourbaix diagrams can be plotted for Al, Sn, Zn, and Cu–H_2O system using Software HSC Chemistry 7.1.

At the first stage, the leaching of nonmagnetic fraction in H_2SO_4 media was carried out in order to remove impurity metals (i.e., Al, Zn (in brass alloys), and Sn) in the Cu extraction. 90% Al, 9% Sn, and 40% Zn were extracted. At the second stage, H_2SO_4 leaching on H_2O_2 oxidant media was performed to completely extract Cu along with Zn (60%). At this stage, the solution pH increased until 4. Cu is insoluble in dilute HCl and H_2SO_4 acids due to its standard electrode potential which is positive (+0.34 to Cu/Cu^{2+}), although any solubilization can occur in the presence of oxygen [41].

Lisinska et al. (2018) found that 5 M H_2SO_4 and 30% H_2O_2 significantly dissolved Cu at 70 °C for 6 h leaching; but, Fe, Zn, and Al dissolution were not substantially influenced by H_2O_2 [43]. Kinoshita et al. (2003) used two-step leaching and solvent extraction for Ni and Cu extraction from WPCBs. Ni was selectively leached with 0.1 M HNO_3 solution in the first-step leaching at 80 °C for 72 h with 98% recovery. Remaining Cu was leached with 1.0 M HNO_3 solution in the second-step leaching at 90 °C for 6 h. Au was recovered in flakes at 99% purity in solids. Solvent extraction with LIX984 and stripping with 4.0 M HNO_3 were used to recover Cu [43].

Hydrometallurgical extraction of Cu has been well-established with regard to Cu ore; however, there is no flowsheet that could be applicable to recycle Cu from WPCBs on the industrial scale. Typically, Cu is dissolved in H_2SO_4 to produce impure Cu-bearing pregnant solution, and then the pregnant solution goes through solvent extraction to upgrade Cu purity. Eventually, pure Cu is obtained by further electrowinning. A H_2SO_4 leaching with H_2O_2 is shown in Eqs. (10.17) and (10.18):

$$Cu^0 + H_2O_2 + H_2SO_4 = Cu^{2+} + SO_4^{2-} + 2H_2O \quad (10.17)$$

$$Zn^0 + H_2O_2 + H_2SO_4 = Zn^{2+} + SO_4^{2-} + 2H_2O \quad (10.18)$$

In the study of selective leaching of valuable metals from WPCBs, it was found that 100% of the Cu and Zn were leached out within 8 h by employing 2 M H_2SO_4 and 0.2 M H_2O_2 at 85 °C; meanwhile, 95% of the Fe, Ni, and Al were dissolved within 12 h. Whereas Pb had low dissolution rate in the H_2SO_4–H_2O_2 and $(NH_4)_2S_2O_3$–$CuSO_4$–NH_4OH systems, thus, the solid residue, mainly $PbSO_4$, undertook NaCl leaching solution to form $PbCl_2$, followed by a solid-liquid separation [44]. Deveci et al. (2010) found that concentration of H_2O_2 and temperature were the most significant factors through a 2^3 full factorial design [45]. Yang et al. (2011) studied the effects of particle size, temperature, and initial Cu concentration with regard to Cu leaching. The results indicated that the fine particle (−1 mm) was efficiently treated, and temperature and initial Cu concentration had insignificant effects on Cu leaching [42]. Moreover, a negative effect was mentioned that the high stirring rate decreased the Cu extraction, which could be attributed to the fact that the increasing stirring speed caused H_2O_2 degradation [46].

In acidic sulfate solution, even though H_2O_2 was extensively used as the oxidant as mentioned above, it suffers from its remarkably high consumption due to its decomposition and high temperature. Thus, ferric ion is regarded as an oxidant alternative because of its low cost and regeneration [45]. The downstream can be purified by goethite and jarosite precipitation before electrowinning [47, 48].

An environmental assessment was investigated by Fogarasi et al. (2013) to evaluate two Cu processes which were the direct electrochemical oxidation and mediated electrochemical oxidation using the Fe^{3+}/Fe^{2+} redox couples [13]. Unfortunately, there was no leaching data reported. Furthermore, Yazici and Deveci (2014) investigated the effect of ferric ions on sulfate leaching of metals from PCBs. The study indicated that high extractions of Cu and Ni were obtained, and

10.2 Base Metal Leaching

increasing temperature, ferric concentration, and acid concentration positively affected the metal extraction, while the increase of solid ratio was an adverse effect on leaching of metals. The addition of air or H_2O_2 could maintain the high ratio of Fe^{3+}/Fe^{2+} [49].

H_2SO_4–$CuSO_4$–$NaCl$ system has attracted attention not only because of the fact that chloride solution has higher solubility and activity of metal compound compared with in sulfate phase but also due to the advantage of Cu ions, which means the usage of Cu can avoid the contamination of Fe^{3+} and decrease the high cost when H_2O_2 is used [50, 51]. The mechanism in H_2SO_4 system is shown in Eqs. (10.19)–(10.23). Ping et al. showed that in the electrooxidation conditions, the recovery of Cu was 100% within 3.5 h [50].

$$Cu + Cu^{2+} = 2Cu^+ \tag{10.19}$$

$$Cu^+ + Cl^- = CuCl; \quad CuCl + Cl^- = CuCl_2^- \tag{10.20}$$

$$4CuCl_2^- + O_2 + 2H_2O = 2[Cu(OH)_2 \cdot CuCl_2] + 4Cl^- \tag{10.21}$$

$$2[Cu(OH)_2 \cdot CuCl_2] + H_2SO_4 = CuSO_4 + CuCl_2 + 2H_2O \tag{10.22}$$

$$CuSO_4 + 5H_2O = CuSO_4 \cdot 5H_2O \tag{10.23}$$

Further investigation was performed with regard to other metals such as Fe, Ni, Ag, and Au by Yazıcı and Deveci (2013) [51]. The experimental results indicated that the Cl^-/Cu^{2+} ratio is the influential factor for the extraction process. At 1% solid ratio, more than 90% Fe, Ni, and Ag could be recovered with 58%, under the condition of a Cl^-/Cu^{2+} ratio of 21 and 80 °C. Furthermore, the addition of oxygen could increase the recovery of the metals even at a higher solid ratio [12].

In addition to H_2SO_4 treatment of Cu, other acids are also used. Kasper et al. (2011) studied Cu extraction by using AR, followed by electrowinning. The material fed to the acid leaching was a metal concentrate by the mechanical process containing 60% Cu. Ninety-five percent of the Cu purity was obtained on the cathodes [52]. Havlik et al. (2010) proposed a method that used HCl acid to leach thermally treated by pyrolyed or burned samples. An improvement of the Cu extraction was accomplished, when the burning temperature increased, which could be explained by the fact that Cu is released from the WPCBs and is oxidized. At the pyrolysis temperature of 900 °C, the highest extraction of Cu was achieved because copper oxide was more efficiently leached out in a nonoxidizing environment than pure copper metal [53]. Kim et al. (2011) were highly interested in electrogenerated chlorine in a HCl acid solution. The effect of cuprous ions and leaching kinetics of Cu from WPCBs were investigated. It appears that depression of cuprous ions is helpful for increasing the leaching rate of Cu, which is due to the fact that the chloride generation preferably occurs in the anode [10]. Moreover, a comparison was made to differentiate the performances in a combined reactor, which could be used as both the chlorine generator and metal leaching, and in two separated reactors. The results showed that current and temperature were of importance in terms of the kinetics of Cu dissolution and metal leaching. The generation of cuprous ions has a negative effect on the leaching efficiency of the metals. Furthermore, it was observed that the surface layer diffusion was the kinetics law of Cu dissolution, which meant that from the viewpoint of kinetics, the rate control step of Cu leaching was the diffusion of the lixiviant through the porous product layer [54].

In a recent study, Yazici and Deveci (2015) further studied cupric chloride leaching of Cu as well as other metals (Fe, Ni, Ag, Pd, and Au) from PCBs. The study showed that almost complete extraction of Cu, Ni, and Fe was achieved over a leaching period of 2 h at 79 mM initial Cu^{2+}. Increasing the initial Cu concentration remarkably enhanced the metal extraction except for Au. Increasing temperature and oxygen supply could also increase the extraction of Pd and Ag to 90% and 98%, respectively, which could be attributed to maintaining the high ratio of Cu^{2+}/Cu^+ and thermodynamically favorable reaction between Pd/Ag and dissolved oxygen [55].

Pressure leaching of Cu has been extensively investigated in recent years due to two benefits: high concentration of oxygen in solution and fast kinetics [56]. However, a limited amount of literature mentioned the application of pressure oxidation leaching in recycling WPCBs. In the study of Jha et al. (2011), it was found that at 150 °C with 2 M H_2SO_4 and 15% H_2O_2 under the oxygen pressure of 20 bar, about 97% Cu could be recovered from the liberated metal sheets, which was pretreated by organic swelling [57].

A chemical preliminary treatment using inorganic acids (HCl, HNO_3, and H_2SO_4) and organic EDTA and citrate was found necessary to remove Cu before the Au can be extracted from WPCBs by Torres and Lapidus (2016). The effect of auxiliary oxidants such as air, ozone, and H_2O_2 was studied. HCl/citrate + H_2O_2 pretreatment achieved Cu extraction more than 90%. In the second stage for Au recovery with thiourea resulted in greater than 90% Au removal in 1 h. In the absence of air, for higher Cu extraction in 24 h, HCl was the best (1 M HCl > 1 M HNO_3 > 1 M H_2SO_4). Air sparging did not significantly improve H_2SO_4 to leach Cu [36]. Torres and Lapidus (2016) suggested two routes: Cu pretreatment with 0.5 M HCl and air sparging during 6 h, followed by Au leaching (83%) with 0.4 M thiourea solution at a pH of 1.5 for 3 h and Cu pretreatment with 0.5 M citrate and 0.1 M peroxide for 6 h, followed by Au leach (93%) with 0.4 M thiourea for 3 h [36].

Selective Cu removal using cupric chloride ($CuCl_2$) or ammonium sulfate (($NH_4)_2SO_4$), solder dissolution with HNO_3, and PM dissolution with AR can be used. Nonselective dissolution can be carried out with AR or chlorine-based chemistry. In hydrometallurgical methods, large amounts of chemicals are used, generating huge amounts of waste acids and sludge, which have to be disposed as hazardous waste. Furthermore, the overall process is very long and complicated. Processes based on strong inorganic acids and H_2O_2 have been developed for WEEE with fluoroboric acid providing useful extraction from mixed stream, including products from pyrolytic processes [58]. Due to oxidation and corrosion of equipment by acids, thiosulfate can be used as a nontoxic and low-cost reagent [59, 60]. Liu et al. (2009) firstly leached the crushed WPCB scraps in the NH_3/NH_5CO_3 solution to dissolve Cu [61]. After the solution was distilled and the $CuCO_3$ residue was converted to copper oxide by heating. After Cu removal, the remaining solid residue was then leached with HCl to remove Sn and Pb. The last residue was used as a filler in PVC plastics which were found to have the same tensile strength as unfilled plastic but had higher elastic modulus and higher abrasion resistance and were cheap.

Some efforts were made in recycling Cu by employing the Cu(I)-ammine complex [22, 62]. On the basis of the thermodynamic prediction, the oxidation-reduction potential of $Cu(NH_3)_4^{2+}/Cu(NH_3)^{2+}$ was greater than that of $Cu(NH_3)^{2+}/Cu$, and the oxidation-reduction potential of Cu(I)/Cu was greater than hydrogen potential, which meant, in this case, that $Cu(NH_3)_4^{2+}$ could be regarded as an oxidant and Cu(I) could be reduced to metallic Cu. The theoretical power consumption was lower than the conventional electrowinning process. The results indicated that the Cu(II)-amine complex had the positive effect on the leaching rate of Cu [22]. Oishi et al. (2007) found that the Cu(I) amine contained the impurities, such as Zn, P, Mn, and some Ni, in both ammonium sulfate and chloride systems. However, Zn, Pb, Ni, and Mn could be removed in the sequent purification process by using LIX-26 (alkyl substituted 8-hydroxy-quinoline). The main impurity for Cu concentrate was Pb, which was due to similar potential between Cu and Pb [62]. In addition, the application of ionic liquid was studied to recycle Cu from WPCBs because of its outstanding properties, such as negligible volatility and high conductivity.

Dong et al. (2009) investigated the behaviors of [bmim]HSO$_4$ in leaching of chalcopyrite. The results suggested that [bmim]HSO$_4$ acted as an acid and catalyst to facilitate the dissolution of chalcopyrite [63]. In the study by Huang et al. (2014), [bmim]HSO$_4$ was further studied to recycle Cu from WPCBs. Particle size, ionic liquid concentration, H$_2$O$_2$ dosage, and solid-liquid radio, as well as temperature, were considered as the parameters. It was observed that 99.9% Cu could be reached under the conditions of 25 mL 80% (v/v) ionic liquid, 10 mL 30% H$_2$O$_2$, S/L ratio of 1/25 at 70 °C within 2 retention hours. Moreover, diffusion through a product layer was the controlling step, using the shrink core model [64].

Table 10.1 summarizes some of the previous BM leach test results and conditions along with metallurgical performances.

10.3 Extraction of Precious Metal by Leaching

PMs have been used in electric and electronic industries due to their excellent electrical conductivity, low contact electrical resistance, and corrosion resistance, even though rare earths have started partially replacing PMs in electronic industry. Therefore, a large number of e-waste contain significant amount of PMs, particularly Au, Ag, and Pd. It is of importance to recycle PMs from e-waste. For instance, the Au content in WPCBs is 35–50 times higher than natural Au ore [73], even though it has been noticed that the Au content in PCBs is decreasing. Extraction of PMs from WPCBs, including leaching, purification, and recovery, is the second stage after the recovery of BMs.

WPCBs have an inherent value because of the PM content. Eighty percent of WPCB's intrinsic value comes from PMs, which is less than 1 wt%. Pretreatment is

Table 10.1 Previous base metal leach results and conditions

Sample	Reagent media	Conditions	Extraction (%)	References
WPCBs	$H_2SO_4 + H_2O_2$	20 mL H_2O_2 + 20 mL H_2SO_4 S/L: 1/2	99.4% Cu 98.8% Au	Zhu-Gu [65]
WPCBs and ECs 0.25–1.0 mm	AR	Leach + electrowinning		Veit et al. [35]
Computer PCBs	H_2SO_4	t: 60 min leach	99.5% Cu	
	H_2SO_4	2.18 N H_2SO_4	2.7% Sn, <0.01% Cu	Castro-Martins [30]
	HCl	3.0 N HCl S/L: 1/10, 60 °C		
	HNO_3	2.18 N H_2SO_4 + 3.0 N HCl		
	Mixture	3.0 N HCl + 1.0 N HNO_3	98.1% Sn, 93.2% Cu	
PCBs	$H_2SO_4 + H_2O_2$ + Cu ion	(15%) H_2SO_4, (30%) H_2O_2, 10 g/L Cu ions, 27 °C, 3 h	>98% Cu	Yang et al. [42]
WPCBs with high Cu content 177–500 mm	HCl + air	1 M acid, 25 °C, 125 rpm, 24 h, S/L: 1/50, 1 L/min air, 30% H_2O_2	85% Cu	Torres-Lapidus [36]
	HNO_3			
	H_2SO_4			
	EDTA + H_2O_2	0.5 M Na-citrate + 0.1 M H_2O_2, 25 °C, 125 rpm, 6 h	30% Cu	
	Citrate + H_2O_2	0.1 M EDTA + 0.1 M H_2O_2, 25 °C, 125 rpm, 6 h	75% Cu	
PCBs (d < 1.0 mm)	$H_2SO_4 + H_2O_2$	85 °C, 12 h, 2 M H_2SO_4	100% Cu and Zn; 95% Fe, Ni, Al	Oh et al. [44]
TV PCBs (d < 3.35 mm)	$H_2SO_4 + H_2O_2$	68 °C, 4 h, 1.6 M H_2SO_4	98.2% Cu	Deveci [45]
PCBs (d < 3.0 mm)	$H_2SO_4 + H_2O_2$	50 °C, 3 h, pH: 1.48	46.3% Cu; 5.7% Fe; 21.1% Sn; 51.1% Zn	Birloaga et al. [66]
		2 M H_2SO_4, 20 mL H_2O_2/100 mL H_2SO_4		

10.3 Extraction of Precious Metal by Leaching

PCBs ($d < 3.0$ mm)	$H_2SO_4 + H_2O_2$	50 °C, 3 h, pH: 1.52	11.5% Cu; 62.2% Fe; 26.4% Sn; 31.8% Zn	Birloaga et al. [66]
Computer PCBs (−300 μm)	$H_2SO_4 + H_2O_2$	2 M H_2SO_4, 20 mL H_2O_2/100 mL H_2SO_4		Behnamfard et al. [67]
		2 M H_2SO_4, 35% H_2O_2 (first step)	85.8% Cu (first step) Total: 99% Cu	
		2 M H_2SO_4, 35% H_2O_2 (second step)	14% Cu (second step)	
Printer PCBs (nonmagnetic) (−1 mm)	H_2SO_4	1 M H_2SO_4, 1/10 w/v, 75 °C, 4 h (first step)	90% Al; 8.6% Sn, 40% Zn	Silvas et al. [41]
	$H_2SO_4 + H_2O_2$	1 M H_2SO_4, 1/10 w/v, 75 °C, 4 h, 240 mL H_2O_2 (second step)	100% Cu; 60% Zn	
Mobile phone PCBs (−2 mm)	$H_2SO_4 + H_2O_2 + CO_2$	2.5 M H_2SO_4, 20% H_2O_2 + supercritical CO_2	89% Cu	Calgaro et al. [68]
		S/L: 1/20, 600 rpm, 35 °C, 20 min supercritical leach	$CO_{2(g)}$ fastens kinetics, reusable and environmentally acceptable	
Mobile phone WPCBs (2–4 cm) not crushed/ground	$H_2SO_4 + H_2O_2$	2–5 M H_2SO_4, 25 < T < 70 °C, 340 rpm, t: 72 h,		Lisinska et al. [43]
		10% < H_2O_2 < 30%, S/L: 1/10		
Computer PCBs (−1 mm)	$H_2SO_4 + CuSO_4 + NaCl$	0.5 M H_2SO_4, 4 g/L Cu^{2+}, 80 °C, 1% w/v solid, 2 h	100% Cu	Yazıcı-Deveci [51]
		0.5 M H_2SO_4, 4 g/L Cu^{2+}, 46.6 g/L $Cl^−$, 80 °C, 1% w/v solid, 2 h	98.7% Cu, 91.9% Ni, 91.8% Ag, 88.8% Fe	
	+ air/oxygen	0.5 M H_2SO_4, 0.5 g/L Cu^{2+}, 25.6 g/L $Cl^−$, 80 °C, 1% w/v solid, 2 h	100% Cu	
Mobile phone PCBs	$H_2SO_4 + FeSO_4$	Cu, Sn, Ag and magnet material dissolution $Fe^{3+} + H_2SO_4$	97% Cu	Lister et al. [11]
	$HCl + Cl_2$	Pd and Au using Cl_2 generated in HCl		
Mobile phone PCBs (non-ferrous fraction)	$H_2SO_4 + Fe_2(SO_4)_3$	Preextraction of BMs in the increase in PM extraction	>90% BMs	Diaz et al. [12]

(continued)

Table 10.1 (continued)

Sample	Reagent media	Conditions	Extraction (%)	References
PCBs	HNO$_3$	1–6 M HNO$_3$, 23–80 °C, 15 min to 6 h, 1 g/3 cm^3		Mecucci-Scott [69]
		1 M HNO$_3$, 2 h, 23 or 80 °C (no temp. effect)	50% Pb	
		1 M HNO$_3$, 6 h, 23 or 80 °C	90% Pb	
		6 M HNO$_3$, 15 min, 80 °C (temp. increases dissolution)	65% Cu	
		6 M HNO$_3$, 6 h, 80 °C	100% Cu	
		Up to 4 M HNO$_3$ H$_2$SnO$_3$ ↓		
Scrap TV boards (−250 mm)	HNO$_3$	2–5 M HNO$_3$, 30–70 °C, 2 h, pulp density: 2–10% w/v, 350 rpm	88.5–99.9% Cu, 16–68% Ag	Bas et al. [70]
WPCBs	HNO$_3$	3 mol/L HNO$_3$, 800 rpm, 60 °C, 5 h, 25 g/L	50% Cu, 80% Fe, 93% Zn	Kumar et al. [71]
WPCBs	HCl + Cl$_2$ electro-generated	Ion exchange (97% efficiency)	94.91% Cu 93.06% Au	Kim et al. [10]
WPCBs	HCl + FeCl$_3$ Fe^{3+}/Fe^{2+}	FR: 20 mL/min, N: 30 rpm, S/L: 1/8, 0.37 M FeCl$_3$, 0.3 M HCl, t: 4 h Chemical dissolution + electroextraction	99% Cu	Fogarasi et al. [13]
WPCBs	HCl + FeCl$_3$ Fe^{3+}/Fe^{2+}	Leach and electrochemical process Simultaneous Cu recovery and Au enrichment	99% Cu Au > 25 times than feed material	Fogarasi et al. [13]
WPCBs	HCl + FeCl$_3$ Fe^{3+}/Fe^{2+}	Chemical dissolution with Fe^{3+} combined with EW of Cu Oxidant regeneration	75% Cu dissolution 99.9% Cu deposition	Fogarasi et al. [15]
WPCBs (−8 mm)	Precombustion+ HCl	Precombustion 15–60 min, 500–900 °C 1 M HCl, 80 °C, 3 h	>90% Cu	Havlic et al. [53]

10.3 Extraction of Precious Metal by Leaching

	$NH_3 + (NH_4)_2SO_4 + SX$	$2 M NH_3 + 2 M (NH_4)_2SO_4$ 35 °C, 400 ppm, 2 h, 12 m^3/h air, S/L: 1/10 g/cm^3	96.7% Cu from leach 99.6% Cu from SX (lix 84 + kerosene)	Yang et al. [72]
PCB solder organic swelling (n-methyl-2-pyrrolidone)	HNO_3	0.2 M HNO_3 Pb leach 3.5 M HCl for Sn leach S/L: 1/100 g/mL, 90 °C, 45 min for Pb leach S/L: 1/20 g/mL, 90 °C, 2 h for Sn leach	100% Pb 98.7% Sn	Jha et al. [7]
Solder leach for dismantling	$MSA + H_2O_2$	3.5 M MSA, 0.5 M H_2O_2, 45 min		Zhang et al. [28]
WPCBs	$HCl + SnCl_4$	Leach + purification + membrane electrodeposition to recover Sn S/L: 1/4, 3.5–4 M HCl + SnCl$_4$, 60–90 °C Na$_2$S for sulfate precipitate Membrane electrodeposition: Sn^{2+}: 60 g/L, 3MHCl, 20 g/L NaCl, 35 °C 200 A/m^2	99% Sn leached	Juan-guang et al. [16]
Pb-free solder paste leach methyl ethyl ketone (MEK)	$HCl + Cu^{2+} + Sn^{4+}$	5 M NH$_3$, 1 M ((NH$_4$)$_2$CO$_3$), 0.1 M CuCO$_3$ at 50 °C, 400 rpm, 1% pulp density for Cu leach		Jeon et al. [20]
30 °C, 200 rpm and 5% pulp density (swelling)		0.5 M HCl, 10,000 mg/L Sn^{4+} (SnCl$_4$) at 50 °C and 1% pulp density for Sn and Bi leach		
WPCB dismantling	Br_2/KBr		Cu, Sn, Pb deposited	Dorneanu [19]
	$HBF_4 + H_2O_2$	2.5 M HBF$_4$ + 0.4 M H_2O_2 20 °C, 20 min	Pb and Sn dissolve, Cu precipitate	Zhang et al. [37]
Solder leach + precipitate	$HCl + HNO_3$	5.5 M HCl 165 min, 50 g/L, 90 °C	97.8% Sn, 4.5 M HCl, 1 h, 90 °C 100% Cu Pb, 0.1 M HNO_3, 60 °C, 10 g/L	Syed [73]

of necessity prior to leaching of PMs from WPCBs including physical and chemical pretreatments. Physical pretreatment is to mechanically break and crush PCB scrap after dismantling the ECs and then, according to the differences of physical characteristics of metals and nonmetals, to use one of the methods, including pneumatic separation, magnetic separation, screening, ECS, and electrostatic separation, to obtain the enrichment of metals and nonmetals. Chemical pretreatment is to dissolve BMs such as Cu, Pb, and Sn in solution, leaving PMs in residue, and then to isolate and purify PMs from the residue. The commonly used chemical pretreatment is HNO_3 pretreatment or $H_2O_2 + H_2SO_4$ pretreatment. Figure 10.3 shows typical flowsheet of recovery of PMs from WPCBs.

Hydrometallurgical processes enable selective recovery of PMs from resources using aqueous chemistry based on a variety of lixiviants including cyanide, AR, thiosulfate, thiourea, and halogens, because of the stable metal complex formed. Some of the above lixiviants are less toxic than others. Advantages and disadvantages of above reagents were summarized before in Table 5.4. The search for less hazardous reagents continues today. Thiourea and thiosulfate are candidates, as well as polyhydric alcohol, ketones, polyether, or cyclic lactone [74]. Table 10.2 summarizes alternative lixiviants for Au leaching [73].

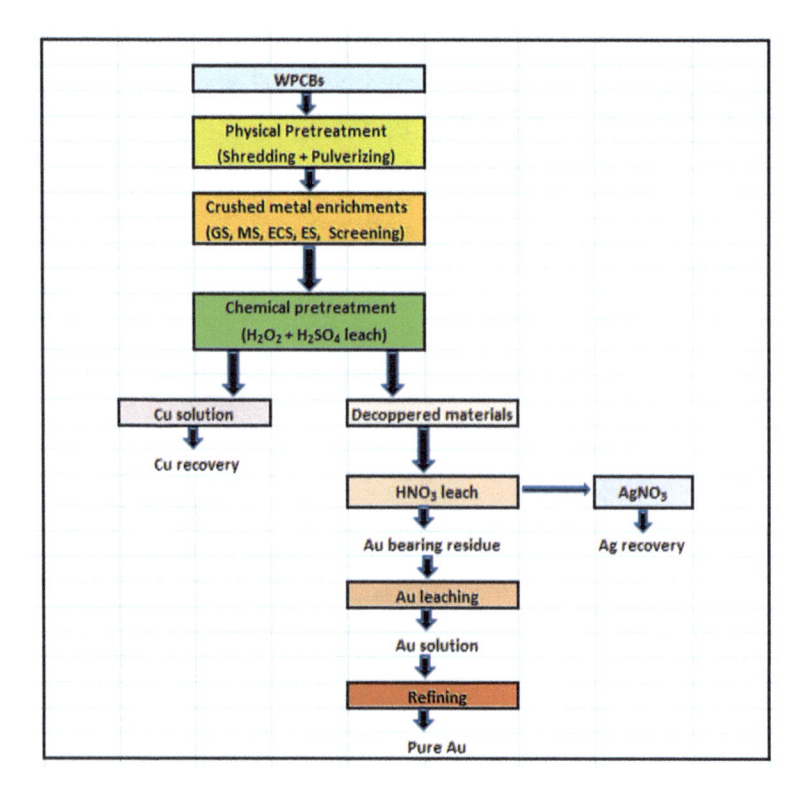

Fig. 10.3 A typical flowsheet of PM recovery from WPCBs

10.3 Extraction of Precious Metal by Leaching

Senanayake (2004) summarized a series of equations to illustrate the mechanism of Au complex formation regarding different lixiviants and also showed the linear correlations of stability constants of Au(I)-complexes, following the order: $CN^- > HS^- > S_2O_3^{2-} > SC(NH_2)_2 > OH^- > I^- > SCN^- > SO_3^{2-} > Br^- > Cl^-$ [75].

$$4Au + O_2 + 2H_2O + 8NaCN = 4NaAu(CN)_2 + 4NaOH \quad (10.24)$$

$$2Au + H_2O_2 + 4L^- = 2Au(I)L_2$$
$$+ 2OH^- \; (L = Cl^-, S_2O_3^{2-}, SC(NH_2)_2) \quad (10.25)$$

$$2Au + L_2 + 2L^- = 2Au(I)L_2 \, (L = Cl, Br, I, SCN, SC(NH_2)_2) \quad (10.26)$$

$$Au + 1.5L_2 + L^- = Au(III)L_4 \, (L = Cl, Br, I) \quad (10.27)$$

$$Au + Cu(II) \text{ or } Fe(III) + 2L = Au(I)L_2 + Cu(I) \text{ or } Fe(II) \quad (10.28)$$

$(L = Cl^-, S_2O_3^{2-}, SC(NH_2)_2, SCN^-, NH_2CH_2COO^-, NH^2CH(CH_3)\,COO^-)$.

Cyanide as lixiviant for Au has dominated in the mining industry for more than one century for its advantages such as has high efficiency, is an inexpensive reagent, is used in lower dosages, is operating in an alkaline solution, etc. Most cyanide leaching processes occur at pH 10, because of the fact that cyanide ion is stable at pH 10.2. Below pH 8.2, cyanide exists as hydrogen cyanide (HCN) that is highly volatile, which results in cyanide loss and harmfulness of operators' health. Recently, the slow cyanidation rate and severe environmental impact of cyanide gold leaching accelerate the development of a substitute that is more effective and environment-friendly. Cyanidation principle is to dissolve Au and Ag of the WPCB surface into the solution by alkali metal cyanide and then to recover Au and Ag by reduction from the cyanide solution. Cyanide leaching is only effective to leach Au/Ag at the surface of WPCBs. Au/Ag found inside WPCBs is hard to dissolve by cyanide process. Although cyanide leaching is still dominant in the Au mining industry, plenty of wastewater containing cyanide is produced in the leaching process, which is harmful for both operators and surrounding environment. Cyanidation has a long residence time for production cycle due to slow leaching rate. Therefore, metallurgical researchers have provided multiple noncyanide leaching processes in recent years, and some of these processes have made significant progress that they are almost to be used in commercial production. Currently, more attention has been paid to the study of several noncyanide leaching process, including thiourea leaching, thiosulfate leaching, and halide leaching.

Thiourea, NH_2CSNH_2, is considered a most promising alternative to cyanide regarding leaching of PMs due to its fast leaching rate and nontoxicity. Thiourea was first synthesized in 1868, also known as sulfurized urea. It is an organic complexing agent with reducibility, being able to form white crystal of complexes with many metal ions. The thiourea leaching is conducted at pH = 1.5 following the reaction shown as Eq. (10.28). The demerits of thiourea leaching are high cost and consumption because of its poor stability. Under acidic conditions, with the presence

Table 10.2 Suggested alternatives to cyanide in Au extraction

Lixiviant	Formula	Concentration	pH	Chemistry	Research level	Commercialization extent
Ammonia	NH_3	High	8–10 alkaline	Simple	Low	Pilot scale
Ammonia/cyanide	NH_3/CN^-	Low	9–11 alkaline	Simple	Extensive	Cu/Au ores
Ammonium thiosulfate	$(NH_4)_2S_2O_3$	High	8.5–9.5 alkaline	Complex	Extensive	Semicommercial scale
Hypochlorite/chloride	ClO^-/Cl^-	High chloride	6–6.5 low acidic	Well-defined	Extensive	Historical and modern
Bacteria		High	7–10	Fairly complex	Low	None
Natural organic acids		High	5–6 low acidic	Fairly complex	Low	None
Aqua regia (AR)	$3HCl + HNO_3$	High	<1 acidic	Well-defined	Low	Analytical and refining purposes
Thiourea	CH_4N_2S	High	1–2 acidic	Well-defined	Fairly popular	Some concentrates
Thiocyanide	SCN^-	Low	1–3 acidic	Well-defined	Low	None
Acidic ferric chloride	$FeCl_3$	High	<1 acidic	Well-defined	Low	Electrolytic Cu slimes

of oxidant (usually Fe^{3+}), thiourea and Au will form soluble cationic complexes. Comparing with other leaching reagents, thiourea shows much poorer stability. It is easy to decompose to sulfide and cyanimide in alkaline solution, and cyanimide can be further converted to urea; and it is easy to be oxidized into disulfide formamidine and elemental S and other various products in acidic solution. Therefore, it's very important to select suitable oxidant and the oxidant concentration so that Au is oxidized as much as possible into the solution; but, thiourea is oxidized as little as possible. Disulfide formamidine, as the product of the initial stage of oxidation, helps to accelerate leaching rate. However, with the increase of disulfide formamidine concentration, it is prone to irreversible decomposition to generate elemental S, which will form a stable passivation layer on the surface of Au particles, hindering Au dissolution. At the same time, a great deal of chemical reagent is consumed. That is the main reason why acidic thiourea leaching has not been used widely and industrially up to now [17, 72].

Stability, process control, reagent recyclability, and economics are the outstanding issues for leaching method. Catalytic processes are being developed for Sn and Pb. Electrochemical methods, microwave-enhanced biodigestion, and ionic liquid extraction for PMs are under investigation. Recently, mechanochemical treatment (MC) technology has been widely applied in extractive metallurgy and waste treatment. In MC treatment, repeated fracturing and cold welding occurred to the reacting particles during collisions with co-milling reactive chemicals. MC treatment technologies include MC leaching, MC sulfidization, and mechanical activation. Mechanical activation involves an increase in the reactivity of target substances, which could promote the subsequent leaching process. MC technology has been applied to recover metals from fluorescent lamps, LCDs, WPCBs, CRTs, and li-ion batteries.

10.3.1 Au Leaching

To our knowledge, the recovery techniques for BMs from WPCBs are mature. However, recovery for PMs is still improper in practice. There is a need for the development of an appropriate technology for Au recovery from secondary resources, which should have salient features like selectivity, high recovery, economical, and eco-friendly even when operated under small-scale operation. The technologies currently available for the recovery of Au include mechanical, pyrometallurgical, hydrometallurgical, electrochemical, or biotechnological processes or their combinations. However, most of these methods suffer from environmental risks and from the need for extensive preprocessing of the material. Hydrometallurgical processes have proven to be successful for leaching of Au from WEEE, but the recovery of Au from solution is often laborious, containing several steps.

The recovery of PM can be accomplished only after the BMs (i.e., Cu, Sn, Pb, etc.) are mainly leached; otherwise, the process involves an excessive use of leaching solutions due to the high BM content (e.g., Cu > 15%). Therefore, preliminary steps are required to separate PM from BMs and other complex materials present in the WPCBs. This is achieved by using efficient hydrometallurgical methods which can extract the metals selectively, as in the case of Cu recovery coupled with the enrichment of Au. However, there are still issues that need to be solved, like increasing the leaching rate and selectivity while reducing the reagent consumption. Moreover, stringent environmental regulations have stimulated further research on the recycling of WPCBs through environment-friendly and energy-saving processes. Presence of Cu, Sn, and Pb has significant effect on the recovery of PMs. It is well known that the extraction of high-purity Au and Ag is not possible if the solution contains Cu. Furthermore, the dissolution of Sn and Pb (solder) is crucial in the separation Au-rich ECs from the other parts of WPCBs. Besides, if the leaching of Cu, Sn, and Pb occurs efficiently in the identified optimal experimental conditions, then there is no doubt that the most reactive BMs like Ni and Fe will be dissolved, from the entire WPCB, with high performances [14].

Hydrometallurgical processes for Au recovery from secondary sources involve three stages: pretreatment, extraction/recovery, and refining. Au leaching can be performed by using a variety of reagents which include HNO_3; mixtures of HNO_3, HCl, and H_2SO_4; $H_2SO_4 + HNO_3 + H_2O_2$; aqua regia; $FeCl_3$; thiourea; K isocyanate; K iodide and iodine; iodide + nitrite mixture; thiosulfate; and cyanides [76]. Au leaching with Cu-catalyzed ammoniacal thiosulfate solution has been extensively studied. Au recovery from thiosulfate solution studied less with cementation with Zn or Fe powder and adsorption on C.

Thiosulfate leaching to recover Au from ores is known for several decades. It is nontoxic, noncorrosive, as well as economical. Oxygen carrying catalyst (Cu $(HN_3)_4^{3+}$) is required. Thiosulfate leaching has many advantages over the cyanidation process including higher leaching rates and less interference from foreign ions. Potassium persulfate ($K_2S_2O_8$) is also nontoxic and strong oxidizing agent. Their decomposition products like CO_2, H_2O, and potassium sulfate are eco-friendly. Halides/chlorides are reliable, safe, nontoxic, noncorrosive, and very selective. Consumption, reagent cost, chlorine gas, and special reactor requirements are disadvantages. Table 5.4 shows some of the acids or caustic solutions like halide, thiourea, thiosulfate, and cyanide to leach PMs (Au, etc.) from WPCBs [77].

10.3.1.1 Cyanide Leaching of Au

Cyanide is the most commonly used conventional leaching reagent for Au and other PMs from complex ores more than a century. Due to toxicity and having serious environmental issues, its use has been prohibited in many countries. Cyanide is largely used due to highly efficient recovery, low cost, and simple operation. The mechanisms of cyanide leaching are the noble metals forming cyanide complexes into the solution during leaching and the reaction for Au is given as

10.3 Extraction of Precious Metal by Leaching

$$4Au + 8CN^- + O_2 + 2H_2O \rightarrow 4[Au(CN)_2]^- + 4OH^- \qquad (10.29)$$

Toxic cyanide leaching of Au has been selectively used in mining industry after the nineteenth century. Leaching by cyanide is dependent on cyanide concentration, oxidant concentration, temperature, pH, particle size, and other physicochemical features of the leaching system. For example, maximum Au leaching rate in cyanide solution can be reached at pH 10–10.5. Cyanide is widely used as lixiviant for PM recovery due to low cost. However, the environmental concerns and toxicity to humans have pushed the development of noncyanide lixiviants.

10.3.1.2 Thiourea Leaching of Au, Ag, and Pd

Thiourea is an organic sulfide, whose crystals can be dissolved in water or acid solution, and then could react with Au/Ag to produce a stable cationic complex or aurous ion. Thiourea is noncyanide, less toxic, and noncorrosive, has less environmental impact, and dissolves Au and Ag at faster leaching rate. But, thiourea is regarded as carcinogenic (due to low chemical stability) and sensitive to the presence of BMs (i.e., Cu, Pb, As, and Sb), which are present in WPCBs. The use of thiourea is not an economical way for Au leaching due to its high consumption and cost.

The kinetics of Au dissolution in acidic thiourea solutions in the presence of a variety of oxidants, namely, H_2O_2, $Fe_2(SO_4)_3$, formamidine disulfide, oxygen, Na_2O_2, and MnO_2 have been studied by many authors [78]. Commercial application of thiourea in Au recovery processing has been hindered due to the higher consumption of reagents and complexity of the leaching system when treating various ores/concentrates and different recycled materials.

At 154 μm, 90% Au and 50% Ag could be reached. Further size reduction achieves full PM recovery. However, thiourea consumption is usually very high and is more expensive compared to cyanide and thiosulfate lixiviants [79]. There are still a lot of challenges existing in PM recovery from WEEE, e.g., low total recovery. In practice, <20% of PMs recycling from WEEE has been recovered. PM loss during pretreatment and multistep leaching should be minimized in WEEE treatment.

Li et al. (2012) examined the thiourea leaching of Au and Ag from WPCBs as functions of particle size, temperature, and thiourea concentration, as well as Fe^{3+} concentration. It appeared that the optimum condition for Au leaching happened when 24 g/L thiourea and 0.6% of Fe^{3+} were used within 2 h. It is also proved that thiourea is less toxic and highly efficient [59]. Birloaga et al. (2014) found that 69% of Au was extracted under the conditions of 20 g L^{-1} thiourea, 6 g L^{-1} Fe^{3+}, and 10 g L^{-1} H_2SO_4, as well as 600 rpm. Furthermore, under the same reagent condition, a multistage crosscurrent leaching was used to reduce the consumption of thiourea and improve the efficiency of Au leaching from WPCBs [80]. Yin et al. (2014) compared thiourea leaching with iodine leaching of Au from WPCBs. It appeared that 93.5% of Au was able to be leached out directly by iodine without pretreatment, while thiourea leaching of Au was carried out after Cu leaching, which resulted in the high consumption of thiourea [81].

Thiourea leaching Au and Ag from WPCBs was also carried out by Xu and Li (2011) [82], and the influence of leaching time, reaction temperature, thiourea concentration, Fe^{3+} concentration, and material particle size on Au leaching rates were investigated. The results showed that Au and Ag leaching rates can reach 90.9% and 59.8%, respectively, under the best conditions. Research by Wu et al. (2009) [83] also showed that selectively leaching Au and Ag from WPCB in acid thiourea solution is possible and effective. The optimal leaching condition was S/L, 1/10; N, 300 rpm; T, 20–25 °C; t, 1 h; thiourea concentration, 12 g L^{-1}; Fe^{3+} concentration, 0.8%; and pH = 1.5. The leaching yields of Au and Ag were 91.4% and 80.2%, respectively. Results of the experiment of leaching Au from WPCBs with thiourea showed that suitable leaching conditions are as follows: S/L, 1/5; T, 35 °C; thiourea concentration, 10 g/L; Fe^{3+} concentration, 0.3%; H_2SO_4, concentration 5%; and t, 1 h. Compared with cyanide leaching, thiourea leaching Au has the advantages of quick leaching kinetics, low toxicity, high efficiency, environmentally friendly, and less interference ions [84]. Commercial application of thiourea leaching has been hindered by the following factors [85]:

- It is more expensive than cyanide.
- Its consumption in Au processing is high because thiourea is readily oxidized in solution.
- The Au recovery step requires more development. However, it was implied that the high costs attached to leaching are likely due to the thiourea process still being in an infancy stage.

Batnasan et al. (2018) used sequential high-pressure oxidative leaching (HPOL) (C: 1 M H_2SO_4, T: 120 °C, P_{total}: 2 MPa, N: 750 rpm, t: 30 min with O_2 gas flow) and thiourea leaching (0.5 M < H_2SO_4 < 4 M; 0.25 g L^{-1} < thiourea < 20 g L^{-1}; 0 g L^{-1} < oxidant < 7.5 g L^{-1}; 25 °C < T < 80 °C; 1 h < t < 10 h at 50 g L^{-1} pulp density and N: 50 rpm). After HPOL, the significant amount of BMs reacted with dilute H_2SO_4 solution (i.e., Cu recovery was 99.3%, Zn 96%, Al 81.6%, Ni 82.7%, Co 84%, Mn 71.4%, and Fe 63.2%). Au, Ag, Pd, Pb, and SiO_2 remained in the solid residue due to HPOL conditions. The use of oxidant ($Fe_2(SO_4)_3$) significantly increases Au and Pd dissolution. Increasing the Fe^{3+} oxidant concentration to 5 g L^{-1} significantly increased Au dissolution (from 18% to 68%) and slightly increased Pd recovery (up to 7.1%) but reduced Ag dissolution (down to 1.6%) [78]. Fe^{+3} oxidant enhances leaching [78]:

$$Au + 2CS(NH_2)_2 + Fe^{3+} \rightarrow Au[CS(NH_2)_2]_2^+ + Fe^{2+} \quad (10.30)$$

$$Pd + 4CS(NH_2)_2 + Fe^{3+} \rightarrow Pd[CS(NH_2)_2]_4^{2+} + Fe^{2+} \quad (10.31)$$

$$Ag + 3CS(NH_2)_2 + Fe^{3+} \rightarrow Ag[CS(NH_2)_2]_3^+ + Fe^{2+} \quad (10.32)$$

Further increase in oxidant in the thiourea solution adversely affects the efficiency of PMs due to oxidation of thiourea to form stable ferric sulfate

10.3 Extraction of Precious Metal by Leaching

complex and formamidine disulfide, which rapidly decomposes to thiourea, cyanimide, and elemental sulfur as follows:

$$Fe^{3+} + SO_4^{2-} + SC(NH_2)_2 = (FeSO_4 \cdot SC(NH_2)_2)^+ \quad (10.33)$$

$$2SC(NH_2)_2 + 2Fe^{3+} = (SCN_2H_3)_2 + 2Fe^{2+} + 2H^+ \quad (10.34)$$

$$(SCN_2H_3)_2 = SC(NH_2)_2 + NH_2CN + S \quad (10.35)$$

Low Ag dissolution is related to the formation of insoluble thiourea complex and silver sulfide (Ag_2S) under the oxidation conditions:

$$2Ag + (FeSO_4 \cdot SC(NH_2)_2)^+ = (Ag_2SO_4 \cdot SC(NH_2)_2)^+ + Fe^{2+} \quad (10.36)$$

$$2Ag + S = Ag_2S \quad (10.37)$$

The effect of thiourea concentration on Au, Ag, and Pd dissolution is important. The maximum dissolution efficiencies of Au, Ag, and Pd were 91%, 81%, and 11.9% at thiourea concentration of 12.5 g L^{-1}. Occurrence of formamidine disulfide (SCN_2H_3)$_2$ is beneficial for leaching PMs [78]. The use of appropriate concentrations of thiourea and formamidine disulfide results in an increase in the leaching of PMs, which can be expressed as follows:

$$2Au + 2SC(NH_2)_2 + (SCN_2H_3)_2 + 2H^+ = 2Au[SC(NH_2)]_2^+ \quad (10.38)$$

$$2Ag + 2SC(NH_2)_2 + 2(SCN_2H_3)_2 + 4H^+ = 2Ag[SC(NH_2)]_3^+ \quad (10.39)$$

$$2Pd + 4SC(NH_2)_2 + 2(SCN_2H_3)_2 + 4H^+ = 2Pd[SC(NH_2)_2]_4^+ \quad (10.40)$$

$$Ag_2S + 6SC(NH_2)_2 + 2H^+ = 2Ag[SC(NH_2)_2]_3^+ + H_2S \quad (10.41)$$

These results indicate that the formamidine disulfide (SCN_2H_3)$_2$ produced can be beneficial to the efficiency of leaching of PMs under precisely optimized leaching conditions. It is noteworthy that, with addition of thiourea, the pH of the leaching medium does not change remarkably, whereas the redox potential in the medium falls from 0.67 to 0.43 V. The variations of the redox potential with addition of thiourea and oxidant show completely opposite trends, which are probably related to oxidation of thiourea to form a ferric-thiourea complex and formamidine disulfide, respectively.

Thiourea leaching under the optimized conditions (H_2SO_4, 1 M; thiourea, 12.5 g L^{-1}; oxidant concentration, 5 g L^{-1}; pulp density, 50 g L^{-1}; T, 25 °C; N, 500 rpm; t, 6 h) was several times faster and could take place under ambient conditions, compared with the conventional cyanidation process that is carried out under alkaline conditions (9.5 < pH < 12) at temperature between 60 and 95 °C during 24 h in the presence of oxygen. These results suggest that PMs, especially Au and Ag, present in WPCB ash samples can be effectively recovered using such a sequential leaching procedure [78]. They found the following results:

- Some BMs present in the WPCB ash sample were effectively extracted via HPOL with dilute H_2SO_4 solution, including recovery of Cu, Zn, Ni, Co, Al, and Fe of 99.3%, 96.0%, 82.7%, 84.0%, 81.6%, and 63.2%, respectively.
- The optimized thiourea leaching process dissolved 100% Au, 81% Ag, and 13% Pd from the WPCB residue sample after HPOL in acidic thiourea solution.
- Higher oxidant concentration played a vital role in leaching of Au and Pd but had a deleterious effect on Ag leaching in thiourea solution.
- The efficiency of dissolution of PMs strongly depended on the concentration of both thiourea and oxidant in the leaching medium.

10.3.1.3 Thiosulfate Leaching of Au

Several studies highlighted thiosulfate leaching of Au. Thiosulfate leaching is operated in alkaline condition to prevent thiosulfate decomposition [86]. Thiosulfate has low chemical stability and low metal recovery. A typical noncyanide lixiviant is ammonium thiosulfate which is nontoxic and economical; but, Au recovery rates are low. Cu^{+2} can be used as a catalyst to enhance the recovery:

$$Au + 5S_2O_3^{2-} + Cu(NH_3)_4^{2+} \rightarrow Au(S_2O_3)_2^{3-} + Cu(S_2O_3)_3^{5-} + 4NH_3 \quad (10.42)$$

There are two kinds of thiosulfate commonly used in leaching Au: one is sodium thiosulfate; the other is ammonium thiosulfate. Au and thiosulfate will form a stable complex with the presence of oxygen. Thiosulfate is stable in alkaline medium, because tetrathionate, the oxidation product of thiosulfate, will turn into thiosulfate again under alkaline conditions in about 60%. But, the pH of the solution cannot to be too high; otherwise it is prone to disproportionation reaction turning thiosulfate into sulphurized form. Thiosulfate leaching has the advantages of high selectivity, nontoxic and noncorrosive. The principal problem with thiosulfate leaching is the high consumption of reagent during extraction. It is reported a loss of up to 50% of thiosulfate in ammoniacal thiosulfate solutions containing Cu. High reagent consumption renders most thiosulfate systems uneconomical overall, in spite of their potential environmental benefits.

When thiosulfate is combined with Au, a complex formation occurs, i.e., aurothiosulfate [Au $(S_2O_3)_2^{3-}$] which is accountable for the Au recovery as shown in Eq. (10.43). Copper sulfate was used as catalyst and oxidizing agent in the solution which forms a stable cupric tetra-amine complex when reacts with ammonia, which stabilizes the aurothiosulfate complex [76]:

$$4Au + 8S_2O_3^{2-} + O_2 + 2H_2O \rightarrow 4Au(S_2O_3)_2^{3-} + 4OH^- \quad (10.43)$$

$$Au + 5S_2O_3^{2-} + Cu(NH_3)_4^{2+} \rightarrow Au(S_2O_3)_2^{3-} + 4NH_3 + Cu(S_2O_3)_3^{5-} \quad (10.44)$$

10.3 Extraction of Precious Metal by Leaching

The cupric tetra-amine complex ions get regenerated by reaction between dissolved oxygen and $Cu(S_2O_3)_3^{5-}$ as per reaction (Eq. 10.45):

$$Cu(S_2O_3)_3^{5-} + 4NH_3 + \tfrac{1}{4} O_2 + \tfrac{1}{2} H_2O$$
$$\rightarrow Cu(NH_3)_4^{2+} + 3S_2O_3^{2-} + OH^- \qquad (10.45)$$

In addition to above chemical reactions, some other degradation reactions to tetrathionate occur: Eq. (10.46) shows an oxidation reaction, which is promoted by the Cu^{2+} ion [58, 77]:

$$2Cu(NH_3)_4^{2+} + 8S_2O_3^{2-} = 2Cu(S_2O_3)_3^{5-} + 8NH_3 + S_4O_6^{2-} \qquad (10.46)$$

By maintaining the appropriate concentration of ammonia and thiosulfate, the conversion from the Cu^{2+} to the Cu^+ state can be controlled to obtain efficient leaching of Au. In this case Cu plays the role of catalyst due to the redox reaction between the Cu^{2+} and Cu^+ state.

Ha et al. (2014) optimized the thiosulfate leaching of Au from WPCBs of discarded mobile phones. The optimum condition was identified as 72.771 mM thiosulfate, 10.0 mM Cu^{2+}, and 0.266 ammonia for 2.395×10^{-5} mol.m^{-2}.s^{-1} Au leaching [77]. Tripathi et al. (2012) achieved 56.7% Au recovery with 0.1 M [(NH$_4$)$_2$S$_2$O$_3$] and 40 mM CuSO$_4$ promoter at $10 < pH < 10.5$, 10 g L^{-1} pulp density, N: 250 rpm, and t: 8 h [76]. Ha et al. (2010) also performed ammonia thiosulfate leaching of Au from waste mobile phones in the presence of Cu^{2+}, which oxidizes thiosulfate. They achieved 90% Au recovery in 10 h from PCBs containing 0.12% Au in the feed [59].

An evaluation [87] was conducted to compare thiosulfate with cyanide and HNO$_3$. Nearly 65% of Au was leached in the cyanide solution, and almost 100% of Ag was leached in HNO$_3$ solution. However, in the case of thiosulfate leaching, only around 15% Au could be extracted, which gave a negative indication of thiosulfate leaching of Au. Further, in this study, it was suggested that the presence of Cu ions promoted Au extraction in the sodium thiosulfate system. Ficeriová et al. (2011) found that 98% of Au and 93% of Ag were recovered from pretreated WPCBs under the conditions of 0.5 M (NH$_4$)$_2$S$_2$O$_3$, 0.2 M CuSO$_4$·H$_2$O, and 1 M NH$_3$ at 40 °C after 48 h. Up to 90% of Pd was also extracted by leaching in AR solution with 2 h [88]. Ha et al. (2010) reported that 98% Au could be recovered using a solution containing 20 mM Cu, 0.12 M thiosulfate, and 0.2 M NH$_3$ [89]. The reaction happening in the thiosulfate leaching is shown in Eq. (10.42).

Recovery of Au from WPCBs was carried out heated to boiling with 20% (w/v) aqueous potassium persulfate in 20 min. Peeled Au is settled as sludge and filtered, washed, dried, and melted with borax and KNO$_3$ [90]. Persulfate ion ($S_2O_8^{2-}$) is one of the strongest oxidizing agents. Under mild conditions, persulfate bivalent ions yield radical anion SO_4^{2-} that appears to be very effective e-transfer oxidizing agent and dissolve all BMs (Cu, Ni, Ag, etc.). After the filtration, BMs and sulfate salts are recovered. Recovered Au purity is 99.5%.

Zhang et al. (2017) studied a mechanochemical (MC) process in a planetary ball mill for effective recovery of Cu and PMs (Pd and Ag) from e-waste scraps [91]. Results indicated that the mixture of $K_2S_2O_8$ and NaCl was the most effective co-milling reagents in terms of high recovery rate. Soluble metallic compounds are produced after treatment with dry ball milling and consequently benefit the subsequent leaching process. According to the above discussion, possible reactions that occurred during ball milling process are expressed as the following reaction:

$$K_2S_2O_8 + 2NaCl + nM \rightarrow (\text{ball milling})\ nMCl_{2/n} + K_2SO_4 + Na_2SO_4 \quad (10.47)$$

If M = Cu, n = 1 and 2 (two reactions for Cu).
If M = Ag, n = 2.
If M = Pd, n = 1.

After co-milling with $K_2S_2O_8$/NaCl, soluble metallic compounds were produced and consequently benefit the subsequent leaching process. 99.9% of Cu and 95.5% of Pd in the e-waste particles could be recovered in 0.5 M diluted HCl in 15 min. Ag was concentrated in the leaching residue as AgCl and then recovered in 1 M NH_3 solution.

The behavior of cathodic process of leaching Au with thiosulfate has been investigated by Jiang [92]. Results show that ammonia reacts with Au ions on anodic surface of Au; the formed $Au(NH_3)^{2+}$ is substituted by $S_2O_3^{2-}$ after entering solution to form more stable $Au(S_2O_3)^{2-}$. The kinetics of sodium thiosulfate leaching of Au was studied by Moore [93], and the results show that it has linear relationship between Au solubility and temperature when temperature is between 45 and 85 °C. Take into account that thiosulfate will be prone to decomposition at high temperature; the suitable temperature is between 65 and 75 °C. The kinetics of electrochemical reaction of thiosulfate leaching under ammonia medium was researched by Heath [94]; results show that leaching rate depends on the concentration of thiosulfate, Cu^{2+}, and ammonia. The higher the concentration of reagents, the faster the leaching reaction goes. Yen (2008) [95] found that the dissolution of Cu minerals is the main reason for the increase of thiosulfate consumption. Thiosulfate leaching process, whose solution is an ammoniacal solution, is suitable for handling Au mine rich of alkaline components, especially for Au ore or Au concentrate containing Cu, Mn, and/or As which is sensitive to cyanide leaching [17]. Table 10.3 shows summaries of previous PM recovery study test conditions used and recovery rates.

10.3.1.4 Halide Leaching of Au

Halide leaching includes chloride leaching, bromide leaching, and iodide leaching. Their common merits are high leaching rate. Au forms both Au^+ and Au^{3+} complexes with chloride, bromide, and iodide depending on the solution chemistry conditions. Only chlorine/chloride has been applied industrially on a significant

10.3 Extraction of Precious Metal by Leaching

Table 10.3 Summary of previous precious metal leaching studies

Raw material	Particle size	Solvent reagents	Conditions/chemistry/species	Recovery rate	References
WPCBs particles	0.43–3.33 mm	NaCN [CN$^-$, Air (O$_2$)]	4 g/L cyanide; flux 20 L/d kg WPCBs day, column leaching 10.5 < pH < 11; t: 15 days $4Au + 8CN^- + O_2 + 2H_2O = 4Au(CN)^- + 4OH^-$/Au $(CN)_{2-}$ (log K = 38.3)/E_0: −0.67 V; pH > 10; 25 °C	Au: 46.6% Ag: 51.3%	Montero et al. [96]
WPCBs particles	<2 mm	H$_2$SO$_4$; thiourea	Cu removal: 2 M H$_2$SO$_4$ + 5% H$_2$O$_2$; 25 < T < 30 °C; N: 200 rpm, t: 3 h Au leach: 20 g/L thiourea, 6 g/L Fe^{3+} and 0.5 M H$_2$SO$_4$; 1/10 S/L; pH: 1, T: room; N: 500 rpm	79.1 mg/L Au 121.1 mg/L Ag	Birloaga et al. [80]
WPCBs particles	53–57 mm	Thiourea	Cu removal: 0.5 M thiourea; 0.03–1.0 M conc.; T: 30 °C, t: 24 h Au leaching: 0.05 M H$_2$SO$_4$; T: 45 °C; t: 6 h Ag leaching: 05 M thiourea in 0.05 M H$_2$SO$_4$; T: 60 °C; t 2 h	3.2 mg/g Au 6.8 mg/g Ag	Gurung et al. [97]
WPCBs particles	<2 mm	Thiourea [CSN(NH$_2$)$_2$, Fe^{3+}]	20 g/L thiourea; 6 g/L Fe^{3+}; 10 g/L H$_2$SO$_4$, N: 600 rpm; pH: 1,4; T: room $2Au + 4CS(NH_2)_2 + Fe^{3+} = 2Au(CS(NH_2)_2)_2^+ + 2Fe^{2+}$/Au $(CS(NH_2)_2)_2$ + (log K = 22) E^0: 0.038; pH: 1–2; 25 °C	69% Au	Birloaga et al. [80]
WPCBs		Thiourea	24 g/L thiourea and 0.6% Fe^{3+}; room temp.; 2 h Mobile phone PCBs	90% Au 50% Ag	Li et al. [60]
WPCBs particles	74–180 mm	Iodine	1.0–2.0% iodine conc.; $n(I_2)$:$n(I^-)$ = 1:8; 1–2% H$_2$O$_2$ conc.; t: 4 h; S/L: 1/10; T: 25 °C; pH: 7	95%	Xu et al. [98]
WPCBs		Iodine H$_2$O$_2$	3% iodine, 1% H$_2$O$_2$ S/L: %15	100% Au	Sahin et al. [99]
WPCBs particles	<4 mm	Iodine-iodide	S/L: 1:2–1:4, N: 300 rpm; T: room; T: 40–180 °C	98.5% Au, 99% Ag 97.2% Pd	Xiu et al. [100]

(continued)

Table 10.3 (continued)

Raw material	Particle size	Solvent reagents	Conditions/chemistry/species	Recovery rate	References
WPCBs particles	<2 mm	NaClO-HCl-H$_2$O$_2$	Cu removal: 2 M H$_2$SO$_4$; 35% H$_2$O$_2$; N: 200 rpm; t: 3 h; T: 25 °C	71.4% Ag	Behnamfard et al. [67]
			Au & Ag leach: 20 g/L thiourea (CS(NH$_2$)$_2$); 6 g/L Fe^{3+}; 10 g/L H$_2$SO$_4$; N: 200 rpm; t: 3 h; T: 25 °C	100% Au	
			Pd leach: 1% v H$_2$O$_2$; 10% v NaClO, 5 M HCl, t: 3 h; T: 90 °C	100% Pd	
WPCBs	−800 μm	H$_2$SO$_4$; NaCl; AR	1. Cu, Zn, Fe, Al, Ni dissolution: oxidative 0.2MH$_2$O$_2$ + 2MH$_2$SO$_4$ leach, −800 μm, pH:6; t: 6 h; T: 80 °C	95% Au	Quinet et al. [101]
		Ammonium thiosulfate	2. Pb and Sn dissolution: 2MNaCl leaching; pH: 6; t: 2 h; T: 25 °C; N: 500 rpm	93% Ag	
			3. Au, Ag leach: 0.5 M (NH$_4$)$_2$S$_2$O$_3$ + 0.2 MCuSO$_4$.5H$_2$O + NH$_3$; pH: 9; t: 48 h; T: 40 °C; N: 500 rpm	99% Pd	
			4. Pd leach: 0.5 M Aqua regia (AR); pH: 2; t: 2 h; T: 25 °C; N: 500 rpm		
WPCB	Crushed	K-Persulfate	Formic acid epoxy resin peeling 20% (v/v): t: 20 min	99.5% Au	Syed [90]
Au-coated bangles-mirrors			20% (w/v) aqueous persulfate heated to boiling: t: 20 min		
WPCB particles	0.5–3.0 mm	Ammonium thiosulfate	0.1 M [(NH$_4$)$_2$S$_2$O$_3$]; 40 mM CuSO$_4$ promoter; 10 < pH < 10.5; 10 g/L pulp density; N: 250 rpm; t: 8 h	56.7% Au	Tripathi et al. [76]
WPCB whole			Only mobile phone particles or whole unit	78.8% Au	
WPCB				98% Au	Ha et al. [89]

10.3 Extraction of Precious Metal by Leaching 251

WPCB	Ammonium thiosulfate	20 mM Cupric, 0.12 M thiosulfate and 0.2 M ammonia; t: 2 h; $10 < \text{pH} < 10.5$			Ha et al. [77]
		N: 200 rpm; T: 25 °C.			
	Ammonium thiosulfate	72.71 mM thiosulfate; 10 nM Cu^{2+}; 0.266 M ammonia; N: 400 rpm	2395×10^{-5} mol/$m^2 \cdot s$		
		Mobile phone PCB			
	$S_2O_3^{2-}$, NH_3, Cu^2+	$4Au + 8S_2O_3^{2-} + O_2 + 2H_2O = 4[Au(S_2O_3)_2]^{3-} + 4OH^-$ / $[Au(S_2O_3)_2]^{3-}$/(log K = 28.7)/E^0: −0.274 to 0.38 V; pH: 8–11; 25 °C			
WPCB	Ammonium thiosulfate	0.5 M(NH_4)$_2S_2O_3$ + 0.2 M$CuSO_4 \cdot 5H_2O$ + 1 MNH_3; pH: 9; t: 48 h; T: 40 °C; N: 500 rpm	98% Au, 93% Ag, 90% Pd	Ficeriova et al. [102]	
WPCBs	Halide			Zhang et al. [103]	
	Cl/Cl_2, Br^-/Br_2				
	I^-/I_2	$Au + 11HCl + HNO_3 = 2HAuCl_4 + 3NOCl + 6H_2O$/ $AuCl_4^-$/(log K: 29.6) E^0: 0.038; pH: 1–2; 25 °C			

scale of the halides. But, chloride leaching requires special stainless steel and rubber-lined equipment to resist the highly corrosive conditions. The vapor pressure of Br is 10 kPa at 0 °C and 28 kPa at 35 °C. Special equipment is required for safety and health risks in Br leaching process, and that restricts it from industrial application [101]. Iodide leaching has the following advantages:

- Quick leaching kinetics
- Good selectivity, less leaching of BMs
- Easy to regenerate iodide, iodine being reduced while recovering Au in anode region
- No corrosion, for iodide leaching in a weakly alkaline medium
- No toxicity

Moreover, the complexes formed by Au and iodine are the most stable complexes formed by Au and halogen [103]. Research on iodide leaching Au from discard PCBs was carried out by Xu et al. (2010) [98]. According to the result, the rational parameters for iodide (I^-) leaching Au are 1.0–1.2% iodide concentration, $n(I_2)$: $n(I^-)$ = 1:8 ~ 1:10, 1–2% H_2O_2 concentration, leaching time 4 h, S/L: 1/10, leaching Au at normal temperature (25 °C), and solution pH 7. Au leaching rate can reach at about 95%. Iodide leaching is a promising technology in noncyanide leaching process. However, iodide leaching consumes a great deal of reagent, and iodine is relatively expensive, and efficiency of electrolytic deposition of Au needs to be improved. All these problems must be solved before iodide leaching Au is applied in industry.

10.3.1.5 Au Leaching by Electro-generated Chlorine

Chlorine generation from HCl acid solution can be represented by the following reaction [10]:

$$\text{Anode}: 2Cl^- \leftrightarrow Cl_2(\text{electrode surface}) + 2e^- \quad E^0 = 1.35\,V_{SHE} \quad (10.48)$$

The Cl_2 gas by Eq. (10.48) dissolves in water as follows (where Log K_{1-3} being the equilibrium constants at 25 °C):

$$Cl_2(g) \leftrightarrow Cl_2(aq) \quad \text{Log } K_{1(25\,°C)} = -1.21 \quad (10.49)$$

$$Cl_2(aq) + H_2O \leftrightarrow HCl + HOCl \quad \text{Log } K_{2(25\,°C)} = -3.40 \quad (10.50)$$

$$Cl_2(aq) + Cl^- \leftrightarrow Cl_3^- \quad \text{Log } K_{3(25\,°C)} = -0.71 \quad (10.51)$$

The dissolution of gold takes place as

$$Au + Cl^- + 3/2 Cl_2(aq) \leftrightarrow AuCl_4^-(aq) \quad \Delta G_{25\,°C} = -27.04\,\text{kcal/mol} \quad (10.52)$$

10.3 Extraction of Precious Metal by Leaching

$$Au + 3/2HClO + 3/2H^+ + 5/2Cl^- = AuCl_4^-(aq)$$
$$+ 3/2H_2O \quad \Delta G_{25\ °C}$$
$$= -33.29\,kcal/mol \quad (10.53)$$

where G is the Gibbs free energy change at 25 °C. The dissolution of Cu can be expressed as follows in Cl_2 media:

$$Cu + Cl_2(aq) \leftrightarrow Cu_2^+ + 2Cl^- \quad \Delta G_{25\ °C} = -46.69\,kcal/mol \quad (10.54)$$

$$Cu^{2+} + Cu + 2Cl^- \leftrightarrow 2CuCl \quad \Delta G_{25\ °C} = -4.87\,kcal/mol \quad (10.55)$$

Thus, thermodynamically the reaction of Cl_2 with Cu (Eq. 10.54) is more favorable (lower ΔG value) than that of Au (Eq. 10.52), whereas the feasibility of reaction of hypochlorous acid (HClO) with Au is higher (Eq. 10.54) than the reaction with Cl_2 (Eq. 10.52) under the same condition. In the presence of Cl_2, Au and Cu dissolve to form Au(III) and Cu(I), Cu(II) chloride complexes, $AuCl_4^-$, $CuCl^+$, $CuCl_3^{2-}$, and CuCl which is evident from the potential Eh–pH [10].

Kim et al. (2011) investigated the selective HCl leaching of Au, Cu, and Ni from mobile phone WPCBs by electro-generated chlorine gas as an oxidant and its recovery by ion exchange process using Amberlite. Leaching was carried out at a separate mixed leaching reactor connected with anode compartment of an electrolytic cell for Cl_2 generation. Au leaching increased with increase in temperature and initial concentration of Cl_2 and was favorable even at low concentration of acid, whereas Cu leaching increased with increase in concentration of acid and decrease in temperature. In a two-stage leaching process, Cu was mostly dissolved (97%) in 165 min at 25 °C during the first-stage leaching in 2.0 mol L^{-1} HCl by electro-generated Cl_2 at a current density of 714 A m^{-2} along with a minor Au recovery (5%). In the second-stage Au was mostly leached out (R: 93%, ~67 mg L^{-1}) from the residue of the first stage by the electro-generated Cl_2 in 0.1 mol L^{-1} HCl. Au recovery from the leach liquor by ion exchange using Amberlite XAD-7HP resin was found to be 95% with the maximum amount of Au adsorbed as 46.03 mg g^{-1} resin. A concentrated Au solution, 6034 mg L^{-1} with 99.9% purity was obtained in the ion exchange process [10].

10.3.1.6 Aqua Regia Leaching of Precious Metals (Au, Ag, and Pd)

Park and Fray (2009) proposed a process of recycling high-purity PMs (Au, Ag, and Pd) from WPCBs using AR. Optimum S/L ratio was 1/20 g/mL and leach time 3 h. Au, Ag, and Pd recoveries were 97%, 98%, and 93%, respectively. In this study Au was recovered as nanoparticles to improve the value. Ag is relatively stable in AR; so, it remained unreacted due to the formation of AgCl black surface which prevents Ag dissolution in AR. Pd forms a red cubic structural precipitate ($Pd(NH_4)_2Cl_6$) with AR. Zn dissolved in AR helped Pd precipitation. A solvent extraction was employed to selectively recover Au, where toluene, dodecanethiol, and sodium borohydride were used. Extracted Au was converted to nanoparticles (round shaped: 20 nm) [104]. However, the application of AR in extraction of PMs is limited in a lab scale because AR is strongly oxidative and corrosive, and the waste water from leaching is too acidic to be dealt with [102].

10.3.1.7 Comparison of Various Leaching Processes

As one of the most popular methods for multi-criteria decision-making, the three-scale analytic hierarchy process (AHP) has been successfully applied to determine the weight of evaluation indexes of different leaching methods by Zhang et al. (2012) [17]. Table 10.4 shows basic scores of different evaluation indexes of various leaching methods along with final AHP leaching scores. Seven leaching technologies mentioned above are technically feasible. However, in order to develop an environment-friendly technique for recovery of PMs from WPCBs, we should pay more attention to evaluating the environmental impact of techniques. A critical comparison of above leaching methods is analyzed based on three-scale AHP. The analysis shows that cyanide leaching gains the highest points for its good economic feasibility and research level. But, when the toxicity of reagents is taken into account, it is an inevitable trend for noncyanide leaching to replace cyanide leaching. With further research, thiourea leaching and iodide leaching are both likely to reduce the cost of the process and to be more reliable in technology, so as to replace cyanide leaching [17].

10.3.1.8 Silver (Ag) Leaching

An oxidative H_2SO_4 leach dissolves Cu and part of the Ag; an oxidative chloride leach dissolves Pd and Cu; and cyanidation recovers the Au, Ag, Pd, and a small amount of the Cu. To recover the metals from each leaching solution, precipitation with NaCl was preferred to recuperate Ag from the sulfate medium; Pd was extracted

Table 10.4 Basic and final scores of different evaluation indexes of various leaching methods

Leaching method	Economic feasibility			Economic impact	Research level	Leaching core
	Leaching rate	Reagent cost	Corrosivity	Toxicity	Reliability	
Cyanide	3	5	5	0	5	4.46
Aqua regia	4	4	0	3	5	3.48
Thiourea	4	4	4	4	4	4.00
Thiosulfate	2	2	5	4	2	2.71
Chloride	5	4	0	3	4	3.25
Bromide	5	2	2	3	2	2.25
Iodide	5	3	5	5	3	3.64

10.3 Extraction of Precious Metal by Leaching

from the chloride solution by cementation on Al; and Au, Ag, and Pd were recovered from the cyanide solution by adsorption on activated carbon. The optimized flowsheet permitted the recovery of 93% of the Ag, 95% of the Au, and 99% of the Pd [103].

In a less concentrated H_2SO_4, Ag metal is rather inert when oxidant is absent. The possibility of leaching Ag increases in a H_2SO_4 solution with higher acid concentration at elevated temperature, e.g., 80 °C, and H_2SO_4 may react with Ag as covalent molecules according to

$$2Ag + H_2O_2 \rightarrow Ag_2O + H_2O \quad (10.56)$$

The redox reaction is more easily happening in concentrated H_2SO_4 (such as 96%). When Ag is already oxidized or oxidant is present during leaching, Ag is reacting with H_2SO_4 following alternative procedures as [24]

$$2Ag + H_2O_2 \rightarrow Ag_2O + H_2O \quad (10.57)$$

$$Ag_2O + H_2SO_4 \rightarrow Ag_2SO_4 + H_2O$$
$$\left(K_{sp} : 1.2 \times 10^{-5} \text{ and solubility at } 25°C\ 8.3\ g\ L^{-1} \text{ in water}\right) \quad (10.58)$$

Zhang and Zhang (2017) used MC process for effective recovery of Cu, Pd, and Ag from e-waste. They used planetary ball milling with the mixture of $K_2S_2O_8$ (oxidant) and NaCl (1:1 ratio) at 600 rpm for 3 h. Using diluted HCl, 99.5% of Cu and 95.5% Pd was leached in 15 min. Then Pd(II) was separated from BMs by solvent extraction. 94.7% of Ag concentrated in the residue as AgCl was leached with 1 M NH_3 and recovered after hydrazine hydrate reduction [91].

10.3.1.9 Palladium Leaching with $CuSO_4$ + NaCl and Extraction with Diisoamyl Sulfide (S201)

Other than Au and Ag, Pd is also considered as a valuable metal to be recycled. Zhang and Zhang (2014) proposed the process to recover Pd, including enrichment and dissolution of Pd and extraction and stripping of Pd(II). Cu was leached out in the solution of $CuSO_4$ and NaCl. An environmentally benign, non-acid process was successfully developed for selective recovery of Pd from WPCBs [105]. The overall reaction was shown as Eq. (10.59):

$$Cu + Cu^{2+} + 2Cl^- = 2CuCl_{(s)} \quad (10.59)$$

When the ratio of $[Cu]/[Cu^{2+}]$ was more than 1.4, Pd was leached out along with CuCl. Thereafter, Pd was further dissolved in the solution of $CuSO_4$ and NaCl, where the ratio of $[Cu]/[Cu^{2+}]$ was less than 0.9. The reaction of Pd dissociation was shown as Eq. (10.60). Diisoamyl sulfide (S201) was applied to extract 99.4% Pd from leaching solution; then a two-step stripping was accomplished using dodecane with 0.1 M NH_3. Therefore, 96.9% Pd was obtained with negligible effect of Cu ions.

$$0.5Pd + Cu^{2+} + 4Cl^- = CuCl_2^- + 0.5PdCl_4^{2-} \qquad (10.60)$$

Process has three stages: enrichment, dissolution, and extraction and stripping steps. In the enrichment step, $Cu/Cu^{2+} \geq 1.4$ and in the dissolution stage, $Cu/Cu^{2+} \geq 0.9$. Cu is removed as CuCl(s) after filtration. The dissolved Pd was then extracted by diisoamyl sulfide (S201). It was found that 99.4% of Pd^{2+} could be extracted from the solution under the optimum conditions (10% S201, aqueous/organic (A/O) ratio, 5 and 2 min extraction). In the whole extraction process, the influence of BMs was negligible due to the relatively weak nucleophilic substitution of S201 with BMl irons and the strong steric hindrance of S201 molecular. Around 99.5% of the extracted Pd^{2+} could be stripped from S201/dodecane with 0.1 mol L^{-1} NH_3 after a two-stage stripping at A/O ratio of 1. The total recovery percentage of Pd was 96.9% during the dissolution-extraction-stripping process. Therefore, this study established a novel, benign, green, and effective process for selective recovery of Pd from WPCBs without corrosive acid or strong oxidant utilization [105].

10.4 Full PM Recovery

In the gold industry, Au is recovered from Au-rich leaching solution by carbon adsorption, following precipitation with Zn dust or electrowinning [106, 107]. Behnamfard et al. (2013) proposed the flowsheet involving a two-step H_2SO_4 leaching, acidic thiourea leaching with Fe^{3+} ions, and $HCl-NaClO-H_2O_2$ leaching, as well as precipitation with sodium borohydride (SBH). Approximately 99% of Cu could be recovered. It was found that 84.3% of Au and 71.4% of Ag could dissolve in acidic thiourea with ferric ions, while Pd could not be dissolved in thiourea solution. In this study, SBH was used as a reducing reagent to selectively precipitate Au from Ag. The optimal precipitation of Au and Ag occurred at 8 g/L SBH in 15 min [67].

10.5 Brominated Epoxy Resin (BER) Leaching

In the last years, the recycling of WPCBs has increasingly been taken seriously for two main reasons. The first is that WPCBs contain considerable amount of metallic and nonmetallic (i.e., fiberglass encapsulated in resins) components. Recently, the separation of WPCBs using organic solvents has become more prevalent because it

10.5 Brominated Epoxy Resin (BER) Leaching

is an environmentally friendly and efficient technique. However, the relatively high temperatures (~135 °C) used during the separation process lead to higher energy consumption, faster solvent degradation, and possibly higher emissions of toxic fumes. BER, which contains about 18% of Br, leaching with organic solvents is claimed to be promising to physico-mechanical treatments in terms of recycling rate and zero dust generation. The economic analysis showed that the application of the developed leach technique on an industrial scale can bring a huge economic return up to 90,000 $ t^{-1} on of WPCBs [108]. Organic solvent dissolution can be supported by ultrasonic and mechanical treatment to accelerate the separation process. If the spent solvent regenerated concurrently with dissolution process by evaporator, this process becomes very economical and environment-friendly. The separation of WPCBs using solvent has three main steps: penetration of organic solvent inside the layers, dissolution of BER between the layers, and delamination of the layers themselves. Contact areas between solvent and WPCB layers can be considered as starting points of the separation process, and their increase leads to acceleration of the separation. The sizes of the contact areas are affected by solder mask layer, Cu tracks, and solder, which hinder progress of solvent penetration inside the matrix. However, separation time does not count as conclusive evidence for whether each element inside the matrix is fully separated or not for the particular BER type. Organic solvent dissolution at low temperature using ultrasonification and vibration process significantly decreases the emission of toxic gases and power consumption during the separation process.

Generally, WPCBs including waste motherboards (MBs) have a similar sandwich structure and are composed of three main layers: substrate (several layers of woven fiberglass and two Cu foils), upper, and lower compound units (which can be defined as an assembly of isolating fiberglass layers, solder joints, conductive tracks, contacts, and solder mask; all these elements are adhered by static friction and epoxy resin (ER)) (Fig. 2.2). Organic solvent fully penetrates the sandwich structure of the substrate firstly swelling and then cracking and finally the separation of fiberglass layers and metal foils [109]. WPCBs comprise six layers of fiberglass (representing WPCB substrate), two Cu foil layers (upper and lower), Cu tracks, through-hole pads, and two solder mask layers (upper and lower). Cu content/pureness varies between 86% and 93%. BFR content changes from 9% to 29%. Br starts to evaporate intensively, and weight loss increases dramatically in the temperature range of 310–360 °C (65–70% mass loss). ER decomposition range is 260–400 °C and weight loss 75%.

In the previous works, dimethylacetamide (DMA) was used as an organic solvent due to its high boiling point, relatively high viscosity, and high thermal stability with an effective ability temperature comparing to other solvents like N-methyl-2-pyrrolidone (NMP), dimethyl sulfoxide (DMSO), and dimethylformamide (DMF). The results showed that RAM was composed of seven layers of fiberglass, epoxy resin, two Cu foil layers, two internal conductive track layers, contacts (Au-coated), through-hole pads, solder mask, and solder joints of integrated circuits. Dimethyl sulfoxide (DMSO) is an organosulfur compound with the formula $(CH_3)_2SO$. DMSO is a nonaqueous highly polar organic solvent and a key dipolar aprotic solvent. It is clear liquid, less toxic, colorless, hygroscopic, essentially odorless, and readily miscible in a wide range of organic solvents as well as water. DMSO effectively dissolves numerous organic and inorganic chemicals but does not corrode metals. DMSO's boiling point is 189 °C, melting point is 18.5 °C, and it has slightly high viscosity (2.2 mPa s at 20 °C). DMSO is highly stable at temperature 150 °C and remarkably stable in the presence of most neutral or basic salts and bases, but acids promote the decomposition. DMSO can be recycled many times [110]. Zhu et al. (2013) tried to dissolve bromine epoxy resin (WPCBs) using DMSO at low temperatures before leaching BMs and PMs. The optimum conditions were 145 °C, 1:7 S/L ratio (μg mL^{-1}), 60 min, and WPCBs area of 16 mm^2 [110].

Yousef et al. (2018) described a new industrial technology for recovery of all metallic and nonmetallic components of full-sized bare waste motherboards using combined chemical-ultrasonic and mechanical treatment techniques. They found that MB was composed of five fiberglass layers adhered by BER (67 wt%), two Cu layers and tracks (19.4 wt%), and through-hole pads (12 wt%). In addition, it was noted that through-hole pads contained Pd coating. MB separation was carried out using MB/DMA ratio of 300 g L^{-1} at 50 °C to dissolve BER at 80 h. The recycling rate was about 99%, and the percentage of nonmetal (7 plain woven fiberglass and epoxy resin) in waste MB was 67%, while metallic part was 33%. Obtained Cu purity was 88%. Separation at low temperature may lead to a significant reduction in the emission of toxic gases and dust comparing to physico-mechanical conventional methods. Ultrasonic treatment accelerates the breakage of the internal van der Waals bonds of BER, thus reducing the separation time [111].

The BFR leach technology using ultrasonification seems to be promising in terms of being economical and environment-friendly. Furthermore, it can be applied at the industrial level, since the developed technique does not require a significant amount of facilities; in particular, only sources of sound wave (transducers) and heating (50 °C) are needed, while other recycling techniques often require much higher temperatures and pressures, e.g., supercritical fluid treatment, pyrolysis, and gasification. Also, the recovered metal and nonmetals can be used to address the shortage of mineral resources for the electronics industry as well as for composite material applications [112].

10.6 Purification Technologies for Precious Metals from Leachates

A number of different refining methods, such as precipitation, cementation, solvent extraction, adsorption, ion exchange, and electrowinning, have been used to purify leaching solutions, which are obtained from metal leaching processes, to obtain pure PM products. However, most of them are still on laboratory scale at present, which could not be used in industrial application due to several obstacles. Hence, developing new refining technologies, which are efficient, low cost, and

10.6 Purification Technologies for Precious Metals from Leachates

Table 10.5 Purification technologies for PMs from leachates

Reagent	Refining method	Conditions	Recovery rate	References
Chloride	Ion exchange	3.0 BV long; 1.0 M NaCl washing; 0.2 M EDTA scrubbing; 2.7 M thiosulfate eluting	72% Ag	Virolainen et al. [114]
HCl	Ion exchange	24 h leach; S/L: 2 g/L; N: 150 rpm; $10 < T < 40$ °C	99% Au	Navarro et al. [115]
Aqueous	Adsorption	0.5 M thiourea; 2 M HCl; pH: 4; N: 300 rpm	99.23% Ag	Wang et al. [116]
HCl	Ion exchange	0.05 M aqueous Na_2SO_3; 30% isoamyl alcohol in micro emulsion	99% Au	Lu et al. [117]
		v/v: 8/1; N: 2500 rpm; t: 5 min; T: 25 °C		
Cl_2	1. Electrowinning	0.1 M HCl; 714 A/m^2; T: 25 °C; t: 165 min	99% Au	Kim et al. [118]
	2. Ion exchange	6034 mg/L amberlite XAD-7HP resin		
$Fe_2(SO_4)_3$	Extraction	0.05–1.0 M H_2SO_4; 0.05–0.5 M Fe^{3+}; $20 < T < 80$ °C; 1% < S/L < 15%	99% Ag	Yazıcı and Deveci [45]
		25.6 < chloride < 116.5 g/L	99% Pd	

environment-friendly, are urgently required. Lu and Xu (2016) reviewed recently purification of PM technology [113]. Chloride, HCl, chlorine, and ferric sulfate were used as a reagent to obtain above 99% Au, Ag, or Pd recoveries. Table 10.5 summarizes purification technologies for PMs from leachates [113].

Cu leaching solution after solid-liquid separation can be either treated by purification or directly taken to the final process, where precipitation or electrowinning is involved to recover Cu or its compounds [56]. Solvent extraction is considered to concentrate Cu from dilute acidic leaching liquors. The common extractants are hydroxyl oximes (such as LIX984N [119]) and alkyl phosphinic acids (such as CYANEX 301 and CYANEX 302 [120]). Fouad investigated solvent extraction of Cu by the mixture of Cyanex 301 (RH) and LIX 984N (LH) [121]. The extraction could be expressed by Eq. (10.61). The results indicated that the mixture of RH and LH with the radio of 1:1 had the higher extraction efficiency, compared with using RH or LH, individually, which could be attributed to the formation of $CuRL_2H$ complex. Moreover, 90.7% of stripping percentage of Cu^{2+} from the organic phase was obtained by the addition of 6 M HCl. Other than solvent extraction, cementation is employed to extract Cu as well. Behnamfard et al. (2013) added metallic Fe to Cu solution and then metallic Cu precipitated [67]. Kumari et al. also reported that in a process of pyrolysis-beneficiation-leaching-solvent extraction, after recovering H_2SO_4 and Fe from the leach liquor using 70% TEHA in kerosene and air sparging, respectively, 99.99% of Cu could be extracted using 10% LIX 841C in two stages at pH 2.5 and O/A ratio 1/1. Similarly, Ni could be completely recovered from the raffinate in two stages when 1% LIX 841C was used at pH 4.58 and O/A ratio 2/1 [122, 123].

$$Cu^{2+} + mRH + n(LH)_2 = Cu(R_mL_2H_{m+2n-2}) + 2H^+ \tag{10.61}$$

Supercritical water (SCW) technology at temperature above 374 °C and pressure 22.1 MPa is generally regarded as green process. SCW is an ideal pretreatment method for metal recovery from WPCBs. Li and Xu (2015) used SCW to decompose brominated epoxy resin and recover metals (99.80%) from WPCBs [124]. However, PM separation needs further studies. Supercritical fluid extraction is faster, efficient, inexpensive, environmentally friendly, readily available, and recyclable. Water or liquefied CO_2 can be used as solvent. Critical point of CO_2 is 31.1 °C and 7.38 MPa. CO_2 has high diffusivity, low viscosity, high solubility, and no surface tension.

It is important to find reliable and cost-effective heavy metal recovery techniques that do not produce any secondary pollution threats to the environment and human health during the processing of WPCBs [3]. One of the possible solutions is provided by electrochemical processes which have high environmental compatibility, due to the fact that the main reagent, the electron, is a "clean reagent." By using direct (concentrated H_2SO_4) or mediated (HCl + Fe^{3+}) electrochemical oxidation for the dissolution of metals, the reagent consumptions can be significantly reduced [125]. Furthermore, in parallel with the oxidation of the metals, electroextraction can be an alternative method to obtain the metals from the leaching solutions [13, 34, 126]. The environmental assessment of two new Cu recovery process (i.e., direct and mediated electrochemical oxidation with simultaneous cathodic deposition of Cu, but instead of Cl or other hazardous reagents) Fogarasi et al. (2013) used electrolytes (H_2SO_4) and less harmful mediators like Fe^{3+}/Fe^{2+} redox couple [13]. Environmental assessment was successfully performed by Biwer-Heinzle method using general environmental indices (GEIs) for inputs and outputs of the two Cu recovery (i.e., direct and mediated) electrochemical processes. Comparing the processes, it was evident that the process using mediated dissolution of Cu had a lower environmental impact than the direct one, because the generated leaching solution could be used for further processing of WPCBs without the addition of fresh reagents. It was also found that, for both processes, the highest risks of toxicity could be attributed to the acidic solutions.

The leaching of metals from the WPCB samples and the electroextraction of Cu and regeneration of $FeCl_3$ involve complex chemical and electrochemical reactions. The dissolution of metals (Me) in the chemical leaching reactor takes place by the following reaction [14]:

$$Me + nFe^{3+} \leftrightarrow Me^{n+} + nFe^{2+} \tag{10.62}$$

However, depending on the chloride concentration, the dissolution process can also involve the simultaneous formation of several metal-chloro complexes:

$$Me + nFe^{3+} + mCl^- \leftrightarrow [MeCl_m]^{(m-n)-} + nFe^{2+} \tag{10.63}$$

10.6 Purification Technologies for Precious Metals from Leachates

In the case of the electrochemical reactor, there are two main processes [14]:

- The electroextraction of metals, mainly Cu, from the leaching solution resulted in the chemical reactor.
- The regeneration of the leaching agent through the anodic oxidation of ferrous ions to ferric ions. The main cathodic process is the electrodeposition of Cu according to the following reactions:

$$Cu^{2+} + 2e^- \leftrightarrow Cu \quad E^0 = 0.337\,V \tag{10.64}$$

$$Cu^+ + e^- \leftrightarrow Cu \quad E^0 = 0.506\,V \tag{10.65}$$

It is also accompanied by secondary reactions:

$$Cu^{2+} + e^- \leftrightarrow Cu^+ \quad E^0 = 0.167\,V \tag{10.66}$$

$$Sn^{2+} + 2e^- \rightarrow Sn \quad E^0 = -0.137\,V \tag{10.67}$$

$$Sn^{4+} + 2e^- \rightarrow Sn^{2+} \quad E^0 = 0.154\,V \tag{10.68}$$

$$Fe^{3+} + e^- \rightarrow Fe^{2+} \quad E^0 = 0.77\,V \tag{10.69}$$

The electrolyte, with high Fe^{2+} content, is passed from the cathodic chamber to the anodic chambers where the main electrochemical process is the oxidation of Fe^{2+} to Fe^{3+}:

$$Fe^{2+} \leftrightarrow Fe^{3+} + e^- \tag{10.70}$$

The anodic process also involves secondary reactions, of which the most important are the following:

$$Sn^{2+} \rightarrow Sn^{4+} + 2e^- \quad E^0 = 0.151\,V \tag{10.71}$$

$$2H_2O \rightarrow 4H^+ + 4e^- + O_2 \quad E^0 = 1.24\,V \tag{10.72}$$

$$2Cl^- \rightarrow Cl_2 + 2e^- \quad E^0 = 1.36\,V \tag{10.73}$$

With the exception of oxygen evolution (Eq. 10.72), the other secondary reactions are also generating species that can dissolve the metals from the WPCB samples, although less efficient than ferric ions. The complex electrolyte composition can involve several side reactions at the cathode [15]:

$$CuCl_2^- + e^- \rightarrow Cu + 2Cl^- \quad E^0 = 0.169\,V/NHE \tag{10.74}$$

$$Cu^{2+} + 2Cl^- + e^- \rightarrow CuCl_2^- \quad E^0 = 0.504\,V/NHE \tag{10.75}$$

$$FeCl_2^+ + e^- \rightarrow Fe^{2+} + 2Cl^- \quad E^0 = 0.508 \text{ V/NHE} \quad (10.76)$$

$$Fe^{2+} + 2Cl^- \rightarrow FeCl_2^+ + e^- \quad E^0 = 0.508 \text{ V/NHE} \quad (10.77)$$

$$CuCl_2^- \rightarrow Cu^{2+} + 2Cl^- + e^- \quad E^0 = 0.504 \text{ V/NHE} \quad (10.78)$$

The dissolution of metals with $FeCl_2^+$ follows the order of Zn, Fe, Ni, Pb, Sn, Cu, and Ag. However, in the case of Ag and Pb, even if the dissolution process is thermodynamically favorited, it will be inhibited by the low solubility of the obtained chlorides. Chemical dissolution of Au is not possible under above conditions [15].

Fogarasi et al. (2014) found that the use of mediated electrochemical oxidation involving $FeCl_3$ solution proved to be an efficient method for the simultaneous recovery of Cu and separation of Au-rich residue from WPCBs [14]. The optimum conditions were 4 mA cm^{-2} and 0.37 M initial $FeCl_3$ concentration to obtain 99.04% Cu deposit at a current efficiency of about 64% and specific energy consumption of 1.75 kWh kg^{-1} Cu. The solid residue obtained simultaneously with the recovery of Cu had an Au concentration 25 times higher than the Au concentration in the initial WPCB samples.

There is a scenario that leaching is not necessary if metals obtained from WPCBs have been concentrated to high purity (60–80%) [127] from physical separation. Under this circumstance, electrometallurgy could be superior to produce Cu. Xiu et al. (2009) reported that after the strong SCWO process, recoveries of Cu and Pb approached to 100% [128]. In the electrokinetic process, the solid residue was suspended in 1 M HCl solution. The anode and cathode, two Pt-coated plates, were isolated by two porous glass frits. The results indicated that the increase of current density promoted the rise of Cu recovery; however, the excessively high current density led to high potential gradient that resulted in more side reactions. Thus, 97.6% of Cu concentration with recovery of 84.2% was obtained at 20 mA/cm^2 current density and 11 h reaction time. More than 74% Cu recovery was attained by cathode deposition as two phases: Cu and Cu_2O [128]. The same electrokinetic setup was also applied in the extraction of heavy metals (Cd, Cr, As, Ni, Zn, and Mn) with different acids [129]. It appeared that HCl could recover 70% of Cd, which could be attributed to the conjugation of low pH and Cd–Cl complexes. The high extraction of Cr, Zn, and Mn was found in the presence of citric acid [129]. Chu et al. (2015) also investigated the electrolysis process as functions of the concentrations of $CuSO_4 \cdot 5H_2O$, H_2SO_4, and NaCl, current density, and time with regard to the concentrated metal scraps containing 83.4% of Cu. It was found that the increase in the concentration of $CuSO_4 \cdot 5H_2O$ simultaneously increased the current efficiency and particle size of Cu powder in the cathode. Increasing the concentration of H_2SO_4 and current density were effective at the increase of current efficiency. The optimization indicated the copper purity of 98.1% was obtained under the optimal conditions of 50 g/L $CuSO_4 \cdot 5H_2O$, 40 g/L NaCl, 118 g/L H_2SO_4, and 80 mA/cm^2 with 3 h [127].

10.6.1 Gold Recovery Using Nylon-12 3D-Printed Scavenger

After leaching, pregnant solutions (PLS) that contained PMs and BMs are subjected to the course of separation and purification, such as precipitation of impurities, solvent extraction, absorption, and ion exchange to isolate and concentrate metals of interest. In the purification of PMs from leachates, chloride and HCl can be used in ion exchange, chlorine in electrowinning, amberlite in ion exchange (IX), and ferric sulfate in the extraction processes [113]. More than 99% of Au, Ag, or Pd can be obtained.

After the leaching, ion-exchange resins are often being used as an adsorbent for the Au, as has been widely reported [130–135]. There have also been reports of polymers, such as polypyrrole and polyaniline, being used to recover Au [136–138]. Because of lack of selectivity of the adsorbents and various other recovery methods, extensive preprocessing of the sample is often required [139, 140]. Typical scavenger materials used to adsorb the dissolved Au consist of small particles, and hence a filtration system is often required, either to recover the used adsorbent or to purify the solution stream of remaining particles. The above difficulties make recycling the adsorbent challenging or even impossible [141].

These problems can be avoided by using 3D printing [142]. The scavenger material can be printed in a form of a column for continuous tests or a cube-shaped mesh for batch process, and the ion-containing solution is either flowed through it or the object is simply dipped into the solution. The captured metal ions can then be recovered by elution with suitable acidic or organic solutions (HNO_3 or glycol dibutyl ether), after which the scavenger is reusable for metal capturing. Cheap 50 μm nylon-12 (N12) powder can be used for printing scavengers and tested these scavengers to selectively adsorb Au from aqua regia solutions containing up to 500 times higher concentrations of other metal ions. Au selectively binds as $[AuCl_4]^-$ on nylon [143]. The binding mechanism between the amide group and $[AuCl_4]^-$ has been predicted to be related to hydrogen bonding between the hydrogen of the protonated amide group and the chlorides of the Au complex [144]. The cause for selectivity toward Au could potentially be because of -1 charge of the complex and square planar geometry of the $[AuCl_4]^-$. The only other species possessing square planar geometries in the solution would be chloride complexes of Pt(II) and Pd(II). These, however, would have a charge of -2. The selectivity of the material should be investigated further in the future by using molecular modeling.

Soaking the nylon-12 $[(CH_2)_{11}C(O)NH]_n$ cubic mesh into synthetic test solution for 4 h yielded 90.4% Au recovery with no noticeable amount of other metals adsorbed into metal mesh. Continuous flow of synthetic test solution with a column of 10 scavenger units achieved Au recovery of 82.7% in less than 30 s residence time [141]. The main advantage of the use of 3D-printed scavengers is that the Au can be separated directly from the diluted leachate with no tedious preprocessing steps. Scavengers can be easily scaled and modified accordingly to meet the users' requirements even up to the industrial scale. It is expected that by using other printable polymers with other functional groups or hybrid materials, other metals can also be captured by using 3D-printed scavengers. In such a case, scavenger modules with different selectivities could be linked together to obtain multi-metal scavengers with detachable ion-specific modules. Because the 3D-printed scavengers are highly selective, they can be used for recovering Au from leachates originating from sources other than e-waste as well. In general, the use of chemically functional printing materials can extend considerably the use of 3D printing in manufacturing of chemically active devices [143].

10.6.2 Solvent Extraction (SX)

In the solvent extraction phase of the process, a metal in the PLS is concentrated and purified into electrolyte. The pure electrolyte is then used in electrowinning to produce a chemically and physically high-quality metal product. Solvent extraction is applied to the WPCB leaching solution. SX using 200 g L^{-1} LIX984 (Henkel) which consists of equal amounts of LIX860IC (5-dodecylsalicylaldoxime) and LIX84I (2-hydroxy-5-nonylacetophenone oxime) diluted with kerosene at 100 g L^{-1} S/L was applied to the Cu metal-loaded (11,180 mg L^{-1}) leaching solution. Ni remained in 4 M HNO_3 stripping solution (backward extraction). Mixing time was 1 h for both forward and backward extractions, and pH values of aqueous feed were adjusted to about 1.0 by adding a small amount of NaOH solution. After attainment of equilibrium, the metal concentrations in the aqueous phases were measured (11,154 mg L^{-1} Cu and 0 mg L^{-1} Ni) [145]. Cu was preferentially extracted 98% while leaving the greater part of Ni in the raffinate (4.3%). Complete back extraction of Cu was attained with 4.0 M HNO_3. This concentrated Cu solution can be sent to electrolytic refining process for Cu recovery.

Outotec has been developing a wide range of industrial SX technologies for decades. Vertical Smooth Flow (VSF) X mixer-settler modular plants offer lower operating and investment costs, significantly shorter lead times, and more reliable and stable production where health and safety issues are given high importance. Outotec technology has industrial Cu, Ni, Co, Li, etc. solvent extraction capacities for decades [144].

The SX equipment (DOP units, SPIROK mixers, settlers, loaded organic tanks, and after settlers) come together to enhance the stability of the process against organic phase oxidation and improve work safety by minimizing organic evaporation (Fig. 10.4). Fire safety is also maximized due to the low oxygen presence and limited combustion space. There are no carbon steel parts inside the equipment, which also reduces fire risks during maintenance.

10.6 Purification Technologies for Precious Metals from Leachates

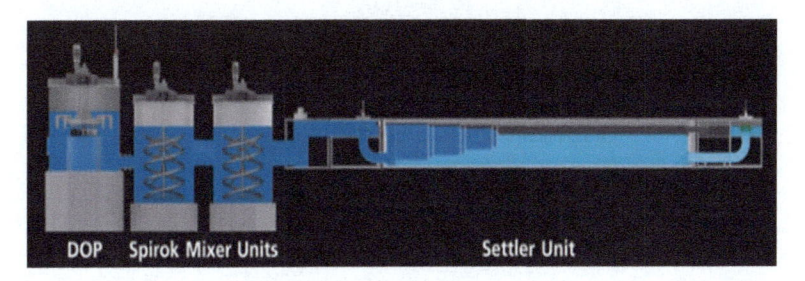

Fig. 10.4 Outotec's DOP, SPIROK mixer, and settler units

VSF X DOP Unit Pumping and mixing are separated in the SX technology. This ensures low entrainment losses even if flows have to be increased above the design values for operational reasons. The DOP unit consists of a suction tank with a conical overflow rim, a turbine, a flow stabilizer, and a baffled cylindrical outer tank.

VSF X SPIROK Mixer Unit The mixer unit is designed to ensure the gentle and uniform mixing needed for proper dispersion mixing. Even at low rotation speed, a vertical flow is maintained throughout the whole mixer. The SPIROK mixer covers the entire volume of the mixer tank, which avoids local high shear forces and contributes to optimizing residual organic and aqueous entrainment levels.

VSF X Settler The settler design is very important for SX plant operation. The VSF X settler has a deep, dense, and packed dispersion layer at the feed side of the settler. The incoming dispersion flow is evenly distributed with a feed launder. Dispersion is then kept at a compressed state by means of DDG separation fences. The dense dispersion layer filters small droplets and enhances coalescence. More flexibility to handle higher feed flows is achieved when the dispersion and the separated organic phase are not allowed to spread over the whole settling length. The separated organic and aqueous phases are finally collected into separate launders. The aqueous inner circulation is taken from the settler bulk without disturbing the flows inside the settler.

Health, safety, and environmental (HSE) aspects are an inseparable part of the plant concept. They are taken into consideration during the product definition phase, providing a safe and sustainable plant that is easy to operate and maintain. The plant concept offers:

- Possibility to use sea water as process water
- Waterlocks in the inspection/maintenance hatches
- Minimized VOC emissions
- Minimized energy consumption
- Reduced plant area with reduced excavation costs
- Minimized environmental impact and carbon footprint

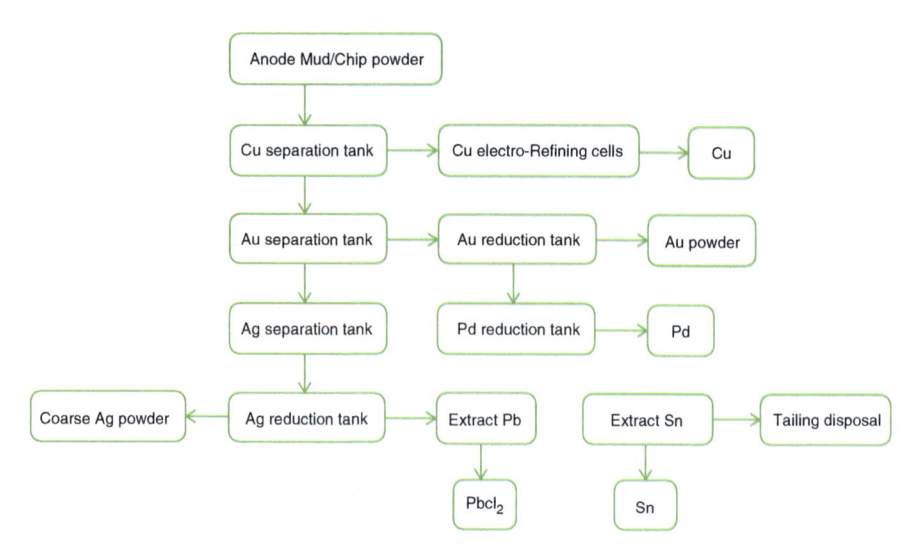

Fig. 10.5 Flowsheet of PM recycling and refining process from WPCBs

- The safest available technology for operators
- Mixer-settlers with a high residual value

10.7 Industrial-Scale E-Waste Precious Metal Separation and Refining Solutions

With increase in the production of electronic appliances and diversity-seeking consumer, hazardous e-waste recycling has challenged field of waste treatment and environmental management because WPCBs have caused environmental pollution without proper treatment. High-quality WPCB has much Au, Ag, Pd, Pt, etc. on them like mobile phones and computer boards.

Two-step pulverization produces mixtures of metal and nonmetal powder. The air separator completely separates metal and nonmetals from each other. The whole process is performed in fully closed operation in an integrated production line without creating any pollution. Figure 10.5 shows the flowsheet of PM recycling and refining process from WPCBs. The recycling rates of PMs can reach up to 97–99% with a purity of 99.99%. High-value WPCB metal powder can be roasted with ozone, which can thoroughly destroy the harmful flue gas caused by burning, and Cu is leached and recovered. Then, roasting residue is leached by HCl to extract Au and Pd. Ag, Pb, and Sn are separately leached after Au removal. This type of plant has a capacity of 100 or 300 kg d^{-1} and power consumption of 20 KW or 30 KW, respectively. Such a plant flowsheet includes three sets of electrolytic tanks, three sets of enamel reactors, one set of vacuum filter unit, one set of flue gas control system, and one set of centralized control system (Fig. 10.6) [146].

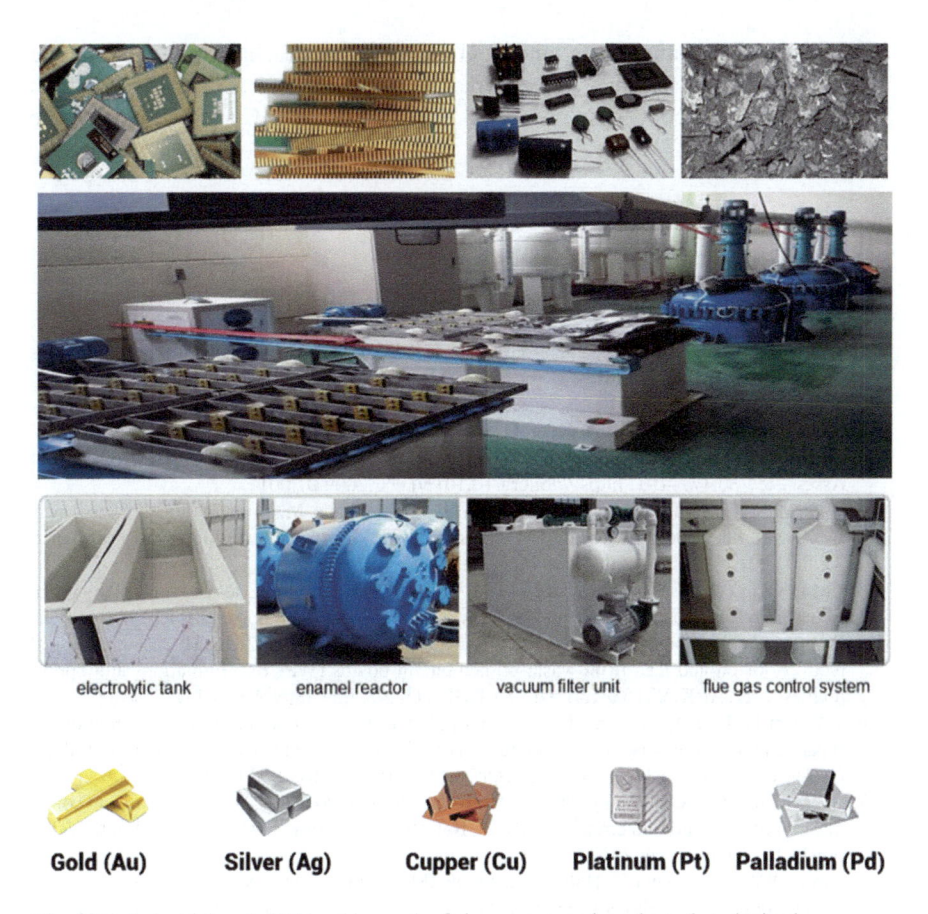

Fig. 10.6 Industrial-scale PM leaching and refining system and products that obtained

10.8 Occupational, Health, and Safety Hazardous Characteristic Determination Tests for PCBs

Different types of leaching tests have been proposed to evaluate leachability of metals in solid wastes from landfills and hazardous characteristics of these wastes. TCLP (toxicity characteristic leaching procedure) test is a widely used test in order to determine whether a solid waste is hazardous. According to TLCP tests, PCBs can be classified as hazardous waste based on the concentration of Pb which is more than 40-fold higher than the US EPA regulatory limits. Pb in PCBs comes from solder (Sn–Pb alloy). PCBs with Pb-free solders are not hazardous according to TLCP tests. Synthetic Precipitation Leaching Procedure (SPLP) test can be implemented to simulate the leachability of metals by the rainwater under atmospheric conditions. Batches ASTM D3987 and EN 12457-2 were also developed to determine the leachability of inorganic constituents and to simulate actual environmental conditions of a landfill [147–149]. The leachates of PCBs can be classified as highly polluted water (type IV) for SPLP and ASTM D-3987 tests with regard to the concentrations of Al, As, Cu, Pb, and Cd which exceed the regulatory limits [147–149].

References

1. Tuncuk A, Atazi V, Akcil A, Yazici EY, Deveci H (2012) Aqueous metal recovery techniques from e-scrap: hydrometallurgy in recycling. Miner Eng 25:28–37. https://doi.org/10.1016/j.mineng.2011.09.019
2. Moberg A, Johansson M, Finnveden G, Jonsson A (2010) Printed and tablet e-paper newspaper from environmental perspective – a screening life cycle assessment. Environ Impact Assess Rev 30:177–191. https://doi.org/10.1016/j.eiar.2009.07.001
3. Widmer R, Oswald KH, Sinha-Kheeetriwal D, Schnelmann M, Boni H (2005) Global perspectives on e-waste. Environ Impact Assess Rev 25:436–458. https://doi.org/10.1016/j.eiar.2005.04.001
4. Cheng CQ, Yang F, Zhao J, Wang LH, Li XG (2011) Leaching of heavy metal elements in solder alloys. Corros Sci 53:1738–1747. https://doi.org/10.1016/j.corsci.2011.01.049
5. Ranitovic M, Kamberovic Ž, Korac M, Jovanovic N, Mihjalovic A (2015) Hydrometallurgical recovery of tin and lead from waste printed circuit boards (WPCBs): limitations and opportunities. METABK 55:153–156. UDC – UDK 669.2/8.:669.48:669.6.4=111
6. Takanori H, Ryuichi A, Youichi M, Minoru N, Yasuhiro T, Takao A (2009) Techniques to separate metal from waste printed circuit boards from discarded personal computers. J Mater Cycles Waste Manag 11:42–54. https://doi.org/10.1007/s10163-008-0218-0
7. Jha MK, Kumari A, Choubey PK, Lee JC, Kumar V et al (2012) Leaching of lead from solder material of waste printed circuit boards (PCBs). Hydrometallurgy 121–124:28–34. https://doi.org/10.1016/j.hydromet.2012.04.010
8. Gibson RW, Goodman PD, Holt L, Dalrymple IM, Fray DJ (2003) Process for the recovery of tin, tin alloys or lead alloys from printed circuit boards. Patent No US6641712 B
9. Park YJ, Fray DJ (2009) Separation of zinc and nickel ions in a strong acid through liquid-liquid extraction. J Hazard Mater 163:259–265. https://doi.org/10.1016/j.jhazmat.2008.06.085
10. Kim EY, Kim MS, Lee JC, Jeong J, Pandey BD (2011) Leaching kinetics of copper from waste printed circuit boards by electro-generated chlorine in HCl solution. Hydrometallurgy 107:124–132. https://doi.org/10.1016/j.hydromet.2011.02.009
11. Lister TE, Wang P, Anderko A (2014) Recovery of critical and value metals from mobile electronics enabled by electrochemical processing. Hydrometallurgy 149:228–237. https://doi.org/10.1016/j.hydromet.2014.08.011
12. Diaz LA, Lister TE, Parkman JA, Clark GG (2016) Comprehensive process for the recovery of value and critical materials from electronic waste. J Clean Prod 125:236–244. https://doi.org/10.1016/j.jclepro.2016.03.061
13. Fogarasi S, Imre-Lucaci F, Ilea P, Imre-Lucaci A (2013) The environmental assessment of two new copper recovery processes from waste printed circuit boards. J Clean Prod 54:264. https://doi.org/10.1016/j.jclepro.2013.04.044
14. Fogarasi S, Imre-Lucaci F, Ilea P, Imre-Lucaci A (2014) Copper recovery and gold enrichment from waste printed circuit boards by mediated electrochemical oxidation. J Hazard Mater 273:215. https://doi.org/10.1016/j.jhazmat.2014.03.043
15. Fogarasi S, Imre-Lucaci F, Egedy A, Imre-Lucaci A, Ilea P (2015) Eco-friendly copper recovery process from waste printed circuit boards using Fe^{3+}/Fe^{2+} redox system. Waste Manag 40:136. https://doi.org/10.1016/j.wasman.2015.02.030

16. Jian-guang Y, Jie L, Si-yao P, Yuan-lu L, Wei-gaing S (2016) A new membrane electro-deposition based process for tin recovery from waste printed circuit boards. J Hazard Mater 304:409–416. https://doi.org/10.1016/j.jhazmat.2015.11.007
17. Zhang Y, Liu S, Xie H, Zeng X, Li J (2012) Current status on leaching precious metals from waste printed circuit boards. Procedia Environ Sci 16:560–568. https://doi.org/10.1016/j.proenv.2012.10.077
18. Mpinga CN, Eksteen JJ, Aldrich A, Dyer L (2015) Direct leach approaches to Platinum Group Metal (PGM) ores and concentrates: a review. Miner Eng 78:93. https://doi.org/10.1016/j.mineng.2015.04.015
19. Dorneanu SA (2017) Electrochemical recycling of waste printed circuit boards in bromide media. Part 1: preliminary leaching and dismantling tests. Stud Univ Babes Bolyai Chem LXII (3):177–186. https://doi.org/10.24193/subbchem.2017.3.14
20. Jeon SH, Yoo K, Allorro RD (2017) Separation of Sn, Bi, Cu from Pb free solder paste by ammonia leaching followed by hydrochloric acid leaching. Hydrometallurgy 169:26–30. https://doi.org/10.1016/j.hydromet.2016.12.004
21. Bari F, Begum N, Jamaludin SB, Hussin K (2009) Selective leaching for the recovery of copper from PCB. In: Proceed. Malaysian metallurgical conference '09 (MMC09), pp 1–4
22. Koyama K, Tanaka M, Lee J (2006) Copper leaching behavior from waste printed circuit board in ammoniacal alkaline solution. Mater Trans 47:1788–1792. https://doi.org/10.2320/matertrans.47.1788
23. Oishi T, Koyama K, Alam S, Tanaka M, Lee J (2007) Recovery of high purity copper cathode from printed circuit boards using ammoniacal sulfate or chloride solution. Hydrometallurgy 89:82–88. https://doi.org/10.1016/j.hydromet.2007.05.010
24. Sun Z, Xiao Y, Sietsma J, Agterhuis H, Yang Y (2015) A cleaner process for selective recovery of valuable metals from electronic waste of complex mixtures of end-of-life electronic products. Environ Sci Technol 49:7981–7988. https://doi.org/10.1021/acs.est5b0123
25. Kim SL, Lee JC, Yoo K (2016) Leaching of tin from waste Pb-free solder in hydrochloric acid solution with stannic chloride. Hydrometallurgy 165:143–147. https://doi.org/10.1016/j.hydromet.2015.09.018
26. Yang J, Lei J, Peng S, Lv Y, Shi W (2016) A new membrane electro-deposition based process for tin recovery from waste printed circuit boards. J Hazard Mater 304:409–416. https://doi.org/10.1016/j.jhazmat.2015.11.007
27. Yang C, Li J, Tan Q, Liu L, Dong Q (2017) Green process of metal recycling: coprocessing waste printed circuit boards and spent tin stripping solution. ACS Sustain Chem Eng 5:3524–3535. https://doi.org/10.1021/acssuschemeng.7b00245
28. Zhang X, Guan J, Gua Y, Cao Y, Gua J, Yuan H, Su R, Liang B, Gao G, Zhou Y, Xu J, Guo Z (2017) Effective dismantling of waste PCB assembly with methanesulfonic acid containing hydrogen peroxide, AIChE. Environ Prog Sustain Energy 36(3):873. https://doi.org/10.1002/ep.12527
29. Barakat MA (1998) Recovery of lead, tin and indium from alloy wire scrap. Hydrometallurgy 49:63–73. https://doi.org/10.1016/S0304-386X(98)00003-6
30. Castro LA, Martins AH (2009) Recovery of tin and copper by recycling of PCBs from obsolete computers. Braz J Chem Eng 26(04):649–657. ISSN 0104-6632
31. Lee MS, Ahn JG, Ahn JW (2003) Recovery of copper, tin and lead from the spent nitric etching solutions of printed circuit board and regeneration of the etching solution. Hydrometallurgy 70:23–29. https://doi.org/10.1016/S0304-386X(03)00045-8
32. Menetti RP, Chaves AP, Tenório JAS (1996) Obtenção de concentrados metálicos não ferrosos a partir de sucata eletrônica (Non-ferrous metals recovery from electronic scrap). In: 51th Annual Meeting of Associação Brasileira de Materiais e Metalurgia, Porto Alegre, Brazil, vol 4, pp 205–216. (In Portuguese)
33. Menetti RP, Chaves AP, Tenório JAS (1996) Recuperação de Au e Ag de concentrados obtidos a partir de sucata eletrônica (Gold and silver recovery from electronic scrap). In: 51th Annual Meeting of Associação Brasileira de Materiais e Metalurgia, Porto Alegre, Brazil, vol 4, pp 217–224. (In Portuguese)

34. Veit HM, Diehl TR, Salami AP, Rodrigues JS, Bernardes AM, Tenório JAS (2005) Utilization of magnetic and electrostatic separation in the recycling of printed circuit boards scrap. Waste Manag 25:67–74. https://doi.org/10.1016/j.wasman.2004.09.009
35. Veit HM, Bernardes AM, Ferreira JZ, Tenório JAS, Malfatti CF (2006) Recovery of copper from printed circuit boards scraps by mechanical processing and electrometallurgy. J Hazard Mater B137:1704–1709. https://doi.org/10.1016/j.jhazmat.2006.05.010
36. Torres R, Lapidus GT (2016) Copper leaching from electronic waste for the improvement of gold recycling. Waste Manag 57:131–139. https://doi.org/10.1016/j.wasman.2016.03.010
37. Zhang X, Guan J, Guo Y, Yan X, Yuan H, Xu J, Guo J, Zhou Y, Su R, Guo Z (2015) Selective desoldering separation tin-lead alloy for dismantling of electronic components from printed circuit boards. ACS Sustain Chem Eng 3:1696–1700. https://doi.org/10.1021/acssuschemeng.5b00136
38. Jiang P, Harney M, Song Y, Chen B, Chen Q, Chen T, Lazarus L, Dubois LH, Korzenski MB (2012) Improving the end-of-life for electronic materials via sustainable recycling methods. Procedia Environ Sci 16:485–490. https://doi.org/10.1016/j.proenv.2012.10.066
39. Zhu P, Chen Y, Wang LY, Zhou M (2012) Treatment of waste printed circuit board by green solvent using ionic liquid. Waste Manag 32:1914–1918. https://doi.org/10.1016/j.wasman.2012.05.025
40. Zeng X, Li J, Xie H, Liu L (2013) A novel dismantling process of waste printed circuit boards using water-soluble ionic liquid. Chemosphere 93:1288–1294. https://doi.org/10.1016/j.chemosphere.2013.06.063
41. Silvas FPC, Correa MMJ, Caldes MPK, Moraes VT, Espinosa DCR, Tenorio JAS (2015) Printed circuit board recycling; physical processing and copper extraction by selective leaching. Waste Manag 46:503–510. https://doi.org/10.1016/j.wasman.2015.08.030
42. Yang JG, Tang CB, He J, Yang SH, Tang MT (2011) A method of extracting valuable metals from electronic wastes, State Intellectual Property Office of the People's Republic of China, ZL 200910303503.5
43. Lisinska M, Saternus M, Willner J (2018) Research of leaching of the printed circuit boards coming from waste mobile phones. Arch Metall Mater 63:143–147. https://doi.org/10.24425/118921
44. Oh CJ, Lee SO, Yang HS, Ha TJ, Kim MJ (2003) Selective leaching of valuable metals from waste printed circuit boards. Air Waste Manag Assoc 53:897–902
45. Deveci H, Yazıcı E et al (2010) Extraction of copper from scrap TV boards by sulphuric acid leaching under oxidizing conditions. In: Going Green – Care Innovation 2010 conference in Vienna, p 45. https://doi.org/10.13140/2.1.2909.7929
46. Yang T, Xu Z, Wen J, Yang L (2009) Factors influencing bioleaching copper from waste printed circuit boards by Acidithiobacillus ferrooxidans. Hydrometallurgy 97:29–32. https://doi.org/10.1016/j.hydromet.2008.12.011
47. Yazici E, Bas A, Deveci H (2014) Removal of iron as goethite from leach solutions of waste of printed circuit boards (WPCB). In: XXVII international mineral processing congress, Santiago, Chile. https://doi.org/10.13140/2.1.2909.7929
48. Habashi F (1999) Textbook of hydrometallurgy, 2nd edn. Laval University, Quebec City, p 610
49. Yazici EY, Deveci H (2014) Ferric sulphate leaching of metals from waste printed circuit boards. Int J Miner Process 133:39–45. https://doi.org/10.1016/j.minpro.2014.09.015
50. Ping Z, ZeYun F, Jie L, Qiang L, Guangren Q et al (2009) Enhancement of leaching copper by electro-oxidation from metal powders of waste printed circuit board. J Hazard Mater 166:746–750. https://doi.org/10.1016/j.jhazmat.2008.11.129
51. Yazici EY, Deveci H (2013) Extraction of metals from waste printed circuit boards (WPCBs) in H_2SO_4-$CuSO_4$-NaCl solutions. Hydrometallurgy 139:30–38. https://doi.org/10.1016/j.hydromet.2013.06.018
52. Kasper AC, Berselli GB, Freitas BD, Tenório JA, Bernardes AM et al (2011) Printed wiring boards for mobile phones: characterization and recycling of copper. Waste Manag 31:2536–2545. https://doi.org/10.1016/j.wasman.2011.08.013

53. Havlik T, Orac D, Petranikova M, Miskufova A, Kukurugya F, Takacova Z (2010) Leaching of copper and tin from used printed circuit boards after thermal treatment. J Hazard Mater 183:866–873. https://doi.org/10.1016/j.jhazmat.2010.07.107
54. Kim EY, Kim MS, Lee JC, Jha JK, Yoo K et al (2008) Effect of cuprous ions on Cu leaching in the recycling of waste PCBs, using electro-generated chlorine in hydrochloric acid solution. Miner Eng 21:121–128. https://doi.org/10.1016/j.mineng.2007.10.008
55. Yazici EY, Deveci H (2015) Cupric chloride leaching (HCl-CuCl$_2$-NaCl) of metals from waste printed circuit boards (WPCBs). Int J Miner Process 134:89–96. https://doi.org/10.1016/j.minpro.2014.10.012
56. Schlesinger ME, King MJ, Sole KC, Davenport WG (2011) Extractive metallurgy of copper. Elsevier, Oxford. ISBN: 9780080967899
57. Jha MK, Lee JC, Kumari A, Choubey PK, Kumar V et al (2011) Pressure leaching of metals from waste printed circuit boards using sulfuric acid. JOM 63:29–32
58. Herrmann S, Landau U (2005) Method for the fine refinement of gold, EP1606424(A1)
59. Ha VH, Lee JC, Jeong J, Hai HT, Jha MK (2010) Thiosulphate leaching of gold from waste mobile phones. J Hazard Mater 178:1115–1119. https://doi.org/10.1016/j.jhazmat.2010.01.099
60. Li J, Xu X, Liu W (2012) Thiourea leaching gold and silver from the printed circuit boards of waste mobile phones. Waste Manag 32(6):1209–1212. https://doi.org/10.1016/j.wasman.2012.01.026
61. Liu R, Shieh RS, Yeh RYL, Lin CH (2009) The general utilization of scrapped PC board. Waste Manag 29:2842–2845. https://doi.org/10.1016/j.wasman.2009.07.007
62. Oishi T, Koyama K, Alam S, Tanaka M, Lee JC (2007) Recovery of high purity copper cathode from printed circuit boards using ammoniacal sulfate or chloride solutions. Hydrometallurgy 89:82–88. https://doi.org/10.1016/j.hydromet.2007.05.010
63. Dong T, Hua Y, Zhang Q, Zhou D (2009) Leaching of chalcopyrite with brønsted acidic ionic liquid. Hydrometallurgy 99:33–38. https://doi.org/10.1016/j.hydromet.2009.06.001
64. Huang J, Chen M, Chen H, Chen S, Sun Q (2014) Leaching behavior of copper from waste printed circuit boards with Brønsted acidic ionic liquid. Waste Manag 34:483–488. https://doi.org/10.1016/j.wasman.2013.10.027
65. Zhu P, Gu GB (2002) Recovery of gold and copper from waste printed circuits. Chin J Rare Metals 26(3):214–216
66. Birloaga I, De Michelis I, Ferella F, Buzatu M, Veglio F (2013) Study on the influence of various factors in the hydrometallurgical processing of waste printed circuit boards for copper and gold recovery. Waste Manag 33:935–941. https://doi.org/10.1016/j.wasman.2013.01.003
67. Behnamfard A, Salarirad MM, Veglio F (2013) Process development for recovery of copper and precious metals from waste printed circuit boards with emphasize on palladium and gold leaching and precipitation. Waste Manag 33:2354–2363. https://doi.org/10.1016/j.wasman.2013.07.017
68. Calgaro CO, Schlemmer DF, da Silva MDCR, Maziero EV, Tanabe EH, Bertuol DA (2015) Fast copper extraction from printed circuit boards using supercritical carbon dioxide. Waste Manag 45:289–297. https://doi.org/10.1016/j.wasman.2015.05.017
69. Mecucci A, Scott K (2002) Leaching and electrochemical recovery of copper, lead and tin from scrap printed circuit boards. J Chem Technol Biotechnol 77:449–457. https://doi.org/10.1002/jctb.575
70. Bas AD, Deveci H, Yazici EY (2014) Treatment of manufacturing scrap TV boards by nitric acid leaching. Sep Purif Technol 130:151–159. https://doi.org/10.1016/j.seppur.2014.04.008
71. Kumar M, Lie JC et al (2010) Leaching of metals from waste printed circuit boards (WPCBs) using sulfuric and nitric acids. Environ Eng Manag J 13(10):2601–2607
72. Yang JG, Wu YT, Li J (2012) Recovery of ultrafine copper particles from metal components of waste printed circuit boards. Hydrometallurgy 121–124:1–6. https://doi.org/10.1016/j.hydromet.2012.04.015
73. Syed S (2012) Recovery of gold from secondary sources-a review. Hydrometallurgy 115–116:30–51. https://doi.org/10.1016/j.hydromet.2011.12.012

74. Kellner D (2009) Recycling and recovery. In: Hester RE, Harrison RM (eds) Electronic waste management, design, analysis and application. RSC Publishing, Cambridge, pp 91–110
75. Senanayake G (2004) Gold leaching in non-cyanide lixiviant systems: critical issues on fundamentals and applications. Miner Eng 17:785–801. https://doi.org/10.1016/j.mineng.2004.01.008
76. Tripathi A, Kumar M, Sau DC, Agrawal A, Chakravarty S, Mankhand TR (2012) Leaching of gold from the waste mobile phone printed circuit boards (PCBs) with ammonium thiosulphate. Int J Metall Eng 2:17–21. https://doi.org/10.5923/j.ijmee.20120102.02
77. Ha VH, Lee JC, Huynh TH, Jeong J, Pandey PD (2014) Optimizing the thiosulfate leaching of gold from printed circuit boards of discarded mobile phones. Hydrometallurgy 149:118–126. https://doi.org/10.1016/j.hydromet.2014.07.007
78. Batnasan A, Haga K, Shibayama A (2018) Recovery of precious and base metals from waste printed circuit boards using a sequential leaching procedure. JOM 70(2):124. https://doi.org/10.1007/s11837-017-2694-y
79. Sun Z, Cao H, Xiao Y, Sietsma J, Jin W, Agterhuis H, Yang Y (2017) Toward sustainability for recovery of critical metals from electronic waste: the hydrochemistry processes. ACS Sustain Chem Eng 5(1):21–40. https://doi.org/10.1021/acssuschemeng.6b00841
80. Birloaga I, Coman V, Kopacek B, Veglio F (2014) An advanced study on the hydrometallurgical processing of waste computer printed circuit boards to extract their valuable content of metals. Waste Manag 34:2581–2586. https://doi.org/10.1016/j.wasman.2014.08.028
81. Yin JF, Zhan SH, Xu H (2014) Comparison of leaching processes of gold and copper from printed circuit boards of waste mobile phone. Adv Mater Res 955–959:2743–2746. https://doi.org/10.4028/www.scientific.net/AMR.955-959.2743
82. Xu XL, Li JY (2011) Experimental study of thiourea leaching gold and silver from waste circuit boards. J Qingdao Univ 26:69–73
83. Wu J, Qiu LJ, Chen L et al (2009) Gold and silver selectively leaching from printed circuit boards scrap with acid thiourea solution. Nonferrous Metals 61:90–93
84. Zhong FW, Li DT, Wei JX (2006) Experimental study on leaching gold in printed circuit boards scrap with thiourea. Nonferrous Metals Recycl Util (6):25–27
85. Brooy SRL, Linge HG, Walker GS (1994) Review of gold extraction from ores. Miner Eng 7 (10):1213–1241. https://doi.org/10.1016/0892-6875(94)90114-7
86. Hilson G, Monhemius AJ (2006) Alternatives to cyanide in the gold mining industry: what prospects for the future? J Clean Prod 14:1158–1167. https://doi.org/10.1016/j.clepro.2004.09.005
87. Petter PM, Veit HM, Bernardes AM (2014) Evaluation of gold and silver leaching from printed circuit board of cellphones. Waste Manag 34:475–482. https://doi.org/10.1016/j.wasman.2013.10.032
88. Ficeriová J, Baláž P, Gock E (2011) Leaching of gold, silver, and accompanying metals from circuit boards (PCBs) waste. Acta Montan Slovaca 16:128–131
89. Ha VH, Lee JC, Jeong J, Hai HT, Jha MK (2010) Thiosulfate leaching of gold from waste mobile phones. J Hazard Mater 178:1115–1119. https://doi.org/10.1016/j.jhazmat.2010.01.099
90. Syed S (2006) A green technology for recovery of gold from nonmetallic secondary sources. Hydrometallurgy 82:48–53. https://doi.org/10.1016/j.hydromet.2006.01.004
91. Zhang ZH, Zhang FS, Yao T (2017) An environmentally friendly ball milling process for recovery of valuable metals from e-waste scraps. Waste Manag 68:490–497. https://doi.org/10.1016/j.wasman.2017.07.029
92. Jiang T, Chen J, Xu S (1993) A kinetic study of gold leaching with thiosuphate. In: Hiskey JB, Warren GW (eds) Hydrometallurgy, fundamentals, technology and innovations. AIME, Littleton, pp 119–126
93. Moore DM, Zhang XR, Li CX (2005) Using thiosulfate as a leach reagent instead of cyanide. Metallic Ore Dressing Abroad, pp 5–12
94. Heath JA, Jeffrey MI, Zhang HG, Rumball JA (2008) Anaerobic thiosulfate leaching: development of in situ gold leaching systems. Miner Eng 21:424–433. https://doi.org/10.1016/j.mineng.2007.12.006

References

95. Yen WT, Xia C (2008) Effects of copper minerals on ammoniacal thiosulfate leaching of gold. In: Proceeding XXIV international mineral processing congresses, Beijing, China
96. Montero R, Guevara A, De la Torre E (2012) Recovery of gold, silver, copper and niobium from printed circuit boards using leaching column. J Earth Sci Eng 2:590–595
97. Gurung M, Adhikari BB, Kawakita H, Ohto K, Inoue K, Alam S (2013) Recovery of gold and silver from spent mobile phones by means of acidothiourea leaching followed by adsorption using biosorbent prepared from persimmon tannin. Hydrometallurgy 133:84–92. https://doi.org/10.1016/j.hydromet.2012.12.003
98. Xu Q, Chen D, Chen L, Huang MH (2010) Gold leaching from waste printed circuit board by iodine process. Nonferrous Metals 62:88–90
99. Sahin M, Akcil A, Erust C, Altınbek S, Gahan CS, Tuncuk A (2015) A potential alternative for precious metal recovery from e-waste: İodine leaching. Sep Sci Technol 50:2587–2595. https://doi.org/10.1080/01496395.2015.1061005
100. Xiu FR, Qi Y, Zhang FS (2015) Leaching of Au, Ag, and Pd from waste printed circuit boards of mobile phone by iodide lixiviant after supercritical water pre-treatment. Waste Manag 41:134–141. https://doi.org/10.1016/j.wasman.2015.02.020
101. Quinet P, Proost J, Van Lierde A (2005) Recovery of precious metals from electronic scrap by hydrometallurgical processing routes. Miner Metall Process 22(1):17–22. ISSN: 07479182
102. Ficeriova J, Bal AZP, Gock E (2011) Leaching of gold, silver and accompanying metals from circuit boards (PCBs) waste. Acta Montan Slovaca 16:128–131
103. Zhang XY, Chen L, Fang ZT (2009) Review on gold leaching from PCB with non-cyanide leach reagents. Nonferrous Metals 61:72–76
104. Park YJ, Frayç DJ (2009) Recovery of high purity precious metals from printed circuit boards. J Hazard Mater 164:1152–1158. https://doi.org/10.1016/j.jhazmat.2008.09.043
105. Zhang Z, Zhang FS (2014) Selective recovery of palladium from waste printed circuit boards by a novel non-acid process. J Hazard Mater 279:46–51. https://doi.org/10.1016/j.jhazmat.2014.06.045
106. Prasad MS, Mensah-Biney R, Pizarro RS (1991) Modern trends in gold processing-overview. Miner Eng 4:1257–1277
107. Kononova ON, Kholmogorov AG, Danilenko NV, Goryaeva NG, Shatnykh KA et al (2007) Recovery of silver from thiosulfate and thiocyanide leaching solutions by adsorption on anion exchange resins and activated carbon. Hydrometallurgy 88:189–195. https://doi.org/10.1016/j.hydromet.2007.03.012
108. Sakunda P (2013) Strategy of e-waste management. http://www.slideshare.net/ketanwadodkar/e-waste-tce-r2?related=2
109. Gulgul A, Szczepaniak W, Zablocka-Malicka M (2017) Incineration, pyrolysis and gasification of electronic waste, E3S web of conferences, 22, 00060. https://doi.org/10.1051/e3sconf/20172000060
110. Zhu P, Chen Y, Wang LY, Zhou M, Zhou J (2013) The separation of waste printed circuit board by dissolving bromine epoxy resin using organic solvent. Waste Manag 33:484–488. https://doi.org/10.1016/j.wasman.2012.10.003
111. Yousef S, Tatariants M, Bendikiene R (2017) Mechanical and thermal characterizations of non-metallic components recycled from waste printed circuit boards. J Clean Prod 167: 271–280. https://doi.org/10.1016/j.jclepro.2017.08.195
112. http://www.stepinitiative.org/files/step/_documents/StEP_WP_One%20Global%20Definition%20of%20E-waste_20140603_amended.pdf
113. Lu Y, Xu Z (2016) Precious metals recovery from WPCBs: a review for current status and perspective. Resour Conserv Recycl 113:28–39. https://doi.org/10.1016/j.resconrec.2016.05.007
114. Virolainen S, Tyster M, Haapalainen M, Sainio T (2015) Ion exchange recovery of silver from concentrated base metal-chloride solutions. Hydrometallurgy 152:100–106. https://doi.org/10.1016/j.hydromet.2014.12.011
115. Navarro R, Saucedo I, Lira MA, Guibal E (2010) Gold(III) recovery from HCl solutions using amberlite XAD-7 impregnated with an ionic liquid (Cyphos IL-101). Sep Sci Technol 45 (12–13):1950–1962. https://doi.org/10.1080/01496395.2010.493116

116. Wang L, Xing R, Liu S, Yu HH, Qin Y, Li KC, Feng JH, Li RF, Li PC (2010) Recovery of silver(I) using a thiourea-modified chitosan resin. J Hazard Mater 180:577–582. https://doi.org/10.1016/j.jhazmat.2010.04.072
117. Lu WJ, Lu YM, Liu F, Shang K, Wang W, Yang YZ (2011) Extraction of gold(III) from hydrochloric acid solutions by CTAB/n-heptane/iso-amyl alcohol/Na$_2$SO$_3$ micro emulsion. J Hazard Mater 186:2166–2170. https://doi.org/10.1016/j.jhazmat.2010.12.059
118. Kim EY, Kim MS, Lee JC, Pandey BD (2011) Selective recovery of gold from waste mobile phone PCBs by hydrometallurgical process. J Hazard Mater 198:206–215. https://doi.org/10.1016/j.jhazmat.2011.10.034
119. Ilias S, Schimmel KA, Yezek PM (1999) Non-dispersive liquid-liquid extraction of copper and zinc from an aqueous solution by DEBPA and LIX 984 in a hollow fiber membrane module. Sep Sci Technol 34(6 and 7):1007–1019
120. Sole KC, Hiskey JB (1995) Solvent extraction of copper by Cyanex 272, Cyanex 302 and Cyanex 301. Hydrometallurgy 37(2):129–147
121. Fouad EA (2009) Separation of copper from aqueous sulfate solutions by mixtures of Cyanex 301 and LIX 984N. J Hazard Mater 166:720–727. https://doi.org/10.1016/j.j.hazmat.2008.11.114
122. Kumari A, Jha MK, Lee JC, Singh RP (2016) Clean process for recovery of metals and recycling of acid from the leach liquor of PCBs. J Clean Prod 11:4826–4834. https://doi.org/10.1016/j.jclepro.2015.08.018
123. Kumari A, Jha MK, Singh RP (2016) Recovery of metals from pyrolysed PCBs by hydro metallurgical techniques. Hydrometallurgy 165:97–105. https://doi.org/10.1016/j.hydromet.2015.10.020
124. Li K, Xu ZM (2015) Application of supercritical water to decompose brominated epoxy resin and environmental friendly recovery of metals from waste memory module. Environ Sci Technol 49(3):1761–1767. https://doi.org/10.1021/es504644b
125. Arslan F, Duby PF (1997) Electro-oxidation of pyrite in sodium chloride solutions. Hydrometallurgy 46:157–169
126. Veglio F, Quaresimaa R, Fornarib P, Ubaldinip S (2003) Recovery of valuable metals from electronic and galvanic industrial wastes by leaching and electrowinning. Waste Manag 23:245–252. https://doi.org/10.1016/S0956-053X(02)00157-5
127. Chu Y, Chen M, Chen S, Wang B, Fu K et al (2015) Micro-copper powders recovered from waste printed circuit boards by electrolysis. Hydrometallurgy 156:152–157. https://doi.org/10.1016/j.hydromet.2015.06.006
128. Xiu FR, Zhang FS (2009) Recovery of copper and lead from waste printed circuit boards by supercritical water oxidation combined with electrokinetic process. J Hazard Mater 165:1002–1007. https://doi.org/10.1016/j.jhazmat.2008.10.088
129. Xiu FR, Zhang FS (2009) Electrokinetic recovery of Cd, Cr, As, Ni, Zn and Mn from waste printed circuit boards: effect of assisting agents. J Hazard Mater 170:191–196. https://doi.org/10.1016/j.jhazmat.2009.04.116
130. Luda MP (2017) Recycling of printed circuit boards. In: Kumar S (ed) Integrated waste management, vol 2. InTech, Shanghai, pp 285–298
131. Trochimczuk AW (2002) Uptake of gold from hydrochloric acid solutions by polymeric resins bearing various phosphorus containing ligands. Sep Sci Technol 37:3201–3210
132. Fujiwara K, Ramesh A, Maki T, Hasegawa H, Ueda K (2007) Adsorption of platinum(IV), palladium(II) and gold(III) from aqueous solutions onto L-lysine modified crosslinked chitosan resin. J Hazard Mater 146:39–50. https://doi.org/10.1016/j.jhazmat.2006.11.049
133. Ramesh A, Hasegawa H, Sugimoto W, Maki T, Ueda K (2008) Adsorption of gold(III), platinum(IV) and palladium(II) onto glycine modified crosslinked chitosan resin. Bioresour Technol 99:3801–3809. https://doi.org/10.1016/j.biortech.2007.07.008
134. Parajuli D, Khunathai K, Adhikari CR, Inoue K, Ohto K, Kawakita H, Funaoka M, Hirota K (2009) Total recovery of gold, palladium, and platinum using lignophenol derivative. Miner Eng 22:1173–1178. https://doi.org/10.1016/j.mineng.2009.06.003

135. Pilsniak M, Trochimczuk AW, Apostoluk W (2009) The uptake of gold(I) from ammonia leaching solution by imidazole containing polymeric resins. Sep Sci Technol 44:1099–1119. https://doi.org/10.1080/01496390902729007
136. Kang ET, Ting YP, Neoh KG, Tan KL (1995) Electroless recovery of precious metals from acid solutions by N-containing electroactive polymers. Synth Met 69:477–478
137. Neoh KG, Tan KK, Goh PL, Huang SW, Kang ET, Tan KL (1999) Electroactive polymer-SiO_2 nanocomposites for metal uptake. Polymer 40:887–893
138. Neoh KG, Young TT, Looi NT, Kang ET, Tan KL (1997) Oxidation–reduction interactions between electroactive polymer thin films and Au(III) ions in acid solutions. Chem Mater 9:2906–2912
139. Alguacil FJ, Adeva P, Alonso M (2005) Processing of residual gold(III) solutions via ion exchange. Gold Bull 38:9–13. ACS Omega Article: https://doi.org/10.1021/acsomega.7b01215. ACS Omega 2017;2:7299−7304, 7303
140. Reck BK, Graedel TE (2012) Challenges in metal recycling. Science 337:690–695. https://doi.org/10.1126/science.1217501
141. Lahtinen E, Kivijarvi L, Tatikonda R, Vaisanen A, Rissanen K, Haukka M (2017) Selective recovery of gold from electronic waste using 3D-printing scavenger. ACS Omega 2:7299–7304. https://doi.org/10.1021/acsomega.7b01215
142. Haukka M, Väisänen A, Rissanen K, Lahtinen E, Kivijärvi L (2017) A porous body, method for manufacturing it and its use for collecting substance from source material. Finland Patent Application no 20175652
143. Cheval N, Gindy N, Flowkes C, Fahmi A (2012) Polyamide 66 microspheres metallised with in situ synthesised gold nanoparticles for a catalytic application. Nanoscale Res Lett 7:182. https://doi.org/10.1186/1556-276X-7-182
144. http://www.outotec.com/products/solvent-extraction/vsf-x-solvent-extraction-plant-unit/
145. Kinoshita T, Akita S, Kobayashi N, Nii S, Kwaizumi F, Takahashi K (2003) Metal recovery from non-mounted printed wiring boards via hydrometallurgical processing. Hydrometallurgy 69:73–79. https://doi.org/10.1016/S0304-386X(03)00031-8
146. https://copper-recycle.com/recycling-plants/e-waste-precious-metal-refining-recycling-solution.html
147. Townsend TG (2002) Evaluation of leaching test results of lead-free solders. Prepared for Abt Associates Inc, April 30
148. Townsend T, Jang Y-C, Tolaymat T (2003) A guide to the use of leaching tests in solid waste management decision making. Prepared for the Florida Center for Solid and Hazardous Waste Management, University of Florida, Report no: 03–01 (A), March
149. Yazici E, Deveci H, Alp I, Yazıcı R (2010) Characterization of computer printed circuit boards for hazardous properties and beneficiation studies. In: XXV International Mineral Processing Congress, B10, Brisbane, Australia

Chapter 11
Hydrometallurgical Recovery of Critical REEs and Special Metals from WEEE

"Keep e-waste, out of trash"
"Cash from trash"

Anonymous

Abstract Rare-earth and/or critical metals are one of the most valuable and neglected part of e-waste. Hydrometallurgical recovery methods of rare earth and critical metals from magnet scraps, fluorescent lamps, CRTs, LCDs, lion batteries, PV cells, and solar panels are broadly presented in this chapter. Industrial-scale recovery techniques of Nd, Pr, Dy, Sm, Co, Zr, etc. from magnet scraps; phosphors from lamps and CRTs; In from LCDs; Cd, Te, Sn, In, Ga, etc. from PV cells/solar panels; Pb, glass, and phosphor from CRTs; and Co from Lion rechargeable batteries are briefly presented in this chapter.

Keywords Rare-earth elements (REE) · Hydrometallurgy · Magnet · CRT · LCD · PV · Lion battery

11.1 Rare-Earth Elements and Special Metals

A rare-earth element (REE) or rare-earth metal (REM) is one of a set of 17 chemical elements in the chemical periodic table, specifically the 15 lanthanides, as well as scandium (Sc) and yttrium (Y). Rare-earth elements are cerium (Ce), dysprosium (Dy), erbium (Er), europium (Eu), gadolinium (Gd), holmium (Ho), lanthanum (La), lutetium (Lu), neodymium (Nd), praseodymium (Pr), promethium (Pm), samarium (Sm), scandium (Sc), terbium (Tb), thulium (Tm), ytterbium (Yb), and yttrium (Y). Table 11.1 shows REEs' atomic numbers, chemical symbols, and application areas in EEE manufacturing. REEs are defined to be the most important critical metals for the world economy, and they are widely used in modern industries for producing various functional materials like permanent magnets, rechargeable batteries, catalysts, and lamp phosphors because of their unique

© The Minerals, Metals & Materials Society 2019
M. Kaya, *Electronic Waste and Printed Circuit Board Recycling Technologies*,
The Minerals, Metals & Materials Series,
https://doi.org/10.1007/978-3-030-26593-9_11

Table 11.1 REE application areas in EEE

Atomic no	Symbol	REE name	EEE applications
21	Sc	Scandium	Additive in metal-halide lamps and mercury-vapor lamps
39	Y	Yttrium	Yttrium aluminum garnet (YAG) laser, yttrium vanadate (YVO$_4$) as host for Eu in television red phosphor, YBCO high-temperature superconductors, yttria-stabilized zirconia (YSZ), microwave filters, energy-efficient light bulbs, gas mantles, and fluorescent lamps
57	La	Lanthanum	Hydrogen storage, battery electrodes, optical glass
59	Pr	Praseodymium	Rare-earth magnets, lasers
60	Nd	Neodymium	Rare-earth magnets, lasers
62	Sm	Samarium	Rare-earth magnets, lasers
63	Eu	Europium	Red and blue phosphors, lasers. Hg-vapor lamps, fluorescent lamps
64	Gd	Gadolinium	Lasers, X-ray tubes, computer memories
65	Tb	Terbium	Additive in Nd-based magnets, lasers, fluorescent lamps
66	Dy	Dysprosium	Additive in Nd-based magnets, lasers
67	Ho	Holmium	Lasers, magnets
68	Er	Erbium	Infrared lasers
69	Tm	Thulium	Metal-halide lamps, lasers
70	Yb	Ytterbium	Infrared lasers

magnetic and electronic properties. REEs are mainly found in WEEE as oxide form such as In$_2$O$_3$, Y$_2$O$_3$, Eu$_2$O$_3$, Ce$_2$O$_3$, Tb$_2$O$_3$, La$_2$O$_3$, etc.

Figure 11.1 shows abundances and scantiness of major industrial metals, PMs and REEs, in the earth's crust according to atomic numbers. PMs are the rarest metals in the earth. REEs are rarer than industrial BMs.

Until 1948, most of the world's rare earths were sourced from placer sand deposits in India and Brazil. Through the 1950s, S. Africa took the status as the world's rare-earth source, after large veins of rare-earth-bearing monazite were discovered there. Through the 1960s until the 1980s, the Mountain Pass Rare Earth Mine in California was the leading producer. Today, the Indian and S. African deposits still produce some rare-earth concentrates, but they are dwarfed by the scale of Chinese production. In 2017, China produced 81% of the world's rare-earth supply, mostly in Inner Mongolia, although it had only 36.7% of reserves. Australia was the second and only other major producer with 15% of world production. All of the world's heavy rare earths (such as Dy) come from Chinese rare-earth sources such as the polymetallic Bayan Obo deposit [1]. Another recently developed source of rare earths is e-waste and other wastes that have significant rare-earth components. New advances in recycling technology have made extraction of rare earths from these materials more feasible, and recycling plants are currently operating in Japan, where there is an estimated 300,000 tons of rare earths stored in unused electronics. In France, the Rhodia group has two factories, in La Rochelle and Saint-Fons, that produce 200 tons of

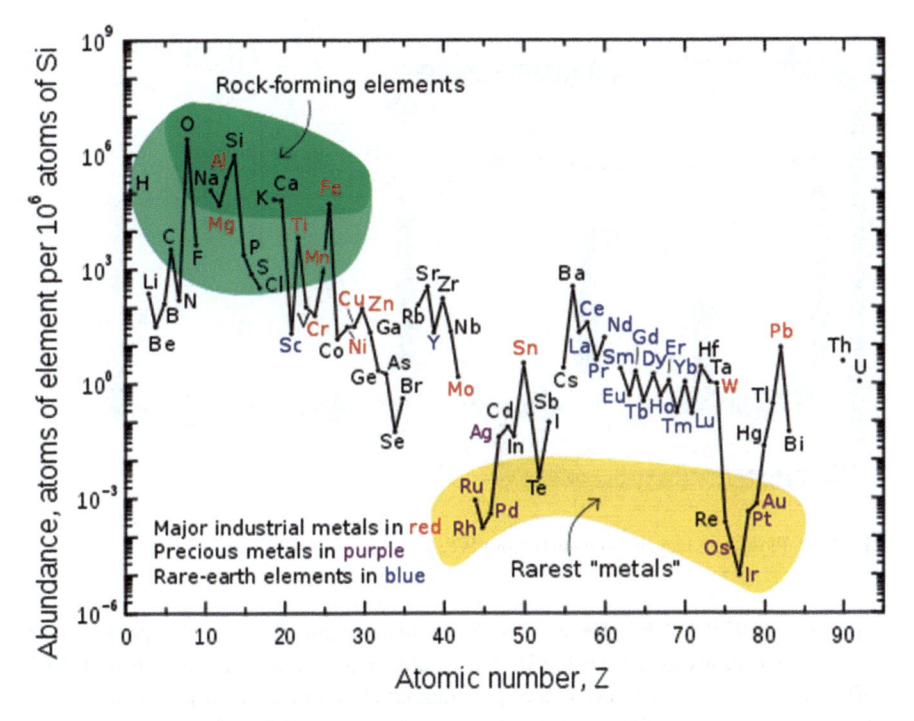

Fig. 11.1 Abundances of BMs, PMs, and REEs in the earth's crust [1]

rare earths a year from used fluorescent lamps, magnets, and batteries [1]. REE recycling ratio was less than 1% in 2011 because of inefficient collection, technological issues, and lack of incentives. Main sources were permanent magnets, NiMH batteries, and lamp phosphors [2]. A mobile phone contains only 0.1–0.25 g of REEs. Challenges in REE recycling are [2]:

- During pyrometallurgical recycling of e-waste and used catalysts, REEs end up as oxides in slags.
- Concentration in oxide slags is too low for recycling.
- Slags used as building material.
- In most applications, the amount of rare earth per item is low (a few grams or less).
- Deep-level dismantling is recommended to recover rare-earth-containing objects.

11.2 Recovery of Magnet Scraps

Pure rare earths are used in two permanent magnets (NdFeB, which includes Nd, Pr, and Dy, and SmCo, which includes Sm). Magnets are 2 wt.% of hard disc drivers (HDD). Rare-earth elements are 0.6 wt.% of HDD. REEs distribute widely in

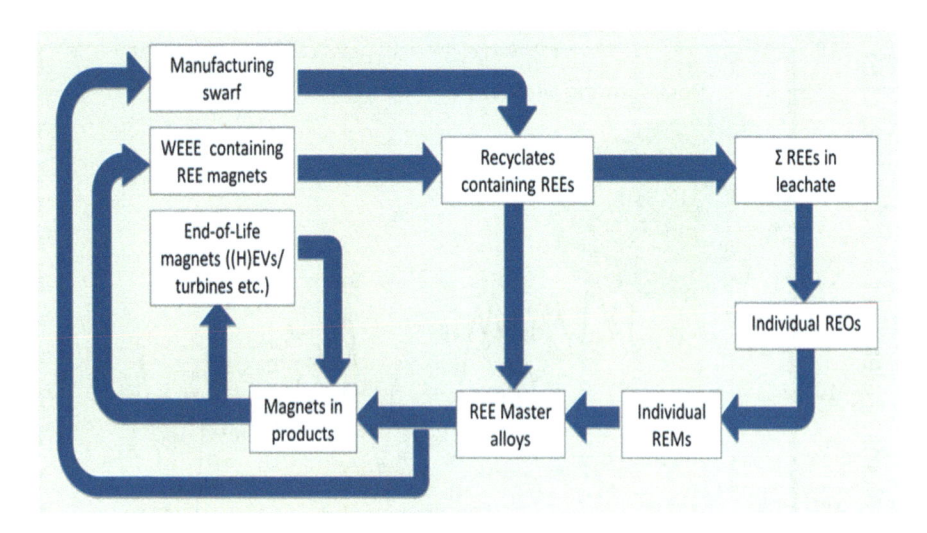

Fig. 11.2 Flowsheet recycling magnets from WEEE

different types of WEEE. Except for the magnet scraps, they usually present with low concentrations in the waste [3]. This reality makes their recovery from WEEE difficult. Since REEs are very reactive compared to other metals, it will be lost in the slag phase, if pyrometallurgical process is applied. Thus, REEs from WEEE should be extracted by leaching. The recycling of SmCo (developed in the early 1970s and has a share of less than 2% in the market) and NdFeB (developed in 1982) spent magnet scrap/swarf by leaching is difficult, but possible. Figure 11.2 shows the flowsheet recycling magnets from WEEE [2].

SmCo brittle and easily crackable magnets are quite low on the market, and recycling of the existing scrap is still important because of high values of Sm and Co. The waste to be recycled is usually $SmCo_5$- or $Sm_2(Co, Fe, Cu, Zr)$-based swarf. Sm/Co scrap is usually leached with H_2SO_4, HCl, HNO_3, or $HClO_4$ that Sm and transition metals (Co, Fe, and Cu) can easily be dissolved into the solution. Zr, Si, O, etc. do not dissolve and separated from the REE-containing solution. Usually double salt of Sm can be obtained if H_2SO_4 is used for leaching, whereas direct solvent extraction is applied in the cases of HCl or HNO_3. In the selective precipitation of Sm, it can be almost completely precipitated as an oxalate or as a sulfate double salt by changing pH of the solution (e.g., addition of oxalic acid or $NaOH/Na_2SO_4/NH_3$). Low solubility of Sm-containing salts (e.g., $Sm_2(SO_4)_3 \cdot 2Na_2SO_4 \cdot 2H_2O$) enables the separation of Sm from the leaching solution. In the solvent extraction process, tributyl phosphate (TBP) or di(2-ethylhenxyl) phosphoric acid is usually used as the extracting reagent [3].

Recycling of Nd-based magnet scrap coming from shredding of HDD is possible with H_2SO_4 leaching at room temperature. Direct leaching, oxidative roasting, and selective leaching can be applied. After purification of the REE-containing solution, they can be precipitated out as double salts [4]. It is also possible to precipitate as oxalates or fluorides depending on the leaching reagents. However, this process needs to completely dissolve the scrap and consume a large amount of acid, and the Fe-containing waste acid after REE precipitation brings difficulties for further treatment. It is unavoidable to leach unwanted elements (Ni, Co, and Cu) into the solution meaning that the purification of the REE-containing solution can be critical for producing high-purity REE products [3]. After oxidative roasting at 900 °C for 6 h, low-acid selective leaching with high REE recovery rate (i.e., 99% REE and 0.5% Fe) and lower reagent and energy consumption is the best and most effective process for magnet scrap treatment. After roasting Fe_2O_3 is formed and leaching of Nd can be more selective by using low acid concentration, and the pH control is important to ensure a high selectivity. Fe is also leached into the solution; it can be precipitated and removed by hydroxide.

$$Fe^{+3} + 3OH^- \rightarrow Fe(OH)_3 \quad (11.1)$$

The assisted techniques, e.g., ultrasounds, microwave, and electrochemistry, may be applied to the above technologies to increase the REEs leaching rate. Furthermore, using ionic liquids both for leaching and extraction of NdFeB magnet scrap is proved more effective than HCl acid, and it is described to be green with minimized wastewater treatment efforts [3].

Santoku Corporation has opened in 2012 a plant in Tsuruga (Japan) for recycling Nd and Dy from magnets. Motor magnets (air conditioners) and magnet production scrap are used. After demagnetization by heating (6 h at 300 °C), magnets are jaw crushed under 75 μm and pulverized. Oxidation by stirring for 12 h in an alkaline solution, selective dissolution in HCl, filtration, rare-earth salts, rare-earth oxide, molten salt, and rare-earth metals for magnet alloys are produced by molten salt electrolysis (Fig. 11.3a, b) [2]. Recovery of REE from magnets has four main steps: milling and roasting, dissolution or selective HCl leaching, removal of transition metals by solvent extraction (with ionic liquids), and precipitation of REEs (oxalate) and calcination to oxides.

In solvent extraction ionic liquid *(trihexyl(tetradecyl)phosphonium chloride)*, Cyphos IL 101 were tested for Nd/Fe and Sm/Co separations. Cyphos IL 101 has negligible vapor pressure, is non-flammable, non-fluorinated, and undiluted. Fe and Co were extracted. Rare earths were not extracted and remained in water phase. Separation factors were Nd/Fe or Sm/Co > 10^6 [2].

11.3 Recovery of Lamp and CRT Phosphors

The reclaimed lamp phosphor mixture is relatively a rich source of REEs, especially heavy REEs (Eu, Tb, Y) which are more critical than light REEs. In fluorescent lightings and CRTs, Y_2O_3, Eu_2O_3, Ce_2O_3, Tb_2O_3, and La_2O_3 are present. Figure 11.4 shows lamp and waste phosphor powder composition and

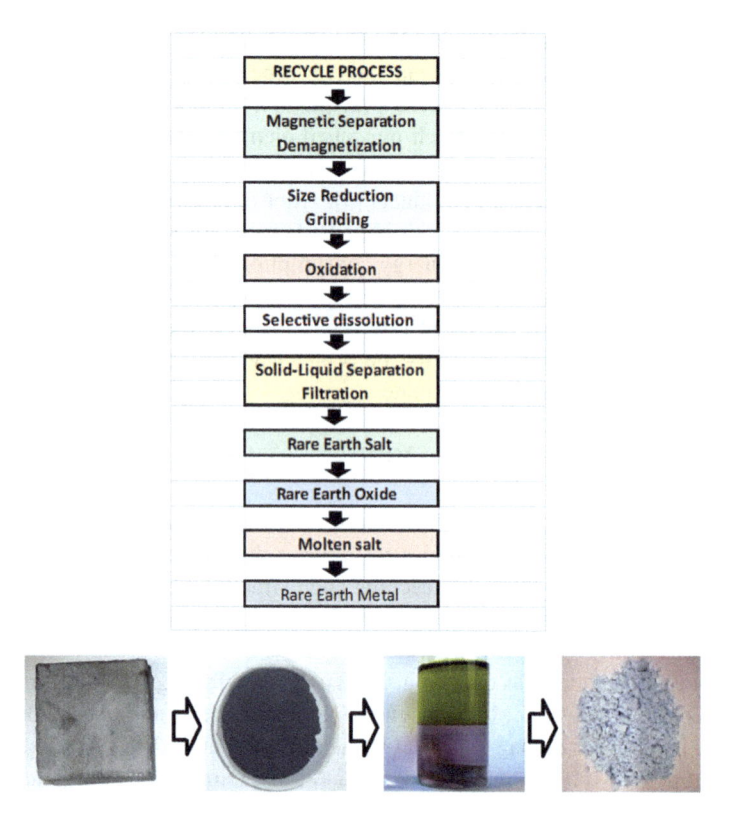

Fig. 11.3 Magnet recycling flowsheet by Santoku in Japan (**a**) and REE recovery steps from magnets (**b**)

REE-based phosphor chemical composition. There are 88% glass, 5% metals, 4% plastics, 3% waste phosphor powder, and 0.005% Hg in an energy-conserving fluorescent lamp. Waste phosphor powder contains 45% halophosphate, 20% fine glass, 10–20% REEs, 12% alumina, and 5% others. REE-based phosphors contain BAM (BaMgAl$_{10}$O$_{17}$:Eu^{2+}), YOK (Y$_2$O$_3$:Eu^{3+}), and LAP (La, Ce)PO$_4$:Tb^{3+}) and CAT (CeMgAl$_{10}$O$_{19}$:Tb^{3+}).

In order to recover REEs from phosphor mixture, leaching and precipitation or solvent extraction can be used [5, 6]. Phosphors have similar features to the REE ores; thus, their acidic dissolution follows halophosphate < yttrium oxide (YOX) < barium aluminate (BAM) < calcium tungstate (CAT) < lanthanum phosphate (LAP). Besides acidic dissolution, molten salts, mechanical activation, pressure leaching, and even supercritical CO$_2$(g) are also used to improve the decomposition of phosphors in waste materials [3]. In practice OSRAM patented process uses multistep leaching treatment for different phosphors [7]:

- Leaching with dilute HCl below 30 °C to leach only halophosphates.
- Increasing temperature to 60–90 °C, the diluted HCl leaches YOX.

11.4 Indium (In) Recovery from LDCs

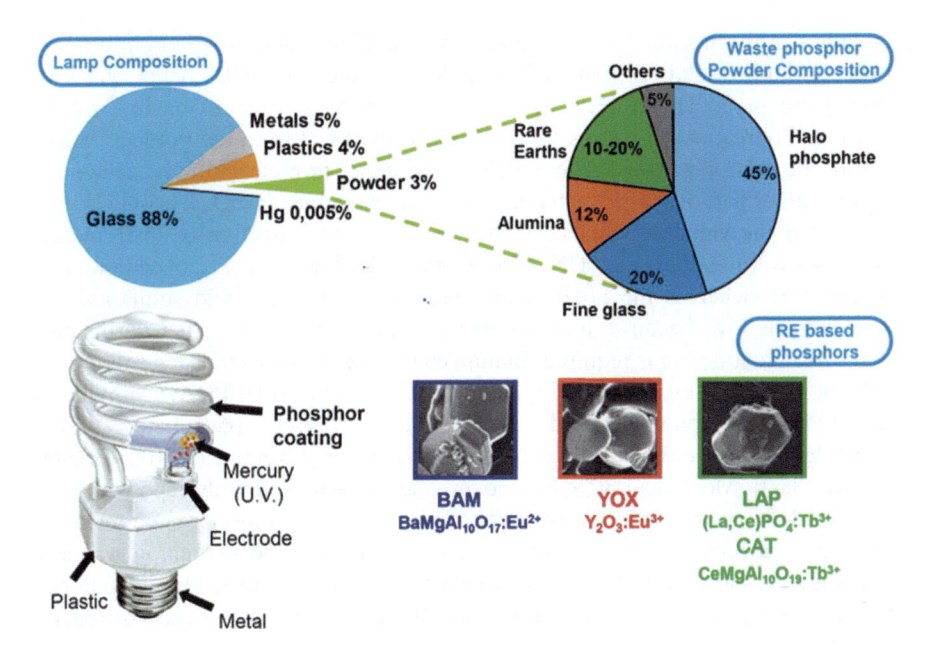

Fig. 11.4 Lamp and waste phosphor powder composition and REE-based phosphor chemical composition

- LAP is then dissolved with concentrated H_2SO_4 above 120 °C.
- CAT and BAM are dissolved in 30% NaOH at 150 °C in autoclave or in molten alkali.

Solvay Group in France has two dedicated facilities for REE recovery from fluorescent lamps in La Rochelle and Saint-Fons using Rhodia process. Industrial demo plant was operational since the beginning of 2012 (capacity: 1000 tons powder/year). In 2013, the full capacity of industrial plant was 2500 tons powder/year. In Rhodia's lamp-recycling project, there are three steps: preparation of REE concentrate from recycled lamp phosphors (La, Ce, Eu, Gd, Tb, Y), separation of REE concentrate in individual REEs, and preparation of new REE phosphors [2].

11.4 Indium (In) Recovery from LDCs

Indium is a rare metal in the earth's crust at a concentration of 0.1 ppm. In the primary ores, In grade is only 10–20 ppm, making the extraction very difficult. In semiconductor compounds are widely used in photovoltaic (PV) devices, IR detectors, and high-speed transistors. The largest application of In is thin film coating in LCDs which are important components of modern electronic components. More than 70% of In is consumed as indium tin oxides (ITO) [1]. Indium content in e-waste can reach more than 500 ppm which is higher than its grade in primary ores. Indium metal content of shredder residue of LCD panel glass is 200 ± 50 ppm. Therefore, it is very important to recycle In or In_2O_3 from e-waste. Pretreatment requires disassembly of the LCD from the product, before rest of the unit operations. After crushing, grinding, or preleaching by acetone to remove all or part of plastic materials or organic materials, H_2SO_4 leaching is used to recover the valuable metals together with In. The extraction of In from acid solutions follows the order $HNO_3 > H_2SO_4 > HCl$ when acid concentration is below 2 M, whereas the order is inversed in concentrated acid solutions. The selectivity is low, because of coleaching of other metals and further purification and solvent extraction is required. Indium can be recovered from leach solution by solvent extraction using di-(2-ethylhexyl) phosphoric acid DEPHA, 2-ethylhexyl 2-ethylhexyl-phosphonic acid (EHEHPA), tributyl phosphate (TBP), trioctylphosphine oxide (TOPO), and methyl isobutyl ketone (MIBK) organic phases [8, 9]. More than 80% In recovery can be achieved with vaporization of In is $InCl_3$ at around 400 °C. Removal of plastic films or other organic materials from In is difficult. Ultrasound-assisted supercritical liquid extraction was also investigated. The liquid crystals can be either separated or extracted from the In-rich phase. The liquid crystal treatment is difficult due to complex structure of organic materials [3].

He et al. (2014) effectively recovered metallic In from LCD panels using coke powder as a reducing agent. Indium oxide (In_2O_3) was reduced to metallic In at 1223 K and 1 Pa with 30% carbon additions for 30 min, and the recovery rate of In reached 90% [10]. Ma et al. (2012) recovered In from waste LCD panels by chlorinated vacuum separation method. NH_4Cl was used as a chlorinating agent at a temperature of 400 °C. NH_4Cl-to-glass ratio was 1:2, and particle size was −0.13 mm [11].

The environmental and financial benefits of vacuum processes for the e-waste recycling have been validated. Gas flow can be efficiently controlled, and furthermore, no wastewater is released or dust emission occurs. This process is a promising and environment-friendly method for metal recycling from e-waste; still, it needs to be improved for further advancement. On the one hand, vacuum processes have many benefits for separating metals having low boiling point and high saturated vapor pressure, such as Zn, Pb, and Cd. In contrast, separation of the valuable and rare metals having low saturated vapor pressure through vacuum condensation method is not so great. Still, pretty much hypothetical issues are waiting for additional clearance before its modern industrial application [12].

11.5 Photovoltaic (PV) Cell/CIGS Solar Panels Recovery

Thin-film PV cells contain various elements such as Cd, Te, Sn, In, and Ga, and their recycling is an excellent example of application of the Recycling Metal Wheel, requiring primary metallurgical industry expertise and infrastructure for

11.6 Cathode Ray Tube (CRT) Recycling at Attero Plant

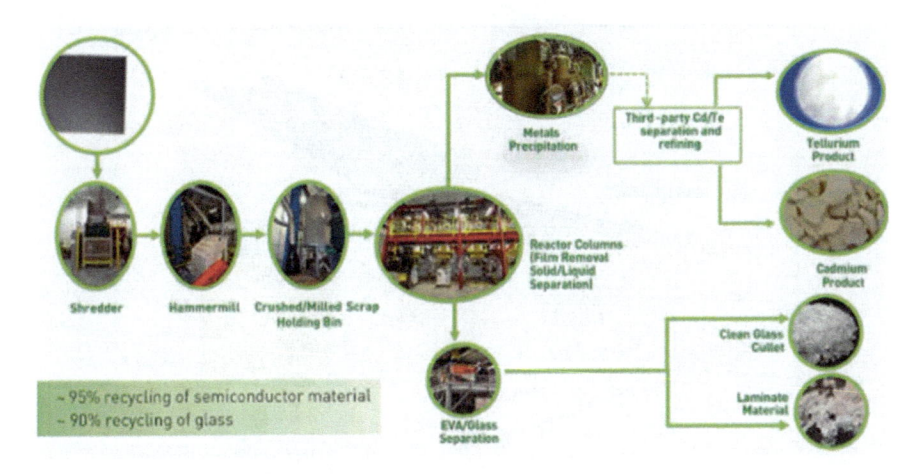

Fig. 11.5 PV cell recovery flowsheet

recovering these metals from the recyclates. GaN, InGaN, GaAs, Cu(In, Ga)Se$_2$, In$_2$O$_3$, Cu(In, Ga)Se$_2$, and Ce can be found in PV cells. The physical separation steps are the same as discussed elsewhere in this book. Figure 11.5 shows a scheme for recycling manufacturing; Te and Cd product recovery are obtained. This can be used for well-sorted EoL scrap as well, but in reality the latter will always be contaminated with other materials and metals. Nevertheless, with a metallurgical infrastructure in place and ensuring that the valuable metals and their compounds are concentrated in suitable phases, the Metal Wheel shows that the recycling system can be "closed." It requires, however, a vision and policy to maintain the metallurgical infrastructure to this end; otherwise, the sustainability-enabling metals used in PV cells will be lost, with a total reliance on mining for further supplies.

11.6 Cathode Ray Tube (CRT) Recycling at Attero Plant

CRT contains significant amounts of Pb and glass which can be recovered and reused. As a first step, CRT display units are unloaded at recycling facility and scanned by barcode reader. The scan details are then uploaded on the system. Next the CRT cutter system separates the glass panel and funnel glass, and fluorescent powder (Pb and phosphor) is collected. The process involves glass cutting, heating through the metal band, and air blowing. Next, they are sent to the vacuum chamber, following which the Pb and phosphor are collected in bags, while the glass is collected in folders. Panel glass cullets can be reused for new panel glass manufacture. For processing one CRT unit, the machine takes roughly 90 s. Circuit boards, chips, and other parts are recycled separately. Figure 11.6 shows optical sorting equipment working principles used in glass separation.

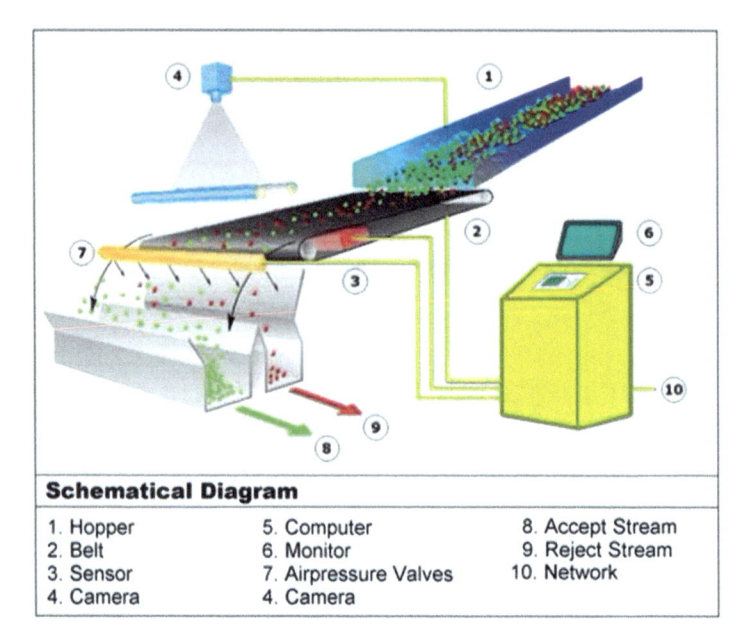

Fig. 11.6 Optical sorting equipment [13]

There is not widespread commercial use of recycled glass. Closed-loop CRT glass recycle is difficult; recycled glass can be used in bricks, building products (such as aggregate), or flux in smelting operations. The plastics stream mixed WEEE processing contains a large number of different polymers that are difficult to separate. Currently, most of mixed plastic is not recycled and may be converted to a liquid diesel fuel. Recycled plastics are melted and regranulated.

11.7 Flat Panel Display Unit Recycling

Recycling of TFT and flat panel display units begins with the segregation of the device, following which the unit is dismantled. During dismantling components like wires, cables, and PCBs are segregated and sent for recycling separately. The display unit then goes through the mechanical shredder, where it is processed. Next the unit is passed through the magnetic separator, where ferrous metals are automatically removed. The next stage involves the separation and collection of nonferrous metals like Al, Cu, etc. in the Eddy current separator. The separated components are then processed individually. The plastic components left behind after the eddy current separator stage is segregated and recycled. The ferrous components collected from the magnetic separator are processed to Fe, while the Cu and Al collected from the eddy current separator are smelted and go through electrorefining, where metals are refined to 99.9% purity.

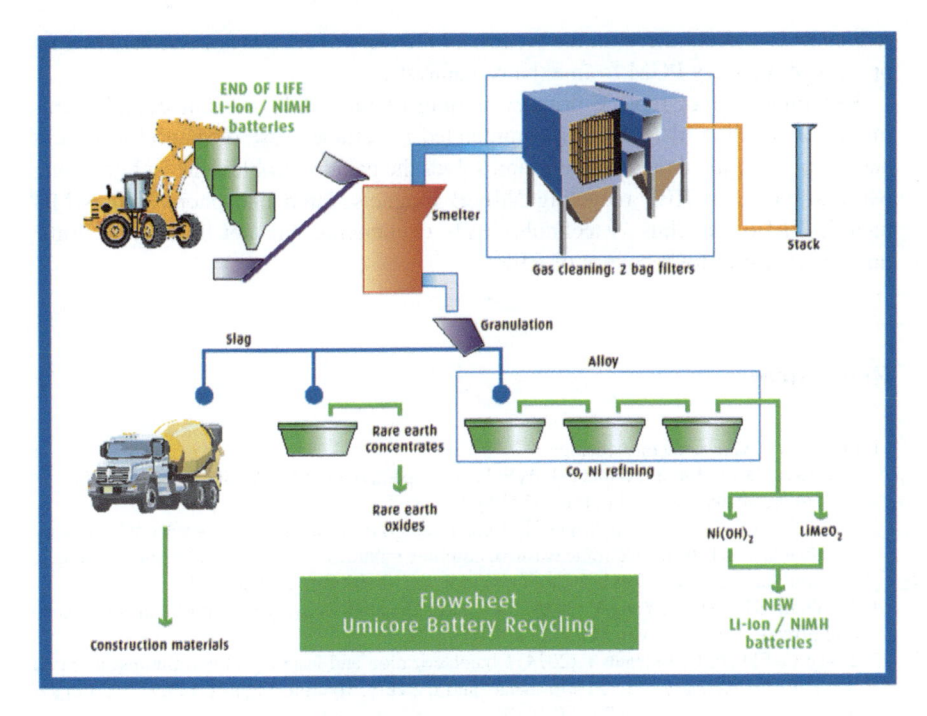

Fig. 11.7 Umicore's battery recycling plant flowsheet

11.8 Cobalt (Co) Recovery from Lion Batteries

Co is mainly found in spent Lion batteries. Lion batteries can be used in electrical vehicles and renewable energy storage devices in the future. Lion battery may typically contain 27.5% LiCoO$_2$. The electrode materials account for around 44% of the whole battery value with Li-Co-O-based cathode materials of 30% and graphite anode materials of 14%. For spent Lion battery recycling, physical processing, e.g., mechanical degradation and thermal treatment, and chemical processing, e.g., leaching (acidic, basic, or bio) and solvent extraction, precipitation, and electrochemical processing, can be used. Li and Co can be leached using inorganic H$_2$SO$_4$, HCl, or HNO$_3$ acids or using organic acids (citric or oxalate) with H$_2$O$_2$ oxidant presence or absence at 60–80 °C. More than 90% Li and Co recoveries were obtained. In spent Lion batteries, the values of Co, Li, and Cu metals are 47.7%, 28.7%, and 19.8%, respectively [2].

Figure 11.7 shows Umicore's battery recycling flowsheet for EoL Lion and NiMH batteries. Smelting and Co-Ni refining sections recover Li(OH)$_2$, LiMeO$_2$, and REOs. Final slags are sent to the construction sector.

The first industrial-scale EoL portable NiMH battery recycling process was developed by Rhodia and Umicore. Cooperation with Rhodia-Umicore produces REE concentrate. Umicore separates REO from harmful elements, while Rhodia refines REE concentrate. But, the process of recovery of REE is not compatible with process of recovery PGM from exhaust catalysts.

Recycling cannot replace primary mining of rare-earth ores but complements mining. Recycling of REEs is recommended for efficient use of natural resources and supply of critical raw materials and solves the balance problem. Most interesting waste streams for REE recycling NdFeB magnets, lamp phosphors, and NiMH batteries. REE recycling is technologically challenging, but not impossible. Ionic liquids are useful for recovery of REEs.

References

1. http://www.wiki-zero.com/index.php?q=aHR0cHM6Ly9lbi53aWtpcGVkaWEub3JnL3dpa2kvUmFyZS1lYXJJ0aF9lbGVtZW50
2. https://core.ac.uk/download/pdf/34595239.pdf
3. Sun Z, Xiao Y, Sietsma J, Agetrhuis H, Yang Y (2015) A cleaner process for selective recovery of valuable metals from electronic waste of complex mixtures of end-of-life electronic products. Environ Sci Technol 49:7981–7988. https://doi.org/10.1021/acs.est.5b01023
4. Abrahami ST, Xiao Y, Yang Y (2015) Rare-earth elements recovery from post-consumer hard-disc drivers. Trans Inst Min Metall C 124:106–115
5. Tunsu C, Ekberg C, Retegen T (2014) Characterization and leaching of real fluorescent lamp waste for the recovery of rear earth metals and mercury. Hydrometallurgy 144:91–98. https://doi.org/10.1016/j.hydromet.2014.01.019
6. Preston J, Cole P, Craig W, Feather A (1996) The recovery of rare earth oxides from a phosphoric acid by-product. Part 1: Leaching of rare earth values and recovery of a mixed rare earth oxide by solvent extraction. Hydrometallurgy 41(1):1–19
7. Otto R, Wojtalewickz-Kasprzak A (2012) Method for recovery of rare earths from fluorescent lamps. U.S. Patent 20120027651 A1, Feb 2
8. Virolainen S, Ibana D, Paatero E (2011) Recovery of indium from indium tin oxide by solvent extraction. Hydrometallurgy 107(1):56–61. https://doi.org/10.1016/j.hydromet.2011.01.005
9. Paiva AP (2001) Recovery of indium from aqueous solutions by solvent extraction. Sep Sci Technol 37(7):1359–1419. https://doi.org/10.1081/SS-100103878
10. He Y, Ma E, Xu Z (2014) Recycling indium from waste liquid crystal display panel by vacuum carbon-reduction. J Hazard Mater 268:185–190. https://doi.org/10.1016/j.jhazmat.2014.01.011
11. Ma E, Lu R, Xu Z (2012) An efficient rough vacuum-chlorinated separation method for the recovery of indium from waste liquid crystal display panels. Green Chem 14:3395–3401. https://doi.org/10.1039/C2GC36241D
12. Zhan L, Xu ZM (2014) State-of-the-art of recycling e-wastes by vacuum metallurgy. Sep Environ Sci Technol 48:14092–14102. https://doi.org/10.1021/es5030383
13. http://www.r-t-g.com/ContentN/E_Developing.asp

Chapter 12
Perspectives of WPCB Recycling

"Think green, Live green, Buy less, Choose well, Use longer"

"Don't trash our future RECYCLE"

Abstract Economic, environmental, and marketability perspectives of WPCB recycling are introduced in this chapter. Comparison of primary ore mining with secondary urban mining (e-waste recycling) shows significant advantages of urban mining. Recycling is a key contributor to sustainable circular economy. Recycling, urban mining, and enhanced landfill mining close the material loop. Market for recycling technologies and eco-design guidelines for manufacturing are presented. Limitations of current e-waste/WPCB recycling technologies and emerging future development perspectives toward sustainability and zero-waste scheme are also explained in detail.

Developments of new separation and extraction methods should consider increase in recovery rates and separation efficiency; better automatic control; lower time consumption, labor force, investment, and operating costs; and environmental impacts simultaneously. Developed new WPCB recycling methods should take into account industrial applicability potential, automation in pretreatment, reducing processes and reagents, and paying more attention to nonmetallic fraction. WPCBs should be regarded as important secondary material sources for achieving higher sustainability under the context of circular economy. There is almost no mature technique for ECs recycling currently.

Keywords Economy perspective · Environmental perspective · Marketability perspective · Eco-design · Future developments

Overall recycling success factors depend on:

- Technical recyclability as basic requirement
- Accessibility of relevant ECs → product design

- Intrinsic and external economic viability
- Completeness of collection system (business models, legislations, and infrastructure)
- Keeping within recycling chain → transparency of flows
- Technical-organizational setup of recycling chain → recycling quality
- Sufficient recycling capacity

Complex products require a systemic optimization and interdisciplinary approaches (product development, process engineering, metallurgy, ecology, and social and economic sciences). In order to focus on sustainable circular economy, innovations and improvements are still needed in every step of product manufacturing, use, EoL, and recycling for secondary raw materials. New product manufacturing step should consider recycling product design, developing business models to close the loop, and recycling production scrap. In the product use step, dissipation should be avoided; residue streams should be minimized and effectively recycled; and holistic system approach should be taken. In the EoL step, collection should be improved; transparency of flows should be increased; instead of mass recycling, more focus should be given to technology metals; and innovative technical recycling challenges should be developed. In the recycling for secondary raw material step, recovered metal yields and ranges should be improved, and energy and water use efficiency should be improved.

12.1 Economy Perspective of Recycling

The economy perspective is always very important or the first priority for the evaluation and implementation of a WPCB recycling technique. The process, which is cost-effective, will have a better chance to be commercialized especially in developing countries due to the lack of sufficient funds for WPCB recycling. One important environmental impact is the loss of usable resources, if materials are not recycled. The extraction of PMs through mining is associated with negative environmental impacts through significant emissions of greenhouse gases and energy, water, and land usage. The environmental impacts of the secondary production in state-of-art operations are much lower than primary production [1]. Besides environmental protection and legislative pressure, recycling is also driven by economic interests. High economic values and limited available reserves of PMs and SMs provide attraction for WPCB recycling.

Large-scale e-waste recycling started in the 1950s for Cu, steel, Al, and PM-graded populated WPCB. In Europe, 1 ton of populated WPCB is about 1000–5000 euros. If medium-graded WPCB costs about 3000 euros, the intrinsic value is more than 3.8 times higher. Therefore, there is a significant economic drivers involved in PCB recycling. Table 12.1 shows typical intrinsic value of populated WPCBs. Au + Pd + Pt + Ag have about 88% of the intrinsic value of WPCB. Cu and

12.1 Economy Perspective of Recycling

Table 12.1 Typical intrinsic value of populated WPCB

Metal	% by wt	Value per kg ($)	Intrinsic value ($/kg)	Intrinsic value (%)	% of value from smelter
Au[b]	0.025	38,772.19	9.693	63.07	98
Pd[b]	0.010	31,723.14	3.172	20.64	92
Pt[b]	0.001	25,698.09	0.257	1.67	
Ag[b]	0.1	459.27	0.459	2.99	95
Cu[a]	16	5.8500	0.936	6.09	96
Sn[a]	3	18.925	0.568	4.28	
Pb[a]	2	2.0670	0.041	0.27	
Ni[a]	1	12.415	0.124	0.81	
Al[a]	5	1.5400	0.077	0.50	
Fe[a]	5	0.317	0.016	0.01	
Zn[a]	1	2.436	0.024	0.16	
Total			15.367		

[a]lme.com
[b]goldprice.com (Sep. 06, 2018)

Sn have approximately 11.6% of the remaining intrinsic value. Commercial smelters typically pay between 92% and 98% of the PM value.

To achieve economic feasibility of metal recycling, it is more and more required that all the valuable metals need to be considered in the recycling of metals instead of only covering the critical metals for improving the sustainability of WEEE recycling. For instance, the production of PMs has been one of the most profitable sections in Umicore instead of recycling of Co, REE, etc. In the world market of metals, secondary Al, Cu, and Pb are part of a large fraction of the metal supply [2]. Critical metals (REEs, Co, Ga, In, PGMs, and Ge) have attracted relatively high attentions recently during metal recycling from WEEE.

Obviously, landfill and incineration have limited economic benefit even without the concern about the treatment of the hazardous substances they emit. The construction and administration costs of landfill sites or incineration plants are large expenses with a scarce return. This is also a reason for the reduction in the current application of these two techniques for WPCB treatment. Pyrometallurgical and hydrometallurgical recycling techniques can achieve partial recycling of WPCB by recovery of MF from the feedstock. However, the intensive energy consumption in pyrometallurgy is attributed for the main expense in the whole process. The energy power in the process is around 0.2–4.0 KW, varying with the different processes, while sufficient data is not available for these techniques since most works were conducted at the laboratory scale and energy analysis is absent for most of the studies. However, the normal pyrolysis temperatures are from 400 to 800 °C, and the next gas emission treatment equipment normally needs the temperature higher than 1200 °C and no doubt consumes a considerable amount of energy. For hydrometallurgical recycling techniques, the chemical reagents used take a considerable and indispensable percentage of the cost. Also, in hydrometallurgy, water is used as leaching media and solvent on a large scale, which also contributed to the operation cost [3].

Currently, percentage of noble metal in the WPCB is falling steadily due to advanced manufacturing technologies. Thus, the driving force for recycling MF from WPCB is decreasing, since previous works focus on the benefit of MF recovery. Recycling of NMF is a potential way to remedy the cost on energy and/or chemical cost. However, the precondition for NMF recycling is the homogenization of the WPCB, which means the pulverization of WPCB feedstock by shredders and hammer mills, which also requires energy and equipment cost. A fine size of NMF (around 0.1 mm) is typical for the recycling of NMF in many studies. Even though in both pyrometallurgy and hydrometallurgy recycling techniques the structure of NMF was totally decomposed, the recycling of NMF is not applicable. In some types of mechanical separation, the structure or composition of NMF was preserved with a scanty effect on the chemical and mechanical properties. Nevertheless, the direct use is not enough to add adequate value to compensate for the cost required to preserve their structure. Therefore, the upgrading and recycling of the value-added nonmetallic fraction are attracting more and more researchers.

12.2 Environmental Perspectives of Recycling

The environmental perspective is another important issue that affects the evaluation of the WPCB recycling techniques. A general rule for emissions/discharges is that the more WPCB recycled, the less will be emitted/discharged. Therefore, the focus of a technique should not only be on the performance of the BM or PM recovery rate but also the fate of the remaining part due to the potential emissions that they may have, especially for gaseous emissions. Environmental emissions can be divided into primary pollution, which comes from the WPCB, and secondary pollution, which is emitted during the treatment or recycling process. For primary pollution, heavy metals including Cu, Pb, Hg, Cd, and Be are of great concern. They can enter the environment in the form of leachate in landfill or vapor in incineration or some pyrometallurgical recycling techniques without proper purification. It can cause the distortion of the human body as well as various other health hazards. The majority of public attention is still paid to the emissions of NO_x, SO_x, or VOCs, while some kinds of pollutants (PBDD/Fs, PCDD/Fs, Br, and HBr) are ignored.

Br concentration in WPCB is relatively high (about 4%) due to the addition of BFRs. The decomposition of BFRs happens in the recycling process, while Br still remains in the form of several kinds of Br or bromorganic compounds including HBr, bromomethane, and bromophenol. However, research shows that Br, especially HBr, has a very bad effect on human health, since it can cause health problems and is accumulated in the human body and is difficult to degrade. Therefore, to achieve complete recycling and elimination, emissions are the ultimate objective of the environmental impact of WPCB recycling techniques [3].

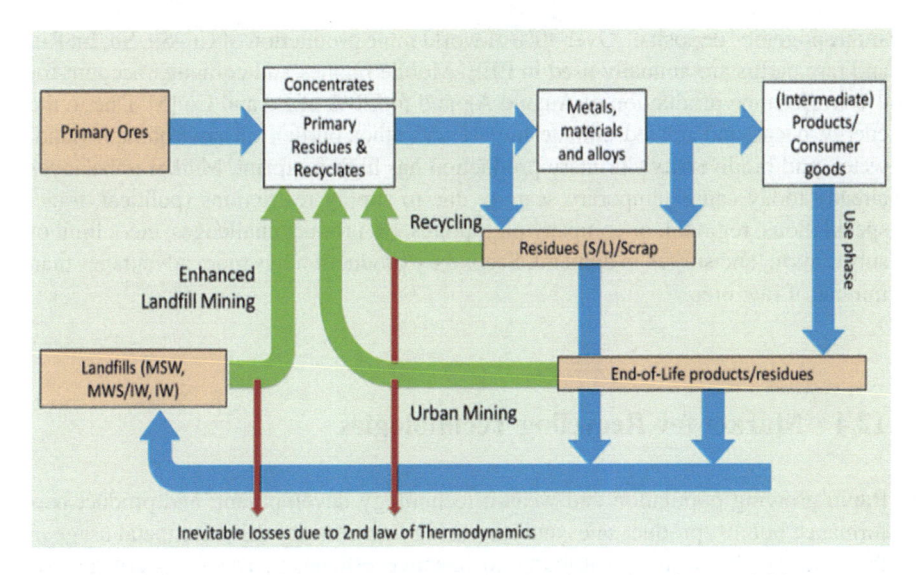

Fig. 12.1 Recycling, urban mining, and enhanced landfill mining close the material loop

12.3 Marketability Perspective

The evaluation of WPCB recycling techniques is based on economic, environmental, and marketability perspectives. Both cost-effective and environmental friendly recycling techniques may not be feasible due to product recovery rate or purity. For some techniques, the product recovery may be very high (such as >90%), while purity is low which prevents the product to be sold or used directly in the industry. The market price of metals is seriously affected by their purity. Small decreases in purity will cause sharp drop in the metal prices. Therefore, gate-to-market ability is a criterion independent of economic and environmental perspective and plays an important role in the evaluation of the feasibility of the process. It is impossible to achieve or satisfy these three criteria simultaneously with one simple recycling technique. Thus, the future research direction of WPCB recycling should focus on the combination of several techniques or in series recycling since the drawback of a process could have a chance to be remedied by one of the techniques, and the evaluation should be conducted on the whole process until the products satisfy the market standard [3].

Comparison of primary ore mining with secondary urban mining (e-waste recycling) shows significant advantages of urban mining. Recycling is a key contributor to circular economy. Recycling, urban mining, and enhanced landfill mining close the material loop like in Fig. 12.1 [4].

Declining ore grades, increasing ore complexities, and needing to mine from greater depths and/or in ecological sensitive areas (arctic regions, oceans, rain forests, etc.) make primary raw materials scarce. Metals can be recycled eternally without loss of properties. There is a massive shift from geological resources to anthropogenic "deposits." Over 40% of world mine production of Cu, Sn, Sb, In, Ru, and rare earths are annually used in EEE. Mobile phones and computer account for 4% world mine production of Au and Ag and for 20% of Pd and Co [5]. Due to the energy needs and related climate impact and other burden on environment (land, water, and biodiversity), primary production has high footprint. Market imbalances already today cause temporary scarcity due to supply restrictions (political, trade, speculations, regional, or company oligopolies, by-product challenges, etc.), limit of substitution, and surges in demand. Secondary production has more advantages than mining of raw ores.

12.4 Market for Recycling Technologies

Rapid growing population and wealth; technology development; and product performance booms; product sales and increasing functionality drive for metal usage in the industry. Recycling contributes to resource efficiency, access to critical raw materials, and innovation. Recycling more e-waste and using better and greener recovery technologies are not a waste of time. Thus, countries should create appropriate legal framework that focuses on high-quality recycling, ensure stringent enforcement of legislations, and support R&D funding and implementation of innovative recycling solutions.

The criterion of markets can be regarded as the basis of technology transfer. The main drivers for the creation of recycling and recycling technologies markets are economic and regulatory factors. However, the market potential of e-waste recycling technologies and the framework conditions vary between countries and regions. Specifically, the most promising technologies for e-waste recycling need to be identified and fostered through relevant instruments. Particularly in many developing countries, tools and instruments are required that promote the finance of collection and transfer of technology innovation in the field of e-waste recycling. This would save costs, energy, and natural resources and could help countries to be less dependent on raw material prices [6]. Technology transfer and capacity building are critical tools to implement e-waste recycling innovation in emerging countries. However, simply copying innovative technologies from post-industrialized to industrializing economies does not necessarily generate the most suitable solution. Thus, capacity building and fostering, coordinating, and strengthening of existing regional capacities are essential for enabling industrialized countries to stimulate local development of sustainable technologies and innovation and to allow them to experience progress and sustainable livelihood.

12.5 Eco-design Guidelines for Manufacturing

In order to achieve viable and effective recycling, product properties and recycling process capabilities must be matched. WPCB products should contain extractible sufficient metallic and nonmetallic values. Composition and concentration of valuable material in WPCB depend on product and technological developments. Current market prices are determined by market developments. Recycling process technology performance and costs are affected by technological efficiency for value recovery (i.e., range, yields, energy, etc.), process robustness and flexibility, environmental and social compliance, available volumes (i.e., economies of scale), factor costs (i.e., labor, energy, and capital), and process chain organization/interface management. Eco-design methodology for PCBs should ensure the minimum consumption of resources (materials, energy, water, etc.) and resource efficiency during the all product life cycle.

Products that are designed with the possibilities and limitations of recycling in mind (design for disassembly (DfD)/design for recycling (DfR)/design for sustainability (DfS)) can further facilitate recycling to become "design for resource efficiency." While designing EEEs or PCBs, material selection guidelines for both manufacture and EoL reuse/recyclability should take the following into consideration:

- Minimize material usage: Use less amount and fewer different types of materials to make new products reduces both the use of natural resources (material, energy, and water) and wastes at the end of the product's life. Use compatible resins which can be recycled together. This facilitates sorting of materials for recycling. Larger amounts of similar materials increase the value of the scrap.
- Avoid using scarce resources: Limit the use of REEs.
- Avoid using dangerous and hazardous substances: Follow legal compliances. Reduce the risk of contact with heavy metals and BFRs during manufacturing, use, and disposal. Avoid using contaminants and toxic additives.
- Use recyclable materials: Increase the possibility of recycling. Reduce natural resource consumption and reduce the amount of waste disposed. Use thermoplastics (PET, PS, HDPE, LDPE, and PP) instead of thermoset plastics. Avoid PVCs and other halogen-containing polymers.
- Reduce/minimize the size: Reduce consumption of materials which results in increased value on disposal. Avoid over-dimensioning.
- Reduce packaging: Reduce consumption of resources.
- Label materials: Stimulate recycling. Define the type of material. But avoid adhesive labels on plastic surfaces which contaminate material on recycling.
- Use recycled material: Reduce consumption of resources and disposal.
- Design simple: Avoid over-designing. Basic and simple design are not only less costly but also easier to disassemble and recycle.
- Extend product life: Delay replacement and conserve natural resources. Products should last longer.

- Disassemble considerations: Ease the disassembly of products by lowering the number of separate parts.
- Thin the walls without strength, stiffness, and toughness: Conserve natural resources.
- Use accessible and removable fasteners and joining: Reduce recycling cost and time.
- Use removable coatings and finishes: Affect the performance of recycled material.
- Material identification: Grade of the material should be known for recycling.

12.6 Limitations of Current WPCB Recycling Technologies

Based on the above analysis and discussions in this book, the following limitations of current WPCB recycling technologies can be summarized:

- Very limited range of materials, especially PMs, is extracted, such as Au and Cu from WPCBs, Co and Li from batteries, and In from display units (LCDs).
- Manual disassembly is still the primary route of acquiring samples. Due to the low processing capacity, potential human health hazards of manual dismantling, high labor cost and amount of WEEE force the development of automatic disassembly.
- Current industrial practices use coarse particle sizes due to energy requirement and time constraints; but a better separation and leaching can be achieved at fine particle sizes and evenly distributed feeds. Particle size is the most influential parameter in separation processes.
- Some portion of nonconductive materials can be entrapped by conductive material aggregations which hinders the effectiveness of electrostatic separation performance.
- Despite the high recovery rates for mass-related elements such as ferrous and Cu, only a quarter of Au and Pd ends up in outputs from which PMs may be recovered.
- For metal extraction inorganic acid leaching is a dominant route. Acid leaching yields higher metal recovery rates than thiosulfate/thiourea leaching but requires a large amount of toxic solution and long leaching times and creates corrosion for equipment. Thiosulfate leaching is more environment friendly but less efficient. Thiourea leaching is more selective for metals over thiosulfate leaching, and its efficiency can be dramatically improved with the addition of acids. However, thiourea is expensive. Bioleaching is an environmentally viable and cost-effective approach, but leaching time is very long and unacceptable [7].

12.7 Future Development Perspectives

The main objective of writing this book is to help to develop a legislative system, formal recycling systems, and advanced recycling techniques in the world. The future development direction of WPCBs recycling should be focused on both higher material recovery rates and more kinds of substances (especially PMs and SMs) recovery.

Developments of new separation and extraction methods could consider increase in recovery rates and separation efficiency; better automatic control; lower time consumption, labor force, investment, and operating costs; and environmental impacts simultaneously. New methods should take into account industrial applicability potential, automation in pretreatment, reduction in processes and reagent costs and pay more attention to nonmetallic fraction as well. WPCBs should be regarded as important secondary material sources for achieving higher sustainability under the context of circular economy. There is almost no mature technique for EC recycling currently. Hence, further studies to develop effective technologies are urgent. Firstly, classification technology and resource recovery techniques for individual ECs should be focused and investigated.

Serious laws and supporting measures should also be enacted to promote regulation and environmental friendliness of e-waste recycling by the governments. Consequently, recycling and resource recovery activities for e-waste should be developed scientifically. Recycling needs a chain, not a single process activity. The volume, quality, and value of the metals produced by recycling are determined by the combined stages in that chain. With each stage dependent for its success on the preceding ones, the final result is profoundly affected by the weakest link in the chain. Figure 12.2 shows the amount of actor changes in WEEE recycling chain. There is an exponential decrease in the number of actors taking part in from collection to endprocessing (global smelting and refining business). Investment significantly increases from collection toward the endprocessing side. Total efficiency is determined by the weakest step in the chain.

The efficiency of collection (about 30%) and dismantling (about 90%) is 60%, preprocessing is 25%, and endprocessing is 95%. Thus, the total recycling efficiency is multiplication of each efficiency of the processes, which is about 15% for formal

Fig. 12.2 WEEE recycling chain activities

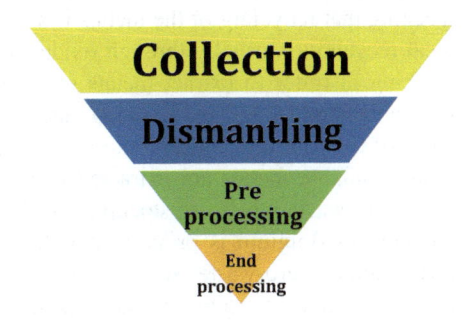

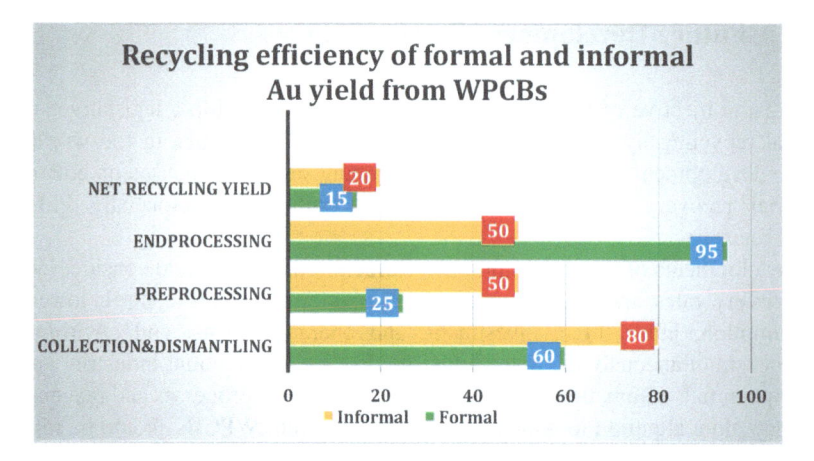

Fig. 12.3 Recycling efficiency between a common formal system in Europe and the informal sector in India for Au yield from PCBs

recycling systems in Europe. In India, the efficiencies of collection and dismantling are 80%, preprocessing is 50%, and endprocessing is 50%. The total net yield is 20% for informal sector for Au recovery from WPCBs (Fig. 12.3) [8]. Technology metals need smart recycling. For mono-substance material without hazards, recycling focuses on mass and costs. For poly-substance materials including hazardous elements, recycling focuses on trace elements and value.

Generally in the developing countries, due to the low WEEE collection rates and illegal dubious exports low-quality WEEE/WPCB recycling processes are used. Technology metals need smart recycling not low-quality recycling. WEEE export tracing and tracking, controls and enforcement, stakeholder responsibility, and transparency are important.

12.8 Toward Sustainability and Zero-Waste Scheme

The processing of WEEE so far largely depends on metal-centric approaches, meaning that recycling of the metals is still the focus. Metal-centric hydrometallurgical recycling is usually through oxidative leaching, direct leaching, or reductive leaching in order to extract metals of interest. In some cases, pressure leaching, mechanochemical-assisted leaching, and chlorination-assisted leaching are applied, e.g., for Ge with low concentration. In metal-centric approach, the most critical or scarce/valuable metals are extracted from WEEE, and other materials are left as secondary waste which are stockpiled or landfilled. After WEEE recycling, slag of Umicore in Belgium is properly handled and used in concrete industry or dyke fortification. Zero-waste schemes for WEEE are requirement of closed-loop recycling circular economy. Zero-waste philosophy can be achieved with a product-centric approach for WEEE treatment. In the product-centric view, it considers the complex chemical processes of all elements at the same time; the extraction and recycling order minimize loss and improve efficiency. The approach demonstrates that zero-waste processing requires more than innovations of technological solutions. An integrated flowsheet considering the zero-waste scheme tailors recovery of both critical and noncritical metals and simultaneously purification and inner-circulation of leach solutions. The circulability value of metals is at the same time improved. Thus, metal recycling, consumer-driven residue matrix production, and zero-waste emission can be achieved in the long term [2].

References

1. Hagelüken C, Buchart M, Ryan P (2009) Materials flow of platinum group metals in Germany. Int J Sustain Manufact 1(13):330–346. https://doi.org/10.1504/IJSM.2009.023978
2. Sun Z, Xiao Y, Sietsma J, Agetrhuis H, Yang Y (2017) A cleaner process for selective recovery of valuable metals from electronic waste of complex mixtures of end-of-life electronic products. Environ Sci Technol ACS 49:7981–7988. https://doi.org/10.1021/acs.est.5b01023
3. Ning C, Lin CSK, Hui DCW (2017) Waste printed circuit board (PCB) recycling techniques. Top Curr Chem (Z) 375:43. https://doi.org/10.1007/s41061-017-0118-7
4. Jones PT, Van Gerven T et al (2011) CR^3: Cornerstone to the sustainable inorganic materials management (SIM^2) partners research program at K.U.Leuven. JOM 63(12):14–15
5. Hagelüken C (2014) High-tech recycling of critical metals: opportunities and challenges, American Association for the Advancement of Science 2015 Annual Meeting (AAAS), Chicago. Feb. High-tech recycling of critical metals: opportunities and challenges. Available from: https://www.researchgate.net/ publication/267541171_High-Tech_Recycling_of_Critical_Metals_Opportunities_and_Challenges. Accessed 19 March 2018
6. http://www.ewasteguide.info/files/UNEP_2009 eW2R.PDF
7. Yang C, Li J, Tan Q, Liu L, Dong Q (2017) Green process of metal recycling: coprocessing waste printed circuit boards and spent tin stripping solution. ACS Sustain Chem Eng 5:3524–3535. https://doi.org/10.1021/acssuschemeng.7b00245
8. http://wedocs.unep.org/bitstream/handle/20.500.11822/8423/-Metal%20Recycling%20Opportunities%2c%20 Limits%2c%20Infrastructure-2013Metal_recycling.pdf?sequence=3&isAllowed=y

Chapter 13
Conclusions

13.1 Summary and Comparison of WPCB Recycling Technologies

The manufacturing of EEE is a major demand sector for metals. E-waste/WEEE are an important secondary source of BMs, PMs, rare metals, and trace elements, and their processing through ecological technologies constitutes a major concern in the world and contributes significantly to the reduction of environmental pollution and preservation of valuable scarce resources of metals. Although state-of-the-art preprocessing facilities are optimized for recovering Fe and Cu, PMs and trace elements are often lost.

Proper management and safe disposal of e-waste has become an emerging issue worldwide. Disposal and incineration can pose threats to the whole environment, from the atmospheric to aquatic and terrestrial compartments. In recent years, recovery of metals from e-waste in the world has become increasingly important due to potential risk of strategic raw material and environmental concerns. WEEE recycling for the production of secondary materials needs to be encouraged. WPCBs are the most complex, hazardous, and valuable components of e-waste. Actually, the problem caused by e-waste increasingly becomes an international issue due to the characteristics of e-waste amount, discarding continuously, high value material content, and most of all migration of pollutants. Treating a more complex type of WEEE, usually mixtures of various kinds of EoL products or wastes from different streams, is still a great challenge. This type of WEEE is one of the main WEEE feedstock today.

In this book, metal extraction and nonmetal recovery processes from e-waste, particularly the existing industrial practices and routes, were reviewed. Industrially, different metallurgical routes are used to extract valuable metals from e-waste. Physico-mechanical and pyrometallurgical processes are used for BM recovery. Combined pyrometallurgical, hydrometallurgical, and electrometallurgical processes can be commonly employed to recover PMs. These routes are described and their advantages and disadvantages are outlined. In the final part of this book, insights into e-waste recycling in the world context are presented. The challenges and barriers associated with recovery of the BMs and PMs are highlighted. Nowadays, WEEE is considered as an important secondary polymetallic resource. Metal recycling from WEEE compensates metal supply and reduces supply risks of critical metals for sustainable resource management.

The treatment of WEEE may be considered as an integration of economic, environmental, legislative, and technology drivers which determine the structure and methodology of waste treatment hierarchy approach. The hierarchy deployed in order of increasing environmental impact may be considered as reduction (extend product life), reuse (refurbishment/repair), recovery, recycle, and disposal (landfill or incinerate). Recycling technology, which aims to take today's waste and turn it into conflict-free, sustainable resources for tomorrow, should achieve resource efficiency and environmental compliance. Although the governments and a number of researchers have made great efforts to improve WPCB recycling, there are still some obstacles that limit the industrial application of WPCB recycling techniques. More suitable approaches, which are lower cost, simple operation, and environmentally friendly, are significant to be investigated and developed to recover noble metals from WPCBs.

From the economic point of view, recycling of valuable materials from WPCBs is extremely attractive. Increasing generation of PCBs and the severe environmental impacts of landfills promote the development of recycling methodologies. The mechanical process is of importance for following chemical processes, because of the need for metal liberation. The hydrometallurgical process has been studied in terms of its advantages, such as simplicity, low capital cost, and less environmental impact. The work done shows the promising future in the world of WPCB recycling. Both ionic liquid and chlorine-based media have the potential for the extraction of BMs and PMs. However, the flowsheets proposed are limited in lab scale. Thus, larger-scale studies should be concentrated on to achieve commercialization.

13.2 Primary Production Versus Secondary Production

The recycling of e-waste contributes to a secondary resource economy, which relates to the practice of sustainable resource management. While developed countries have established high-capital, state-of-the-art plants for the processing of various fractions and components of e-waste, developing countries face environmental and health challenges from a prevalence of artisanal and informal recyclers. Table 13.1 summarizes the advantages and disadvantages of both primary raw material production (mining) and secondary raw material production (waste recycling/urban mining):

Mining provides all metals/nonmetals in huge amounts to the society. Natural ore grades are low and getting lower continuously. Mining destroys the lands and forests. Thus, there is a serious objection for mining operations. Scarce raw

13.3 Manual Dismantling Versus Automated Dismantling

Table 13.1 Comparison of mining with urban mining

Primary production (mining) (about 5 g/t Au or PGM in ore)	Secondary production (e-waste recycling) (urban mining) (150 g/t Au, 40 g/t Pd & Ag, Cu, Sn, Sb in PC PBBs, 300 g/t Au, 50 g/t Pd in cell phones)

Pros
  Supply all metals in large quantities
Cons
  Natural ore grades are low
  Difficulty in mining conditions
  Large amount of lands is used
  Scarce energy, raw materials, and water are consumed/depleted
  $SO_2(g)$ and $CO_2(g)$ emissions are produced
  Environmental pollutions are created
  Requires new residue ponds
	Pros
	  10–40 times more abundant than ores
	  Sustainable
	  Metal shortages
	  Reduces loss of scarce sources
	  Creates new jobs
	  Less land is used for mining
	  Less landfill is required< Preserves depleting resources
	Cons
	  Mostly informal sector without pollution control
	  Complex structure and composition
	  About 60 elements are closely linked
	  Metals are cross-linked to plastics
	  Plastics are toxic (includes BFR)
	  Supply is limited

materials, energy sources, and water are depleted by mining. Environment is polluted by mining operations. Gas emissions and hazardous wastewater ponds are serious problems. However, urban mining from wastes is much richer in metal contents. There is a serious metal shortage for many metals. Secondary production reduces the loss of valuable metals. Recycling of waste material creates new jobs for uneducated people and sustainable economy. Less land is used for mining operations [1, 2].

13.3 Manual Dismantling Versus Automated Dismantling

Table 13.2 shows the comparison of manual and automated/semiautomated dismantling methods for e-waste recycling. Automated dismantling has more advantages and less disadvantages than manual dismantling. Manual dismantling is cheap, utilizes simple tools, creates jobs for low educated people, and is suitable for underdeveloped and developing countries. Manual dismantling is a widely and selectively used technique. But, it takes long times, is expensive due to high labor force, and generates occupational health and safety problems and environmental issues. Automated dismantling is a fast, operational cost-effective, large-scale, environmental friendly, eco-efficient process. However, initial investment cost is high, and recovery efficiency is low due to simultaneous disassembly. Automated disassembly is generally preferred by developed countries.

A disassembly stage is always required to remove dangerous components such as batteries and condensers and valuable and reusable ECs. Disassembly has

Table 13.2 Comparison of manual and automated/semiautomated dismantling methods

Manual dismantling	Automated/semiautomated dismantling
Pros	Pros
Low investment cost	Time saving
Utilize simple tools	Cost-effective/economical
Job creation for low educated workers	Minimum labor force
Can be performed selectively/simultaneously	Large scale
High recovery efficiency	Semi/full automatic
Small scale	Environmental friendly
Suitable for developing countries	Eco-efficient
Cons	Without dust
Takes long times	Suitable for developed countries
Occupational health and safety problems	Cons
Dust exposure	High investment cost
Bed smell and black fumes	Lower recovery efficiency
High labor force	Not generic for all kinds of products
Expensive	
Informal, banned process	
Has its limit at some point in the process	

traditionally been undertaken manually, but newly developed automated systems will impact upon future recycling strategies both to maximize cost-effectiveness for low-value component recovery and as an initial stage for recycling approaches to maximize yield of residual intrinsic material value. WPCB is markedly heterogeneous in nature, and the key to all mechanical treatment methodologies is in the liberation of the component material fractions. Crushing and separation are then key points for improving successful further treatments. Physical recycling is a promising recycling method without environmental pollution and with reasonable equipment investments, low energy cost, and diversified potential applications of products [1, 2].

13.4 Conventional Versus Novel WPCB Recycling Technologies

Table 13.3 shows the advantages and disadvantages of traditional and emerging industrial WPCB recycling technologies. Conventional (incineration/combustion, mechanical separation (gravity separation)) and emerging (pyrometallurgy, hydrometallurgy, electrostatic separation, and magnetic separation) e-waste recycling methods are compared, and advantages and disadvantages are summarized.

The metal value in e-waste is typically extracted by both formal and informal industries. This starts from the collection, sorting, size reduction, and processing. Size reduction is done so as to concentrate and/or liberate metals. Choice of size reduction technology is also influenced by the choice of subsequent processing techniques used to recover the metals. In informal industry or artisanal recycling,

13.4 Conventional Versus Novel WPCB Recycling Technologies

Table 13.3 Advantages and disadvantages of most used e-waste recycling methods

Conventional WPCB recycling processes	Novel WPCB recycling processes

Incineration/combustion
Pros
  Conventional process
  Accepts every e-waste as is
  Short process time
  Heat energy produced
  Change chemical composition and physical phase
  Irreversible process
Cons
  Energy intensive
  High investment cost
  Expensive process
  Large capacity
  Low efficiency
  Low recovery
  Low metal purity
  Low selectivity between metals
  Requires expensive gas purification systems
  Environmentally unfriendly
  Requires corrosion-resistant equipment
  Organic material and glass in WPCBs +ECs are burned and lost
  Fe and Al cannot be recovered
  Only Cu, Au, and Ag recovered
  Cu is purified by H_2SO_4 (hydrometallurgy)
Mechanical separation
Size reduction (shredder/pulverizing)
  Multi-crushing-grinding steps
  Only plastics are liberated
  Metals are not liberated
  Produce toxic gases and dust/particulate matter
  Low recovery rates for PMs
  Produce low-value nonmetallic powder
Separation
  Loss of PMs in dry separation due to liberation problem between plastics and metals
  Simple
  Appropriate
  Environmentally friendly
  Equipment is cheap
  Energy cost is low
  Products can be directly used
Cons
  Significant dust generation
  Metal loss in shredding and grinding
	Pyrometallurgy
	Pros
	  Most commonly used process
	  Integrated smelters and refineries (ISR) are used
	  PCBs+ores are fed directly without size reduction to reduce PMs losses
	  Short process time
	  Remove massive Fe and Al parts for valorization and upgrading Cu and PMs
	  Produces a Cu alloy-containing PMs
	Cons
	  May require sorting, dismantling, and mechanical pretreatment before smelting to maximize energy efficiency
	  Energy intensive (>1200 °C)
	  High-temperature requirement
	  High investment cost
	  Large capacity
	  Requires expensive gas purification systems for VOCs, dioxins, and furans
	  Requires corrosion-resistant equipment
	  Low selectivity between metals
	  Low efficiency
	  Low recovery
	  Low metal purity
	  Environmentally unfriendly
	  Expensive process
	  Only Cu, Au, and Ag recovered
	  Mixed oxide product is obtained
	  Glass and plastics are lost and converted to slag, ash, and heat energy
	  Cu is purified by H_2SO_4 (hydrometallurgy)
	  Subsequent hydrometallurgical and electrometallurgical techniques are required to extract pure metals
	Hydrometallurgy
	Pros
	  Advanced method
	  Easy to apply
	  Easy management
	  Simple to operate
	  Selective
	  Flexible
	  Stable
	  Expectable
	  Faster kinetics
	  More exact
	  Highly predictable
	  Single or multistage (sequential) leaching is

(continued)

Table 13.3 (continued)

Conventional WPCB recycling processes	Novel WPCB recycling processes

Gravity separation
Pros
  Metals can be easily separated from plastics, Al, and glass
  Gravity separation is more economical than electrostatic separation
  Wet separation reduces noise and dust pollution
  Air tables and zigzag three-way classifiers do not have wastewater problem
  Recoveries are more than 95%
Cons
  Plastic separation from Al and glass is not easy
  Dry separators require dust removal systems and do not need water
  Wet shaking tables, heavy media separation, and jigs produce huge wastewater
  Air tables and zigzag three-way classifiers do not have wastewater problem
	possible
	  Less hazard to environment
	  May eliminate secondary wastes
	  Low waste gas emission
	  Low energy consumption
	  Low temperature requirement
	  Easily controlled
	  Low investment cost
	  Reagent recyclability
	  High recovery rates
	  Cu, Au, Ag, Pd, Pb, Sn, etc. can be recovered
	  No slag generation except few plastics
	Cons
	  Strong hazardous inorganic acids/bases are used
	  Generate acidic or basic waste/water solution
	  Leachate discharge
	  Corrosive reactive used
	  Special corrosion-resistant equipment requirement (stainless steel or rubbers)
	  Needs pretreatment (finer size reduction)
	  Energy consumption for comminution
	  Long process time (time-consuming)
	  Tedious
	  Involves large number of steps
	  Easy working condition
	  High cost of recovery
	  NMF cannot be recovered
	  Free heavy metal ions present in effluent
	  Downstream processes are difficult
	  Very specific for substances
	  No one solution for all
	Biohydrometallurgy
	Pros
	  Weak organic acids are used
	  Eco-friendly (green technology)
	  Suitable for both BM and PM extraction
	  Low temperature and energy requirement
	  Clean nonmetal product
	  Low investment/operating cost
	  Cost-effective
	  Selective recovery
	  Promising technology
	  Less gas and water
	Cons
	  Difficulty in microorganism isolation
	  Difficulty in microorganism reproduction/culture
	  Requires nutrients for microorganisms
	  Selective to specific metals
	  Vulnerability to heavy metals

(continued)

13.4 Conventional Versus Novel WPCB Recycling Technologies

Table 13.3 (continued)

Conventional WPCB recycling processes	Novel WPCB recycling processes
	Bacteria toxicity
	Low leaching speed
	Slow leaching kinetics
	Long process time (48–245 h)
	Electrostatic separation
	Pros
	Simplicity
	Metals and nonmetals/nonferrous materials can be separated
	Dry separation
	No wastewater discharge
	No gaseous emission
	Low energy consumption
	Room temperature
	Environment friendly
	Separation efficiency of 95–99% purity
	Cons
	Requires dust extraction system
	High-level voltage requirement
	High roller speed
	Corona electrostatic separation (CES)
	Separates conductive and nonconductive particles (Cu from plastics)
	Separates mixed particles that have similar conductivities
	Separation efficiency is as high as 99%
	Particle size: 0.6–1.2 mm
	Easy operation
	Simple maintenance
	Small area requirement
	Little cross contamination
	Triboelectric separation (TES)
	Contact charged
	Two similar or dissimilar materials are separated
	Eddy current separators (ECS)
	Charged by ion bombardment
	Nonferrous metals (Cu and Al) can be separated from plastics
	Particle size: 3–150 mm
	Magnetic separation
	Separates magnetic particles from nonmagnetic particles
	Wet or dry separation is possible
	Agglomeration problem in dry separation
	Low intensity magnetic separators (LIMS)
	Ferromagnetics can be separated <2 Tesla from Cu, Al, and plastics
	Dry LIMS for course particles (0.5–6 mm)
	Wet LIMS for fine particles (−0.5 mm size)

(continued)

Table 13.3 (continued)

Conventional WPCB recycling processes	Novel WPCB recycling processes
	High intensity magnetic separators (HIMS)
	Paramagnetics can be separated at 10–20 Tesla
	Pyrolysis
	Pros
	Conventional process
	Accepts every e-waste as is
	Short process time
	Gases, oils, and chars (containing metals) are produced
	Change chemical composition and physical phase
	Irreversible process
	Reduce volume of WPCBs
	Cons
	Energy intensive
	High investment cost
	Large capacity
	Requires expensive gas purification systems
	Requires corrosion-resistant equipment
	Low selectivity between metals
	Low efficiency
	Expensive process
	Low recovery rates
	Low metal purity
	Commercial absorbents can be produced for wastewater and Cd uptake
	WPCBs+ECs depolymerize
	Environmentally unfriendly
	Glass fiber and Cu foil are recovered
	Glass and plastics are lost
	Cu and Sn are purified by hydrometallurgy

manual sorting, dismantling, and open burning of WPCBs are performed to separate Cu and Au metals from plastics, followed by metal leaching using acid baths without any safety procedures. Cu is cemented by Fe fillings and Au is cemented by Zn. The main extraction methods in formal industry are either high-temperature smelting with subsequent hydrometallurgical refining of the resulting blister, mattes, and slags or direct leaching of metal values from WPCBs suitably reduced in size or otherwise pretreated and selective recovery of dissolved metals from the leach liquor. Integrated pyrometallurgical and hydrometallurgical processes are carried out at large scale in Cu smelters for the recovery of metals. In such operations, e-waste is used both as a metal and energy source, but these operations are found only in developed countries.

It is clear that pretreatment, e.g., mechanical processing, disassembly, and thermal treatment, is frequently required with the aim to improve the metal recovery rate.

13.4 Conventional Versus Novel WPCB Recycling Technologies

Mechanical treatment approaches would appear to offer significant environmental and operational benefit, and this reflected in the amount of developed work undertaken on such during the past 30 years, with focus having been on improving yield and efficiency. Mechanical and hydrometallurgical recycling approaches have been able to take the advantages of intrinsic material, physical, and chemical property differences, respectively; such would include density, magnetic and electrical conductivity, and chemical reactivity. As the PMs (especially Au) contained are the most valuable component for WPCB recycling, the loss of these metals should be minimum. Mechanical handling of the e-waste to reduce the size required for efficient dissolution consumes long time. Approximately 20% of metals may be mechanically lost during liberation process that may lead to a substantial reduction in the overall metal recovery. Physico-mechanical treatment processes, including crushing, cyclone air separation, and electrostatic separation, have been industrially applied to separate metals and nonmetals, and Cu is the main objective production of this course currently. Actually, the loss of PMs is high during these processes, because the PMs are usually plated onto the surface of WPCBs and generally softer. Globally, pyrolytic/smelting route is generally used with high cost and ecological problems. Second-generation physico-mechanical separation along with hydrometallurgical treatment approaches offers a cost-effective and more sustainable alternative methodology to smelting.

Hydrometallurgy: Hydrometallurgical aqueous approaches offer the opportunity to eliminate material loss from recycling processes but have potentially more significant environmental impact in implementation. Hydrometallurgical treatment of e-waste has been developed on two fronts, namely, the extractions of BMs and PMs. This group of processes is attractive from an economic point of view due to their suitability for small-scale applications and for the treatment of low-grade wastes. Hydrometallurgical processing routes start with the same sorting and dismantling as is done for pyrometallurgical processing, followed by pretreatment mainly for liberation and size reduction, which are then followed by the leaching of metals by a suitable lixiviant, purification of pregnant leach solution, and recovery of metals. Hydrometallurgical treatments need more mechanical pretreatment and finer grinding than pyrometallurgical processes. Hydrometallurgical treatments are based on leaching agents in aqueous solutions, such as strong acids and bases. These are often applied together with other complexing agents, such as oxalic acid, acetic acid, cyanide, halide, thiourea, and thiosulfate. Hydrometallurgical processes are less energy intensive and cost demanding than pyrometallurgical treatments, and they are also applicable in plants with relatively small capacities. Hydrometallurgical processes are tedious and time-consuming and impact recycling economy. There is not much research on downstream processes, such as metal recovery from solution and treatment of effluent streams. Recently hybrid technologies have also been applied, which integrate the chemical approach (more efficient) with bioleaching (more eco-compatible).

The selection of leaching lixiviants depends on the features of the WEEE and also environmental impact of the whole process. Strong corrosive acids and oxidizing conditions require special equipment made of stainless steel or rubbers for leaching.

Also there is a risk of metal losses during subsequent steps affecting the overall recovery of metals. Nontoxic leaching reagents, the concept of green chemistry, and the principles of sustainable economy are important factors to be considered in the next stage of hydrochemical metal recovery from WEEE. Single- or multistage sequential leaching process can be used. Chemical leaching (hydrometallurgy) of WPCBs is faster and more efficient; however, wastewater and waste gas are generated invariably during this process. On the contrary, bioleaching is greener due to less wastewater and waste gas originated; meanwhile, bioleaching (biometallurgy) is selective to specific metals [3]. Nevertheless, the specific microorganism is difficult to select and culture. Simultaneously, microorganisms employed in the leaching process are sensitive to metal ions which would be toxic to the microorganisms at certain concentrations leading to inefficient operations. Therefore, chemical leaching is more suitable and advantageous compared to bioleaching process for WPCBs due to their complex material makeup [4].

Metal recovery can be performed by traditional pyrometallurgical approaches on metal-concentrated PCB scrap fractions. Comparing with the pyrometallurgical processing, the hydrometallurgical method is more exact, more predictable, and more easily controlled. New promising biological processes are now under development. It should be kept in mind however that the chemical composition of e-waste changes with the development of new technologies and pressure from environmental organizations to find alternatives to environmentally damaging materials. A sound methodology must take into account the emerging technologies and new technical developments in electronics. Miniaturization of electronic equipment in principle would reduce the volume of WPCBs but make collection more difficult and repair costlier, so that a large amount of WPCB is still expected in the e-waste in the future. In addition, many of the metal extraction techniques have not been optimized, and thus commercial-scale hydrometallurgical operations in the e-waste recycling industry are still limited.

Summarily, hydrometallurgy recycling techniques are easy to apply and simple to operate. However, the problem that cannot be avoided or ignored is the discharge of leachate as well as the pollutants since these techniques do not incorporate the recycling of NMF. The process is mainly undermining the structure of NMF and emits them, which not only wastes the useful part of NMF but also converts them into pollutants. Another issue that should be highlighted is that the recovery rate for hydrometallurgy recycling techniques is full recovery. Therefore, there will be a certain amount of heavy metal present in the effluent in the form of free ions, which will reinforce the hazardous content of it. New developments, for instance, molten salt extraction, selective leaching, and bioleaching or even combining pyrometallurgy and hydrometallurgy for metals recycling, provide ample possibilities of pushing WEEE treatment to the theme of product-centric philosophies.

Pyrometallurgy: Nowadays, the main technologies used for the recovery of valuable materials from WEEE are based on pyrometallurgical and hydrometallurgical processes. Current industry of base and/or valuable metals recycling is widely (70%) based on pyrometallurgy, e.g., smelting, with a considerable secondary residue/waste being landfilled. But, recovery of REEs from WEEE is not possible by

13.4 Conventional Versus Novel WPCB Recycling Technologies

Table 13.4 Comparison of different types of metal recovery techniques

Recycling technique	Cost	Efficiency	Toxicity effect	Recycling time
Pyrometallurgy	High	**High**	High	**Short**
Hydrometallurgy	**Low**	**Medium**	**Medium**	**Medium**
Biometallurgy	**Medium**	Low	**Low**	Long

pyrometallurgical processes. Pyrometallurgical processes require the heating of WEEE at high temperatures (often greater than 1000 °C) to recovery metals. These thus represent highly energy-consuming treatments that lead to the production of hazardous gases that must be correctly removed from the air with flue gas cleaning systems.

Pyrometallurgy shows the best performance in efficiency and leaching duration since it can significantly remove NMF (i.e., resin) in WPCB by thermo-decomposition. However, the high-energy consumption and initial investment cost also make it the most expensive e-waste recycling technique. Furthermore, the toxic harmful emissions and residues generated in the process cause considerable health effects to the human and environment if performed without proper treatment. Compared to pyrometallurgy, hydrometallurgy and biometallurgy are not comparable to pyrometallurgy, thus making them extremely time-consuming techniques although their cost is several times lower than pyrometallurgy. As can be seen from Table 13.4, the partial recovery of MF from WPCB by the above methods has obvious drawbacks. It is imperative to develop further full recovery techniques for WPCB to improve the current methods of WPCB recycling [5]. From high efficiency and recycling time points of view pyrometallurgy; from medium-cost and low toxicity points of views biometallurgy, and from low-cost and medium efficiency, toxicity, and recycling time points of view hydrometallurgy are better recycling process.

Current recycling methods include both pyrometallurgy and hydrometallurgy, which have a long history and wide application. Biometallurgy as an emerging technology also takes up a certain share of the WPCB recycling market. Many studies have been done to improve the performance or cut the cost of these techniques. However, the inherent drawbacks are very obvious for these techniques, including the irritation to the environmental intensive energy consumption. Therefore, new advanced technologies are in great demand due to the requirement of technology with high safety and economic feasibility. Figure 13.1 shows one of the best e-waste recycling flowsheets.

Pyrolytic approach is attractive because it allows the recovering of valuable products in gases (3–4%), oils (20–28%), and solid char residue (70–76%). Organic (resins), metallic (Cu foil, solder, etc.), and glass fiber can easily be separated. Pyrolysis reduces waste volume seriously. Evolution of toxics PBBD/PBDF can be controlled by appropriate treatments such as the addition of suitable scavengers or dehydrohalogenation, which are still under development. New technologies are proposed such as vacuum pyrolysis or depolymerization in supercritical methanol.

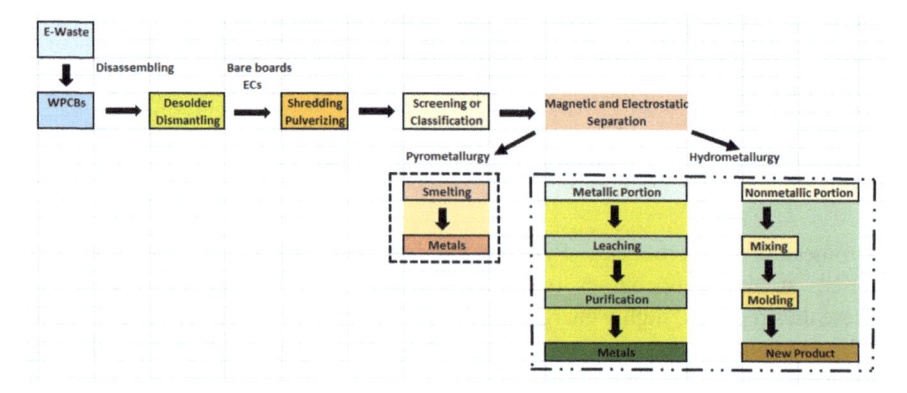

Fig. 13.1 E-waste recycling flowsheet

13.5 Recycling Solution

A successful recycling approach of WPCB should take into consideration the valorization of the recycled items to compensate for recycling costs. Recycling of WEEE, and of WPCB in particular, is still a challenging task due to the complexity of these materials and possible evolution of toxic substances. Traditionally, the recovering of valuable metals by WPCBs was carried out on a large scale for positive economic revenue. Legislation pushes now toward more comprehensive processes which include recovering and recycling of the ceramic and organic fractions in substitution to not-eco-efficient disposal in landfill.

Sustainable recycling technology ensures that e-waste is processed in an environmentally friendly manner, with high efficiency and lowered carbon footprint, at a fraction of the costs involved with setting multibillion dollar smelting facilities. Recycling can offer a potent resource for our planet, transforming millions of tons of e-waste (i.e., electronic asset) into reusable material.

E-waste recycling converts today's serious solid waste problem to a challenge or an opportunity from economic (sustainable profitability), social (generates new jobs and well-being for the society), and environmental (generates superior environmental performance and full utilization of depleted resources) point of view.

Although many endeavors have been attempted to solve e-waste problem, yet until now there is still a huge room to achieve the sustainability of e-waste management. At least in the perspective of academic research, some tasks and questions related to fundamental knowledge, recycling technology, and eco-design have been completed and answered. However, there are still many key questions not to be well concerned. Strict environmental regulations of some countries are changing the landscape of e-waste trafficking. The struggle against illegal imports of e-waste has become one of the major challenges for most developing countries. Practical e-waste management in developing countries is unregulated so that rudimentary techniques are widely used. Although tightening of regulations alone will not solve these problems, regulations should be issued and updated to control local e-waste system, and it should be designed in conjunction with the establishment of formal recycling infrastructure. Meanwhile, consumer participation should be also controlled and reinforced to improve the collection channels. Technology and paradigm of e-waste recycling should be altered toward international standardization. Basically, to extend recycling capacity and develop new technology and to formalize informal recycling sector are urgent to improve e-waste system for developed and developing countries, respectively. In the near future, integrated modular and mobile e-waste recycling process and infrastructure will be effective to handle various types of e-waste to solve local and national problems.

Finally in this book, current metallurgical processes for the extraction of metals and nonmetals from e-waste, including existing industrial routes, are covered. Then, an integrated technological route, including metal enrichment and PM recoveries, is proposed. Finally, in order to promote the development of metal and nonmetal recovery from WPCBs, some improvements and recommendations in techniques and the future trend are also put forward.

Direct hydrometallurgical treatment for recovering PMs from WPCBs in two steps, (1) leaching metals into solutions for separating Cu and other BMs from PMs and (2) PM recovery, could reduce these metal losses. Further, more efficient and flexible techniques for PM purification from leaching solutions ought to be focused on and be drawn more attention to in the future study to obtain high added-value final products, promoting WPCB recycling industry. Hence, recycling of WPCBs is an essential and significant course not only for hazardous waste disposal but also from the PM recovery viewpoint. Simultaneously, in order to recover PMs cost-effectively and environmentally friendly, hydrometallurgical processes should be suitably developed to improve the present disadvantages to provide a new method to dispose and recover WPCBs.

References

1. Kaya M (2018) Chapter 3: Current WEEE recycling solutions. In: Veglio F, Birloaga I (eds) Waste electrical and electronic equipment recycling – aqueous recovery methods, Woodhead Publishing series in electronic and optical materials, 1st edn. Woodhead Publishing, Duxford, UK/Cambridge, MA. https://doi.org/10.1016/B978-0-08-102057-9.00029-9
2. Kaya M (2018) Waste printed circuit board (WPCB) recovery technology: disassembly and desoldering approach. In: Reference module in materials science and materials engineering/ encyclopedia of renewable and sustainable materials, 1st edn. Elsevier. https://doi.org/10.1016/B978-0-12-803581-8.11246-9
3. Kaya M (2018) Waste printed circuit board (WPCB) recycling: conventional and emerging technology approach. In: Encyclopedia of renewable and sustainable materials. Elsevier. https://doi.org/10.1016/B978-0-12-813195-4.11296-9
4. https://www.researchgate.net/publication/278849195_Recycling_-_from_e-waste_to_resources
5. Ning C, Lin CSK, Hui DCW (2017) Waste printed circuit board (PCB) recycling techniques. Top Curr Chem 375:43. https://doi.org/10.1007/s41061-017-0118-7

Index

A
Acid leaching, 228
Acidic sulfate solution, 230
Active disassembly (AD), 80
Active disassembly using smart materials (ADSM), 72
Air/pneumatic gravity separators (air tables), 157–160
Air/water classifiers, 139
Al electrolytic capacitors (AEC), 60
Ammonium sulfate ((NH$_4$)$_2$SO$_4$), 232
Analytic hierarchy process (AHP), 254
Aqua regia leaching, PMs, 253, 254
Attero recycling, Roorkee, India, 206, 207
Au leaching
    biotechnological, 241
    BMs, 242
    catalytic processes, 241
    cyanidation principle, 239
    cyanide, 239, 240, 242, 243
    electrochemical, 241
    electro-generated chlorine, 252, 253
    halide, 248, 252
    hydrometallurgical, 241
    hydrometallurgical processes, 242
    K$_2$S$_2$O$_8$, 242
    lixiviants, 238–240
    mechanical, 241
    mechanism, 239
    pyrometallurgical, 241
    secondary resources, 241
    stringent environmental regulations, 242
    thiosulfate, 242, 243, 246–249, 251
    thiourea, 239, 243

Aurubis recycling center, Lünen (Germany)
    Cu scrap, 203, 205
    KRS, 204–206
    PMs, 205
    pyrometallurgical preparation, 204
    raw materials, 205
Austrian Müller-Guttenbrunn Group (MGG), 185, 186
Automated disassembly, 62
Automatic sorting, 144
Auxiliary oxidants, 232

B
Base metal leach, 228
Base metal operations (BMO), 97, 182
Basel Convention, 15
Best available technique (BAT), 43
BFR leach, 227, 257, 258
Bioleaching, 106–108
Biometallurgical leaching
    A. ferrooxidans, 107
    advantages, 108
    bioleachability of metals, 107
    biosorption process, 109
    Cu from ores, 107
    cyanide-producing bacteria, 107
    e-waste, 108
    extraction of Cu, 108
    Fe source, 108
    ferrous ion, 108
    heavy metals, 106
    limitations, 108
    low extraction rate, 108

Biometallurgical leaching (cont.)
    low-grade ores, 106
    mechanism, 107
    metal ions, 107
    microbes, 106
    microorganisms, 106
    recovery of PMs, 107
    recovery period, 108
Biosorption process, 109
Biwer-Heinzle method, 260
Black Cu, 95
Blade/hammer mill pulverizers, 129–131
Boliden's Rönnskar smelter, 95
Br$_2$-based lixiviants, 223
Brominated epoxy resin (BER) leaching, 256–258
Bromine-based resins, 42
Bromine-bromide (Br$_2$/KBr) media, 223, 224
Bubbling depicts, 115

C
Cathode ray tubes (CRTs)
    and lamp phosphors, 281–283
    recycling at Attero Plant, 285, 286
Cathodic process, 261
Cement solidification, 212
Cementation, 109, 110
Cementation process, 66
Centre for Science and Environment (CSE), 206
Ceramics, 36
Chemical and biological methods, 123
Chemical precipitation of metals, 109
Chemical recycling
    degradation/modification/depolymerization, 213
    and direct, 211, 212
Chinese rare-earth sources, 278
Classification, 137, 139
Clean reagent, 260
Closed-loop CRT glass recycle, 286
CO$_2$ emission, 185
Cobalt (Co) recovery
    lion batteries, 287, 288
Component Removal Machine, 207
Concentration criterion (CC), 147
Conductive materials, 161
Conventional vs. novel WPCB recycling technologies
    advantages and disadvantages, 304–308
    extraction methods, 308
    hydrometallurgy, 309, 310

mechanical treatment, 308
metal value, 304
physico-mechanical treatment processes, 309
PMs, 309
pretreatment, 308
pyrometallurgy, 310, 311
size reduction, 304
Conveyor belts, 140
Corona electrostatic separation (CES), 163–165
Cu smelters
    anodic Cu production, 95
    black Cu, 95
    Boliden's Rönnskar smelter, 95
    converter, 95
    electronic material, 95
    electrorefining process, 94
    flash and bath smelting, 95
    molten mass, 94
    pyrometallurgical production, 93, 94
    smelting, 94
    Umicore's ISR, 96
    WEEE material melts, 95
Cu(II)-amine complex, 233
Cupric chloride (CuCl$_2$) leaching, 232
Cupric ions, 223
Cyanide leaching, 227, 238–240, 242–244, 247, 248, 254
Cyanide-producing bacteria, 107
Cyclone air separation-corona electrostatic separation (CAS-CES), 197

D
Daimler Benz in Ulm, Germany, 187, 188
Damaged ECs, 69
Delamination, 131–134, 136, 137
Depolymerization, 213, 217
Design for disassembly (DfD), 295
Design for recycling (DfR), 295
Design for resource efficiency, 295
Design for sustainability (DfS), 295
Desoldering, 48, 50
    ADSM, 72
    chemical treatment, 72
    heating method (see Thermal treatment)
    leaching, 72
    SMP, 72
    THCs, 72
di(2-ethylhexyl) phosphoric acid, 280
Diisoamyl sulfide, 255
Diisononyl phthalate (DINP), 218
Dimethyl sulfoxide (DMSO), 257, 258

Index

Dimethylacetamide (DMA), 257
Dimethylformamide (DMF), 257
Dip soldering, 51
Direct recycling
    cement solidification, 212
    and chemical, 211, 212
    disposal problem, 212
    epoxy resin matrix, 212
    HDT, 212
    leaching test, 212
    mechanical characterization, 212
    mechanical properties, 213
    PMC, 212
    reinforcing fillers, 212
Direct smelting, 88
Disassembling rate (DR), 73
Disassembly
    AD, 80
    and dismantling (*see* Dismantling)
    characteristics of connection types, 70
    desoldering (*see* Desoldering)
    eco-design/DfD concept, 80
    ECs, 69
    segregate ECs/materials, 69
    WPCBs, 70
Dismantling
    characteristics of connection types, 70
    damaged ECs, 69
    ECs, 69
    environmental-friendly, 70
    functional and usable ECs, 69
    manual informal, 71
    WPCBs, 70, 71
Dismantling/disassembly
    automated, 62
    automatic process, 62
    ECs on WPCBs, 62
    Manuel, 62
    materials, 62
    selective, 62
    simultaneous, 62, 63
Dismantling machine
    semiautomatic PCB electronic component (*see* Semiautomatic PCB electronic component-dismantling machines)
Dissolution-extraction-stripping process, 256
Dodecylamine, 173
Double-roller electrostatic separators, 163
Double-shaft shredders, 128
DOWA Group
    China, 189, 191
    Japan, 189, 190

Drum-/barrel-type dismantling machines 75, 76
Dry/air gravity separation
    air/pneumatic, 157–160
    crushing-pulverizing-classifying, 155
    FBS, 157
    materials and nonmetal materials, 155
    vertical vibration, 155
    zigzag, 155, 156
Dry film photoimageable solder mask (DFSM), 53
Dry LIMSs, 171
Dry mechanical separation processes, 86
Dry processes, 65
Dry separation methods, 65

E

Eco-design
    guidelines for manufacturing, 295, 296
Eco-design/design for disassembly (DfD), 80
Economic analysis, 257
Economy perspective, 290–292
Eddy current separators (ECS), 86, 167–170
Edge trim, 192
Eldan recycling, Spain, 186, 187
Eldan WEEE recycling plants, 186, 187
Electrical and electronic equipment (EEE), 17
Electrochemical reactor, 261
Electrodynamic fragmentation (EDF)
    component detachment, 136
    components, 137, 138
    delamination, 136
    depopulated WPCBs, 137, 138
    depopulation, 136, 137
    depopulation treatment, 137, 138
    electrical and acoustical properties, 135
    energy consumption comparison, 136
    entire fragmentation, 136
    HVP laboratory-scale equipment, 134, 135
    internal structures, 137
    laboratory-scale batch EDF systems, 137
    operating parameters, 136
    preweaking tool, 136
    pulse power generation, 134, 135
    selective fragmentation, 136
    Selfrag technology, 136
    shock wave, 135
    structure opening, 137
Electro-generated chlorine, 231, 252, 253
Electrokinetic process, 262
Electrolysis separation, 173, 174

Electrometallurgical technique, 226
Electrometallurgy, 262
Electronic components, 47
Electrooxidation conditions, 231
Electroplating, 110
Electrorefining process, 94
Electrostatic discharge (ESD), 166
Electrostatic separation (ES)
   CES, 163–165
   conductive materials, 161
   conductor and nonconductor (insulator/
      dielectric) properties, 159, 161
   conductor materials, 162
   conductors, 159, 162
   ECSs, 167–170
   high-voltage, 162
   insulators/nonconductors, 161, 162
   low energy consumption, 162
   nonconductors, 162
   principles, 159, 162
   semiconductors, 162
   TES, 165–168
   triboelectric charging, 162
   tribo-electricity/contact charging, 162
Electrowinning (EW)
   bubbling depicts, 115
   calculated electrolysis data, 115
   complexing agents, 111
   Cu, 114
   electrodes, 113
   electrolyte, 110
   electroplating, 110
   EMF series, 114
   industrial-scale Cu EW plant, 111, 112
   metals, 111
   oxidation, 111, 112
   parameters, 110
   Pourbaix, 111
   reactivity of metals depends, 111
   reactivity order of metals, 113, 114
   recovery of metals, 110
   redox reaction, 111
   reduction, 111, 112
   reduction half-reaction in acidic solutions, 112, 113
   SS, 114
   standard half-cell potential, 114
   waste streams, 115
EMF series, 114
End processes
   purification and refining, 66–68
Energy-saving process, 242
Environmental assessment, 230, 260
Environmental-friendly dismantling process, 70
Environmental perspectives, 292
Environment-friendly process, 242
Epoxy resin (ER), 218, 257
Equilibrium isothermal adsorption tests, 215
E-waste
   advantages and disadvantages, 305
   chemical composition, 310
   classification, 1–3
   complex, hazardous and valuable components, 301
   components, 41, 302
   compounds, 43
   definition, 1–3
   disassembly and upgrading methods, 67
   energy savings, 27
   fundamental issues, 2
   handling, 21
   hydrometallurgical treatment, 309
   indium, 283
   landfill, 21
   life expectancy of electronics, 19
   management, 19, 20, 22, 312
   manual and automated/semiautomated dismantling methods, 303
   mechanical handling, 309
   metal value, 304
   metals and nonmetals, 313
   PM distribution, 36
   postproduction to disposal, 19, 21
   preprocessing, 67
   problems, 10, 13, 15
   processing, 26
   proper management and safe disposal, 301
   rare earths, 278
   recycling
      benefits, 26
      challenges, 30
      efficiency, 10
      metal source, 29, 30
      objectives, 26
      raw ores, 29
   recycling chain, 60
   recycling flowsheets, 311, 312
   recycling industry, 310
   reusable WEEE, 28, 29
   secondary resource economy, 302
   separation and dismantling criteria, 62
   solid waste, 312
   streams, 61
   sustainable recycling technology, 312
   thermal treatment, 213
   3R principle, 26
   toxic emission levels, 28
   types, 313
   vacuum processes, 284
   weight and value distribution, 27, 28

Index

F
Ferric ion, 230
Ferromagnetic substances, 170
Ferromagnetism, 171
Ferrous material, 148
Ferrous scrap, 93
Filtration systems, 140–142
Flame retardants, 42
Flash and bath smelting, 95
Flat panel display unit recycling, 286
Fluidized bed separator operation principles, 133
Fluidized bed separators (FBS), 157
Formamidine disulfide (SCN$_2$H$_3$)$_2$, 245
Four-shaft shredders, 129
FR-2 type, 42
FR-4 type, 42
Fractionation process technology
    delamination, 132
    material preparation, 131
    output fractions, 132, 134
    physical principles, 131
    separation into high-purity output fractions, 132, 133
Free take-back schemes, 15
Froth flotation, 172, 173
Functional and usable ECs, 69
Future development perspectives, 297–298

G
Gaseous phase, 91
Gasification, 90–92
General environmental indices (GEIs), 260
Gibbs free energies, 229, 253
Glass/binder encapsulation, 98
Gold recovery, 263
    nylon-12 3D-printed scavenger, 263
Granulators, 129–131
Granulometric separation, 159
Gravity concentration methods, 147
Gravity/density separation
    CC, 147
    dry/air gravity separation (*see* Dry/air gravity separation)
    metal particles, 147
    mineral processing industry, 147
    particles size and density, 147
    plastics and Al/glasses, 147
    sink-float, 147
    wet (*see* Wet gravity separation)
Green reverse flotation technology, 173
Green technology, 106

H
H$_2$SO$_4$-CuSO$_4$-NaCl system, 231
Halide, 239, 242, 248, 252
Hard disc drivers (HDD), 279
Hard solder, 48
Health, safety and environmental (HSE), 265
Heat deflection temperature (HDT), 212
Heat transfer liquids, 74
Hellatron recycling, Italy, 201, 204
High-intensity magnetic separators (HIMS), 171
High-pressure oxidative leaching (HPOL), 102
High-voltage pulses (HVP)
    EDF, 134–138
Hot fluids, 74
Hydraulic shaking tables, 147
Hydrocyclone separation, 149–152
Hydrometallurgical process, 302
Hydrometallurgical recovery
    cobalt (Co), 287, 288
    CRT, 285, 286
    flat panel display unit recycling, 286
    indium (In), 283, 284
    lamp and CRT phosphors, 281–283
    magnet scraps, 279–281
    PV cell/CIGS solar panels, 284, 285
    REE (*see* Rare-earth elements (REE))
Hydrometallurgical treatment, 215
Hydrometallurgy, 65, 66, 309, 310
    advantages, 221
    biometallurgical leaching, 106–109
    components, 222
    leaching (*see* Leaching)
    mechanical pretreatment step, 222
    production processes, 221
    solvent leaching, 98–106

I
Identification data (ID), 79
Immiscible liquid phases, 115
Incineration
    e-waste, 93
    and mechanical separation, 83
    vs. pyrolysis, 89, 90
    pyrometallurgy, 88, 89
    uncontrolled, 85
    without smelting, 97
Indium (In) recovery
    LCD, 283, 284
Industrial applications, 229
Industrial-scale Cu EW plant, 111, 112
Industrial-scale electrostatic separators, 165

Industrial-scale e-waste, 266, 267
Industrial-scale e-waste/WPCB recycling lines
  academic research *vs.* industrial practices, 208
  Attero recycling, Roorkee, India, 206, 207
  Aurubis recycling center, Lünen (Germany), 203, 204, 206
  Chinese e-waste recycling flowsheets, 198
  Daimler Benz in Ulm, Germany, 187, 188
  DOWA Group, Japan, 189
  Eldan recycling, Spain, 186, 187
  Hellatron recycling, Italy, 201, 203, 204
  ISRs (*see* Umicore's integrated smelters-refineries (ISRs))
  NEC Group, Japan, 188
  Noranda Smelter in Quebec, Canada, 207
  PCB manufacturing waste recycling, Taiwan (*see* PCB manufacturing waste recycling, Taiwan)
  Rönnskar Smelter, Sweden, 207
  Sepro Urban Metal Process, Canada, 195–197
  SwissRTec AG, 198, 200
  unpopulated/populated WPCB, 178 179, 181
  WEEE Metallica, France, 200, 201
Industrial-scale plants, 64
Industrial-scale screen types, 139
Industrial-size fractionator, 132
Industrial waste heat, 74
Informal waste sectors, 60
Infrared (IR) heating, 73, 74
Inorganic acids, 229, 232
Insulators/nonconductors, 161, 162
Integrated smelter and refinery (ISR), 88, 179
Interdisciplinary approaches, 290
Intermetallic compounds (IMCs), 50
Intrinsic value, 290, 291
Ion exchange (IX), 118, 119
IR heater, 77, 79
IsaSmelt process, 184

J
Jigs, 151, 153
Joining components
  soldering (*see* Soldering)
Joint Electron Device Engineering Council (JEDEC), 49

K
Kayser Recycling System (KRS), 203
Kelsey Centrifugal Jig separation, 149–152
Kosaka Smelting and Refining, 189

L
Laboratory-scale analytical pyrolysis, 91
Laboratory-scale batch EDF systems, 137
Lamp phosphors
  and CRT, 281–283
Large-scale e-waste recycling, 290
Laser-induced breakdown spectroscopy (LIBS), 80
LCD
  indium (In) recovery, 283, 284
Leaching
  base metal, 228
  BER, 256–258
  BM and PM recoveries, 221
  PMs (*see* Precious metal (PMs) leach)
  solder stripping (*see* Solder stripping leach)
Leaching test, 212
Life cycle assessment (LCA)
  analysis tools, 25
  approach based on life cycle, 23, 24
  environmental approach, 24
  relative approach, 24
  scientific approach, 24
  stages, 24, 25
  standards, 23
  transparent approach, 24
Life cycle management (LCM), 23
Lion batteries, 287, 288
Liquid photoimageable solder mask (LPSM) inks, 53
Liquid-liquid extraction (LLE), 115
Liquid wastes, 54
Lixiviants, 238–240
Low-intensity magnetic separator (LIMS), 171

M
Magnet scraps
  flowsheet recycling, 280
  HDD, 279
  ionic liquids, 281
  magnet production, 281
  motor magnets, 281
  NdFeB, 279, 280
  oxidative roasting, 281
  recycling flowsheet, 282
  recycling of Nd-based, 280
  REEs, 280
  Santoku Corporation, 281
  SmCo, 280
  solvent extraction ionic liquid, 281
  solvent extraction process, 280
Magnetic dry substances, 171
Magnetic separation (MS), 170–172
Manual dismantling, 12

Index

Manual informal dismantling, 71
Manual *vs.* automated dismantling, 303, 304
Manuel disassembly, 62
Market, 294
Marketability perspectives, 293, 294
Mechanical separation, 85–86
Mechanical sorting, 61
Mechanochemical (MC) treatment, 241
Metal-centric approach, 298
Metal-containing slags/bottom ash, 93
Metal recovery, 310
Metal resources
    BAT, 43
    electronics, 45
    and materials, 43
    PCB manufacture, 43, 44
    phones and PCs, 45, 46
    PMs (*see* Precious metals (PMs))
    postconsumer recycling rates, 43, 46
    product-centric, 43
    recycling of WEEE, 45
    WEEE stream, 43
Metals, 36
Methanesulfonic acid (MSA), 225
Microwave heating, 92
Mineral acid leaching, 226
Mineral processing industry, 147
Mineralogy, 19
Mining process, 18
Molten salt electrolysis, 281
Motherboards (MBs), 257
Mozley separators, 154, 155
Multi-roller electrostatic separators, 163

N
National Strategy for Electronic Stewardship (NSES), 15
NdFeB, 279, 280
NEC Group, Japan, 188
Nernst equation, 113, 114
N-methyl-2-pyrrolidone (NMP), 257
Non-/diamagnetics, 170
Nonaqueous layer, 116
Nonferrous metals, 186
Nonferrous scrap, 93
Nonmetal fraction (NMF)
    chemical recycling (*see* Chemical recycling)
    direct recycling (*see* Direct recycling)
    physical recycling, 42
    pyrolysis (*see* Pyrolysis)
    TGA, 42
    thermal stability, 42
Nonmetallic fraction, 215
Nonselective dissolution, 232

Noranda Smelter in Quebec, Canada, 207
Nylon-12 3D-printed scavenger, 263

O
Occupational Safety and Health Standards (OSHS), 65
Optical sorting equipment, 285, 286
Opto-electronic sorters, 80
Organic acids, 102
Organic compounds, 115
Organic materials, 36
Original equipment manufacturers (OEMs) 37, 79
OSRAM patented process, 282
Outotec technology, 264, 265
Oxidation and reduction potential (ORP), 102
Ozone-depleting substances (ODS), 28

P
Palladium (Pd) leaching, 255
Paramagnetic substances, 170
Pb-free solder, 48, 49
PCB manufacturing waste recycling, Taiwan
    edge trim, PCBs, 191, 192
    PTH process, 193, 194
    rack stripping process, 193
    recovery, Cu and PMs, 190
    Sn/Pb solder dross, 192, 193
    Sn/Pb spent stripping solution, 194, 195
    spent basic etching solution, 193, 194
    wastewater sludge, 192
    WPCBs, 190
Peroxysulfuric acid (H_2SO_5), 228
Phenolic molding compound (PMC), 212
Photovoltaic (PV) cell/CIGS solar panels recovery, 284, 285
Physical methods, 123
Physical separation methods, 145, 146, 226
Physico-mechanical and pyrometallurgical processes, 301
Physico-mechanical process, 61
Plastic/cellulose composite, 72
Polyester, 144
Polymers, 87
Polyvinyl chloride (PVC), 18
Postconsumer recycling rates, 43, 46
Potassium persulfate ($K_2S_2O_8$), 242
Pourbaix diagrams, 229
Precious metal (PMs) leach
    advantages and disadvantages, 238
    Ag, 254, 255
    aqua regia leaching, 253, 254
    Au (*see* Au leaching)

Precious metal (PMs) leach (cont.)
    chemical pretreatment, 238
    comparison, 254
    electric and electronic industries, 233
    extraction, 233
    full recovery, 256
    industrial-scale e-waste, 266, 267
    lixiviants, 238
    MC treatment, 241
    Pd, 255
    physical pretreatment, 238
    purification (see Purification)
    recovery flowsheet, 238
    WPCBs, 233
Precious metal operations (PMO), 96, 182
Precious metals (PMs)
    in notebook PCBs, 43, 45
Pregnant solutions (PLS), 263
Prepreg, 34
Preprocessing, 60, 61, 65, 67
Pressure leaching, 232
Primary vs. secondary productions, 302, 303
Printed circuit boards (PCBs)
    alignment, 35
    average metal content, 38, 39
    by weight, 36
    consumer electronics, 36
    ECs, 33
    electronic components, 47
    epoxy resin and glass fiber, 33
    green/yellow board, 34
    hazardous elements, 35
    joining components (see Joining
        components)
    manufacturing process, 53–55
    material compositions, 34, 36, 37
    metal resources, 43–47
    metals, 38, 40
    packaging, 50
    prepreg, 34
    single- and double-sided, 35
    structure, 35
    substrate, upper and lower compounds, 34
    types, 33, 34
    unpopulated and EC compositions, 37
    unpopulated/bare, 59
Printed wiring board (PWB), 59
Product-centric recycling, 43
Proton exchange membrane (PEM) fuel
    cells, 45
Pulse air separator, 179
Pulse power generation, 134, 135
Pulse-jet bag filter dust collection, 140–142
Pulverizer, 124, 129–131, 141
Purification
    anodic process, 261
    cathodic process, 261
    cementation, 109, 110
    chemical precipitation of metals, 109
    clean reagent, 260
    Cu leaching, 259
    direct/mediated electrochemical
        oxidation, 260
    dissolution of Me, 260
    dissolution of metals, 262
    electrochemical oxidation, 262
    electrochemical reactor, 261
    electrokinetic process, 262
    electrolyte, 261
    electrolyte composition, 261
    electrometallurgy, 262
    environmental assessment, 260
    EW, 110–115
    gold recovery, 263
    IX, 118, 119
    metal recovery processes, 110
    physical separation, 262
    PMs from leachates, 259
    refining technologies, 258
    SCW technology, 260
    solvent extraction, 259
    SX, 115–118, 264–266
Purification and refining endprocesses, 66–68
Pyrolysis, 197
    ANMF, 215
    bromide, 216
    chemical activation, 214
    chemical composition, 89
    depolymerization, 217
    disposal of WPCBs, 215
    electrochemical characterization, 214
    epoxy resins, 218
    formation, 214
    gases, 89
    vs. incineration, 89, 90
    metals, 214
    nonmetallic fraction, 215
    pH, 215
    physical activation, 214
    physical phase, 89
    PO, 89
    properties, 215
    PVC resin, 218
    pyrometallurgical route, 90
    recycling of cross-linked polymers, 218
    SCMO recycling, 216
    SCWD, 215, 216
    SCWO, 215, 216
    simultaneous recovery, 218
    solvents, 217

Index

subcritical fluids, 217
supercritical fluids, 217
supercritical water, 215
temperature, 214
temperature range, 214
thermal behavior of epoxy resins, 214
thermal-alkaline activation process, 215
thermochemical decomposition, 89
vacuum, 90, 214
Pyrolytic oil (PO), 89
Pyrometallurgical processes
   limitations, 97
   recovery of metals
      Cu smelters, 93–96
      ferrous scrap, 93
      metal-containing slags/bottom ash, 93
      nonferrous scrap, 93
      shredder residues, 93
Pyrometallurgy, 65, 66, 310, 311
   direct smelting, 88
   direct treatment, 87
   e-waste/Cu/Pb scraps, 87
   ferrous fractions, 87
   gasification, 90–92
   incineration, 88, 89
   polymers, 87
   pyrolysis, 89, 90
   recycling facilities, 87
   smelting, 87
   technologies, 97, 98
   thermal/thermochemical processes, 87
Pyrometallurgy+supergravity +hydrometallurgy, 173, 174

R
Raffinate, 115
Ramp to spike (RTS), 52, 53
Ramp-soak-spike (RSS), 52, 53
Rare-earth elements (REE)
   abundances, 278, 279
   and calcination to oxides, 281
   application areas in EEE, 277, 278
   chemical elements in chemical periodic table, 277
   concentrates, 278
   definition, 277
   e-waste, 278
   WEEE, 278, 280
   leaching rate, 281
   phosphor chemical composition, 282, 283
   recovery rate, 281
   recycling, 279
   scantiness, 278, 279
Rare-earth metal (REM), 277

Recycle, 15
Recycling chain, 179
   AEC, 60
   classification, 60
   collection programs, 60
   dismantling, 62, 63
   ECs, 59
   end-of-life WEEE/WPCB treatment options, 60, 61
   endprocessing, 66–68
   e-waste, 60
   hazardous components, 61
   informal waste sectors, 60
   mechanical pretreatment, 61
   mechanical sorting, 61
   physico-mechanical process, 61
   postconsumer goods, 60
   separation/upgrading/extraction, 65, 66
   size reduction, 63–65
Recycling lines
   equipment details and technical parameters, 181, 182
   unpopulated/populated WPCB, 177, 181
Recycling rate, 257
Recycling technologies
   market, 294
Redox reaction, 111
Reduce, 15
Refining, 242, 258, 264, 266, 267
Reflow soldering, 52, 53
Reuse, 15
Rising-current separators, 148, 149
Rod- and brush-type EC disassembly apparatus, 77, 79
Rönnskar Smelter, Sweden, 207
Roorkee-based Attero Recycling, 207

S
Santoku Corporation, 281
Scanning and laser desoldering automated component-dismantling machine, 79
SCMO recycling, 216
Screen, 137, 139
Screens and shaking tables, 186
Selective disassembly, 62
Selfrag technology, 136
Semiautomatic PCB electronic component-dismantling machines
   drum-/barrel-type dismantling machines, 75, 76
   labor force, 75
   rod- and brush-type EC disassembly apparatus, 77, 79
   scanning and laser desoldering, 79

Semiautomatic PCB electronic component-
        dismantling machines (cont.)
    tunnel-type dismantling machines, 76–78
Sensing technologies, 80
Separation
    electrolysis separation, 173, 174
    ES (see Electrostatic separation (ES))
    Froth flotation, 172, 173
    gravity/density (see Gravity/density
        separation)
    MS, 170–172
    pyrometallurgy+supergravity
        +hydrometallurgy, 173, 174
Separation preprocessing upgrades/
        concentrates and extraction, 65, 66
Sepro Urban Metal Process, Canada
    Au and heavy metal recovery process, 195
    e-waste/WEEE, 195
    WEEE/e-waste pyrolysis process, 197
    WPCBs, 196
Shanghai Xinjinqiao Environmental Co., Ltd.,
        China, 197, 198
Shape-memory alloys (SMAs), 80
Shape-memory polymers (SMPs), 72, 80
Shredder residues, 93
Shredders
    automatic reverse sensors, 126
    blade wearing, 126
    characteristics and features, 126
    double-shaft shredders, 128
    four-shaft shredders, 129
    high-capacity, 125
    machine details and cutter knife discs/
        blades, 125
    single-shaft shredders, 127, 128
Shredding, 64
Silver (Ag) leaching, 254, 255
Simultaneous disassembly, 62, 63
Single- and double-sided PCBs, 35
Single-shaft shredders, 127, 128
Sink-float methods (heavy media), 150–153
Sink-float separation, 147
Size reduction, 63–65, 123, 124
Smelting, 65, 87
Socket pedestal, 48
Soft solder, 48
Solder, 48
Solder alloys, 50, 51
Solder flux, 49
Solder mask/solder stop mask/solder resist, 53
Solder paste, 49
Solder stripping leach
    ammonia-ammonium carbonate, 225
    ATMI automated WPCB recycling, 227, 228
    Au connectors, 227

BM components, 227
Br_2/KBr media, 223, 224
Br_2-based lixiviants, 223
cupric ions, 223
dismantled and shredded WPCBs, 227
electrochemical regeneration, 223
green chemistry, 227
green engineering methodologies, 227
HNO_3, 222
hydrometallurgical method, 226
ICs, 227
Pb and Sn, 222
Pb-Sn, 225
physical separation methods, 226
scrap material, 225
$SnCl_4$, 224, 225
Sn-Cu-bearing precipitates, 226
stannic ions (Sn^{4+}) oxide tin metal in HCl
        solutions, 224
WPCBs by dismantling, 226
$ZnSO_4$ solution, 225
Soldering
    application types and orders, 51
    dip, 51
    Pb-free solder, 48, 49
    reflow, 52, 53
    solder alloys, 50, 51
    solder flux, 49
    solder mask/stop mask/resist, 53
    solder paste, 49
    wave, 52
Solvent extraction (SX), 115–118
    electrolytic refining process, 264
    equipment, 264
    HSE, 265
    LIX84I, 264
    LIX860IC, 264
    LIX984, 264
    mixing time, 264
    Outotec technology, 264, 265
    plant concept, 265
    VSF X DOP Unitunit, 265
    VSF X settler, 265
    VSF X SPIROK mixer unit, 265
    WPCB leaching solution, 264
Solvent leaching
    advantages and disadvantages, 98, 99
    aqua regia (AR), 101, 104, 105
    Au, 105
    balanced equations, Cu, 103
    corrosion, 98
    heat pretreatment, 102
    higher acid concentrations, 102
    higher temperatures enhance leaching
        rates, 102

HPOL, 102
lixiviants/chemicals, 98
mechanical crushing, 98
metal recoveries, 102
mineral acids, AR and ionic liquids, 98, 101
MSA-H$_2$O$_2$ desoldering leaching, 106
Ni and Zn, 104
organic acids, 102
ORP, 102
oxidants, 102
oxidizing agents, 103
Pb dissolution, 104
Pourbaix, 102
precombustion, 102
S/L ratio, 102
selectivity and effectiveness, valuable metals, 101
Sn, 104
Sn and Pb dissolution, 106
Solvents, 213, 217
Solving the E-waste Problem (StEP), 206
Sorting
   automatic, 143, 144
   manual, 143
   types, 144
Stainless steel (SS), 114
Standard half-cell potential, 114
Steam gasification, 92
Subcritical fluids, 217
Supercritical fluids, 213, 216, 217
Supercritical water (SCW), 215, 260
Supercritical water depolymerization (SCWD), 215, 216
Supercritical water oxidation (SCWO) 215, 216
Surface-mount components (SMCs), 49
Surface-mount technology (SMT), 52
Surface-mounted device (SMD), 48
Sustainability, 298, 299
Sustainable recycling technology, 312
SwissRTec AG, 198, 200–202
Synthetic precipitation leaching procedure (SPLP), 267
Systemic optimization, 290

T
Thermal treatment
   burning, 73
   DR, 73
   explosion, 73
   heat transfer liquids, 74
   industrial waste heat, 74
   IR heating, 73, 74
   IR lamps, 72

   tin melting stoves, 73
   working temperature, 72
Thermogravimetric analysis (TGA), 42
Thermogravimetry brominated epoxy resins, 214
Thermoplastics, 42
Thermoset organic polymers, 213
Thermosetting, 42
Thermosetting polymer, 215
Thiosulfate, 232, 238, 239, 242, 243 246–249, 251
Thiourea, 232, 238, 239, 241–243, 254, 256
3R principle, 26
Through-hole components (THCs), 52
Through-hole device (THD), 48
Through-hole technology (THT), 52
Tin melting stoves, 73
Top-blown rotary converter (TBRC), 204
Toxicity characteristic leaching procedure (TCLP), 267
Traditional processes
   mechanical separation, 85–86
   uncontrolled incineration, 85
Traditional *vs.* advanced WPCB recycling processes
   classification, 83, 84
   disadvantages, 84
   hydrometallurgy (*see* Hydrometallurgy)
   melting points, densities/physical/chemical characteristics, 84
   purification (*see* Purification)
   pyrometallurgical (*see* Pyrometallurgical processes)
Triboelectric charging, 162, 165, 166
Triboelectric cyclone separation, 149–152
Triboelectric separators (TES), 165–168
Triboelectric series, 166, 167
Tributyl phosphate (TBP), 225, 280
Tunnel-type dismantling machines, 76–78

U
Ultrasonification, 258
Umicore's battery recycling flowsheet, 287
Umicore's integrated smelters-refineries (ISRs)
   BMO, 182
   furnace, 183, 184
   leach-electrowinning plant, 184
   off-gas treatment, 184
   operation flowsheet, 183
   PM refinery, 184
   PMO, 182
   recovered metals, 183
   resource efficiency, 185
   vertical *IsaSmelt furnace*, 182

Umicore's ISR, 96
Uncontrolled incineration, 85
Unpopulated PCBs, 37
Unpopulated/bare PCB, 59
Unpopulated/populated WPCB recycling lines
    dismantled ECs, 178
    EC removal, 177
    ISR, 179
    pulse air separator, 179
    separation processes, 178
    stages, 179
Urban mining, 38, 303

V
Vacuum pyrolysis, 90, 214
Value chain, 37–41

W
Waste electrical and electronic equipment (WEEE), 1
    key features, 15
    management/treatment, 4
    manual dismantling, 12
    nonhazardous and hazardous substances, 4–6
    physical properties, 144–146
    recycling process, 10, 11
    significance and characteristics, 3
    treatment, 302
    wet gravity separation, 12
    WPCB, 10
Waste printed circuit board (WPCB), 4
    characterization, 41, 42
    classification, 137, 139
    composition, 18
    conventional *vs.* novel, 304–309, 311
    desktop and laptop PCs, 37
    economic value, 38, 41
    glass materials, 36
    grades, 38, 41
    grinding, 16, 17
    heterogeneous materials, 16
    industrial application, 146
    layers and components, 34, 35
    manual *vs.* automated dismantling, 303, 304
    material compositions, 36, 37
    metallic materials, 36
    metals, 16, 38, 40
    mineralogy, 19
    mining process, 18
    mobile phone, 37
    NMF (*see* Nonmetal fraction (NMF))
    nonmetallic fraction, 144
    OEMs, 37
    primary *vs.* secondary productions, 302, 303
    recycling chain (*see* Recycling chain)
    researchers and enterprises, 35
    resin materials, 36
    shredding, 16, 17
    size reduction, 123, 124
    urban mines, 18
    valuable materials, 302
    value chain, 41
    values, 38
Water treatment, 119
Wave soldering, 52
Wet gravity separation, 12
    ferrous material, 148
    flowsheet of PCB recycling, 148, 149
    hydrocyclone, 149–152
    industrial-scale plants, 148
    jigs, 151, 153
    Kelsey Centrifugal Jig separation, 149–152
    metals and nonmetal materials, 145, 148
    Mozley separators, 154, 155
    rising-current separators, 148, 149
    shaking tables, 154, 155
    sink-float methods (heavy media), 150–153
    triboelectric cyclone, 149–152

Y
Yangzhou Ningda Precious Metal Co., Ltd., China, 197, 198

Z
Zero dust generation, 257
Zero-waste scheme, 298, 299
Zigzag separators, 155, 156

Printed by Printforce, the Netherlands